O9-ABG-295

The Book of Revelation

The Book of Revelation
Apocalypse and Empire

LEONARD L. THOMPSON

New York   Oxford
OXFORD UNIVERSITY PRESS
1990

Oxford University Press

Oxford   New York   Toronto
Delhi   Bombay   Calcutta   Madras   Karachi
Petaling   Jaya   Singapore   Hong Kong   Tokyo
Nairobi   Dar es Salaam   Cape Town
Melbourne   Auckland

and associated companies in
Berlin   Ibadan

Copyright © 1990 by Leonard L. Thompson

Published by Oxford University Press, Inc.,
200 Madison Avenue, New York, NY 10016

Oxford is a registered trademark of Oxford University Press

All rights reserved. No part of this publication may be reproduced,
stored in a retrieval system, or transmitted, in any form or by any means,
electronic, mechanical, photocopying, recording or otherwise,
without the prior permission of the publisher.

Library of Congress Cataloging-in-Publication Data
Thompson, Leonard L., 1934–
The book of Revelation : Apocalypse and empire /
Leonard L. Thompson.
p. cm.   Bibliography: p.   Includes index.
ISBN 0-19-505551-9
1. Bible.   N.T.   Revelation—Criticism, interpretation, etc.
2. Bible.   N.T.   Revelation—Language, style.   I. Title.
BS2825.2.T46   1990
228′.06—dc20   89-3405   CIP

2 4 6 8 9 7 5 3 1
Printed in the United States of America
on acid-free paper

To My Mentors
Robert M. Grant and E. Graham Waring
in their retirement
"For they too are honorable men"

Bookstore

2/96

1/11/91

42507

Preface

Throughout the following chapters there are references to many books, inscriptions, and papyri from the early Christian and classical world of the first centuries of the Common Era (CE). They are listed on pages 211–212. Translations of the Greek New Testament usually follow the *Revised Standard Version*, though I sometimes use my own translation.

There are four parts to the book. I provide an orientation in the Introduction and part 1. Some of the rather complex ideas at the end of chapter 2 are discussed again in chapter 11. Parts 2 and 3 could be read in reverse order, but I recommend following numerical order. The similarities between the arrangement of this book (four parts and twelve chapters) and the number symbolism of the Book of Revelation are purely coincidental.

In the best of circumstances writing involves stretches of isolation punctuated by fruitful conversation with various groups of colleagues. Helpful conversation occurred in the seminar on early Christian apocalypticism of the Society of Biblical Literature and the New Testament seminar of the Upper Midwest Region of the same organization. I express special gratitude to David Aune, William Beardslee, Adela Yarbro Collins, Howard Kee, and Carolyn Osiek. My colleagues at Lawrence University listened patiently to some of these ideas in their roughest form. Robert M. Grant and John J. Collins made observations regarding some of the chapters when they were in not-so-rough a form. I approach the text differently from those two scholars, but their writings provide the foundation. Diane Jeske, my undergraduate assistant, read the text and corrected it in many ways. In the final stages of the manuscript, the comments of Bob Grant, M. L. Ray, and Cynthia Read helped me revise and improve several chapters. Lastly, Michael Lane read the manuscript with great care and helped clarify both language in the manuscript and references to other works.

Stretches of isolation were made possible by a year-long fellowship from the National Endowment for the Humanities and a Summar Research Fellowship from the same endowment. In that connection I note Edwin Good's unflagging support with appreciation. I am grateful for those fellowships which provided an *inclusio* around the research — a most appropriate form for the Book of Revelation. Along the way, research grants from Lawrence University provided opportunities for visits to major research libraries.

Appleton, Wisconsin   L. L. T.
March 1989

Contents

Chronology

Emperors	Non-Christian Authors	Christian Authors
Julius (d. 44 BCE)		
Augustus (27 BCE–14 CE)		
Tiberius (14–37)		Paul (d. 62–68)
Gaius (Caligula) (37–41)		*Letters* (c. 50–60)
Claudius (41–54)		
Nero (54–68)		
Galba (68–69)		
Otho (69)		
Vitellius (69)		
Vespasian (69–79)		
Titus (79–81)	Martial (d. c. 104)	
	Spectacles (c. 80)	
Domitian (81–96)	*Epigrams* (from c. 86–)	John
		Revelation (92–96)
	Statius (45–96)	Didache (110?)
	Silvae (92–)	Ignatius (d. 115)
	Achilleid (95–96)	*Epistles* (c. 110)
	Quintilian (b. c. 30–35)	
	Institutio Oratoria (96)	
	Josephus (b. c. 37)	
	Jewish Antiquities (93?)	
	Against Apion (c. 96)	
	Life (c. 96–97)	
Nerva (96–98)	Silius Italicus (d. c. 101)	
	Punica (c. 98)	
Trajan (98–117)	Tacitus (b. c. 56)	

Emperors	Non-Christian Authors	Christian Authors
	Agricola (98)	
	Germania (98–99)	
	Histories (100–110)	
	Pliny the Younger (c. 61–112)	
	Panegyricus (100)	
	Letters (105–109)	
	Dio Chrysostom (c. 40–112)	
	Discourses	
Hadrian (117–138)	Juvenal (c. 60–127)	Polycarp (d. 155)
	Satires (115–27)	*Letter to the Philippians*
	Suetonius (b. 70)	(c. 112)
	Lives of the Caesars (c. 120)	
Antoninus Pius (138–61)		Justin Martyr (d. 165)
		Apologies (c. 150)
		Dialogue with Trypho (c. 155)
		Hegesippus (fl. 120–80)
Marcus Aurelius (161–80)		Melito (fl. c. 180)
		Irenaeus (140–200)
		Against Heresies (180–90)
Commodus (180–92)		
Septimius Severus (193–211)		Tertullian (160–200)
		Apology (c. 200)
Caracalla (198–217)	Dio Cassius (c. 150–235)	
	Roman History (215–20)	
	Philostratus (c. 170–245)	
	Life of Apollonius of Tyana	
	Lives of the Sophists	
Elagabalus (218–22)		
. . . .		Dionysius of Alexandria (d. c. 265)
		Fragments
. . . .		Eusebius (260–339)
. . . .		*Ecclesiastical History* (c. 300)
. . . .		
Constantine (306–37)		

The Book of Revelation

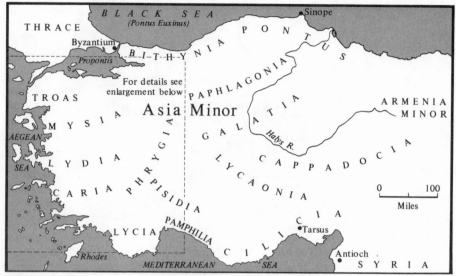

Map of Asia Minor about the time when the Book of Revelation was written.

Introduction

In Robert Coover's *The Origin of the Brunists* (1971) Justin Miller, editor of the local newspaper, reflects on the fact that he "had read Revelations at the age of thirteen and never quite got over it." That happens to a lot of people who read the Book of Revelation—even so-called scholars—and it is from a scholarly point of view (specifically that of a literary and social historian of early Christianity) that the following chapters have been written.

A book like Revelation can grip people in different ways, and it is not easy to explain just how the scholarly "grip" differs from, for example, Coover's Brunists who use the Book of Revelation to illumine an unfinished note of a preacher-miner killed in a mining explosion. Nor is the scholar likely to be gripped by the book as a window into the future of Middle Eastern affairs, with the number of the beast, 666, enigmatically identifying the Soviet Union or Iran. I could say that the scholar is systematic in his or her approach to the Book of Revelation. But the Brunists and other millenarians, as they decode the book, are probably just as systematic as a social or intellectual historian of early Christianity, though in a different way.

Scholars, like millenarians, must also enter into the symbols, syntax, and literary structures of the book. Dell Hymes, writing about the need for linguists to be participants in the language community that they are studying, says, "No amount of acoustic apparatus and sound spectography can crack the phonemic code of a language, and a phonemic analysis . . . is the necessary basis for other studies" (Giglioli 1982, 24). So research and analysis of the Book of Revelation, as of all language, require participation in the structures of meaning and the shared codes of apocalyptic language. Nor do scholars necessarily participate in a "more objective manner" than millenarians. The persona of disinterested, cool, objective observation sometimes slips, and then passions and commitments become visible to everyone. We all have our axes to grind. Nor need the scholarly inquiry be any less meaningful existentially for the life of the scholar than religious inquiry for the

1

millenarian: for both, self-identity becomes intertwined with the pursuit of learning and interpreting systematically the words, sentences, and symbolic representations in "Revelations" — though, again, along different lines.

There may be something to the distinction that the millenarian — I use the term broadly to include all those personally committed to apocalyptic spirituality — believes and lives by the book, whereas the scholar makes inquiry into that belief and commitment; that is, the millenarian can do without the scholar, but the scholar cannot do without the millenarian. In that sense scholarly inquiry is "once removed" from the apocalyptic spirituality of the millenarian. In reflecting upon the Book of Revelation, the scholar will probably "explain" it by comparing it to something else; by showing how it belongs to a broader category (a literary genre or a phenomenon in the history of religion); or by linking its presence to a social class, a psychological type, or a specific time and local place in history. Yet the devout apocalypticist also has moments of reflecting on the faith and "looking in" at the spirituality in which he or she participates. The most faithful can ask, "How did I get mixed up in this? What attracts me to it? Where did these ideas come from? What do I have in common with other millenarians?" That is, "believing" and "living by" do not necessarily bracket out the second-order activity of reflection.

Nonetheless, there is a distinction — at least one of emphasis — to be made here between the "scholarly" and the "faithful" approach. The scholar does not view the Book of Revelation as a self-contained, self-authenticating work. It is authenticated by how it enters into other social, political, psychological, historical, literary, and religious structures. The scholar seeks to uncover the connections that are not apparent on the surface — the latent connections, the hidden structures, and the invisible systems of which the Book of Revelation is a part. But don't the "faithful" do the same, as they uncover connections between the Book of Revelation and other social, historical, and political events? There is at least this difference: for the scholar, the Book of Revelation is not *the* determining force in the invisible systems. The Book of Revelation — which includes any divine or human actor in the book — does not contain the code, the DNA strand, that shapes and unfolds the rest of life and history. It is not *the independent* variable that determines all else. The scholar looks for reciprocity, for some situations in which the book is determinative and for others in which it is determined. The scholar hesitates to draw simple lines of cause and effect and searches rather for feedback loops and reciprocal interaction among inseparable elements. Scholars are mealymouthed, unwilling, or at least hesitant, to state anything as simple and direct. And that can evoke impatience from those who know what makes the world work — whether that be divine or economic determinism — or from those who want a simple and direct answer in all analysis.

But I have been too simple in my analysis. Scholars and the faithful are not two generic brands packaged in plain, white containers. Members of each group come in different sizes, qualities, colors, and shapes. What lies behind these reflections on how a scholar approaches a work like the Book of Revelation is this: I am interested in having conversation with all those who are committed in different ways to apocalyptic spirituality and with those who are not. Further, I should like to converse with both groups at the same time. In brief, I am interested in conversation in a pluralistic context, and I think that a "scholarly" approach to a work such as the Book of Revelation can contribute to fruitful, pluralistic dialogue. Only one group need by definition be left out: those who reject the possibility that the study of a book like Revelation could have any value at all.

In the following chapters I will use a scholarly approach to what is said, but in the packaging process the content will take on distinctive hues, shapes, and qualities that may not be to every scholar's liking. Nonetheless, all scholars will recognize, explicitly or implicitly, how the chapters unfold and they will use the same approach as I have used to show how, in their opinions, I have colored in things incorrectly and shaped things along the wrong designs; that is, there is a common approach in the scholarly world even though it can lead in different directions and can produce different sets of conclusions. That common approach promises fruitful exchange in pluralistic contexts.

Pluralism is limited by the fact that all readers of the Book of Revelation share in a common text: they look at the same words, the same sentences, the same paragraphs. Without that shared focus in the same text there could be no hope for conversation in a pluralistic context. But "common text" is no simple item. Any text that continues to be read in any kind of community — university, church, political party — does not exist simply as a given fact. It does not exist as a *communicating text* except in the act of reading it. Through reading, the text is once again "brought to pass." With every reading, a reader gains the text once again; different readers in reading acquire the same text differently. The Book of Revelation is, then — like all texts — an ongoing accomplishment, an ever-recurring acquisition. Further, none of us reads it fresh in the Garden of Eden. We receive it in a tradition of readers who connect us to the text — often through complicated, unrecognized links — and who shape what we see. One way of recognizing the "ongoingness" of that chain of readers would be to trace out the history of the reading of the Book of Revelation: Who read it, when, in what circumstances, to what end? Bernard McGinn (1979) has done that for the medieval period; Patrides and Wittreich have edited a volume which does that for the Renaissance. Through that process one could begin to write a history of readership of the Book of Revelation and trace out continuities, discontinuities, and explosive revolutions in its reading history. For most readers of the Book of Revelation that is too ambitious a project, but every reader should be as aware as possible of the "tradition of readers" through which he or she reads the Book of Revelation. That awareness can often clarify conversation in a pluralistic context.

My interest in the Book of Revelation is limited to the situation in which it was first read and written — the situation of the author and the situation of those to whom the Book of Revelation was originally addressed. I claim no privileged position for that original situation except that it captures my interest; I am not prepared to argue that you must first understand what it meant to those original readers before you can understand its meaning for readers in the Middle Ages or the Renaissance or contemporary life. Nor would I even argue that the original meaning should control what possible meanings can be drawn from the text (though there are other controls). Language carries too many bits of information in too many dimensions for such claims to be made. The heyday of historical positivism — that authorial intentionality and original context controls all meaning—is over. Augustine made the same point from a different angle centuries ago: "What harm is there if a reader believes what you, the Light of all truthful minds, show him to be the true meaning? It may not even be the meaning which the writer had in mind, and yet he too saw in them a true meaning, different though it may have been from this" (*Conf.* 12.18). Besides, original meanings are never recovered in their pristine purity without the taints of that "reading history" mentioned above. Indeed, without those taints and stains, the pure text would have no point of entry, no place for a mind-hold.

Just as "original meaning" and "original context" are not normative for all subsequent readings, so subsequent readings of the Book of Revelation should not control how we understand its "original context." Its "original meaning" and "original context" may be quite different from later ones, including those of today's reader; inquiry into the situation in which the Book of Revelation arose should not be harnessed quickly into the service of social, political, and/or religious concerns of the reader. Past situations are not necessarily mirrors of the present. Past situation may disclose novel meanings, new kinds of connections among texts and their contexts, and even distinctive categories in which to conceive of relationships among religious, literary, political, and social dimensions of life. Historical inquiry should be allowed to do more than provide examples or counterexamples of how we understand the world—at least, if we want history to disclose its secrets. I am not denying the need for categories, frames, and models from the life of the reader as bridges to the past, nor do I eschew the importance of morality in historical inquiry; but I am arguing that our categories, models, and moral commitments (all of which we bring to historical inquiry) can be chastened by careful inquiry into the past.

In brief, in the process of reconstruction, one adjusts both the situation being reconstructed and one's own understanding of various relationships in the world. Any reconstruction of the original situation of the Book of Revelation—how the seer understood the world, why he wrote to the people he did, what he had in mind in writing to them, how it was to affect their situation—is a process, a feedback loop of continuous readjustment as one moves from reconstructing to the thing being reconstructed and back again to reconstructing. Justin Miller "had read Revelations at the age of thirteen and never quite got over it." That happens to many people, so let the reader beware.

The Book of Revelation seems to evoke excess response. Two monks in Umberto Eco's *Name of the Rose* hold the following exchange as they seek to solve a series of murders in the abbey: "I asked him why he thought the key to the sequence of crimes lay in the Book of Revelation. He looked at me, amazed: 'The book of John offers the key to everything!'" (1984, 303). On the other hand, George Bernard Shaw, in his preface to *The Adventures of the Black Girl in Her Search for God*, dismisses the Book of Revelation as "a curious record of the visions of a drug addict which was absurdly admitted to the canon under the title of Revelation" (quoted in McGinn 1984, 35). In limiting my inquiry to the original situation of the Book of Revelation, I will argue only that a seer named John wrote the Book of Revelation as a means of communicating with other Christians who were living in roughly the same geographical zone and at the same time as John; that is, from this approach the Book of Revelation is a text intended to communicate something to others (contra Shaw), but I limit the language of communication to Christians living in cities of western Asia Minor at the end of the first century of the Common Era.

For many people—those of the Shaw variety—the issue may be whether the Book of Revelation offers a key to anything in normal, ordinary human life. The symbolic, metaphoric, even bizarre language of the seer does not at first glance serve as a window into urban life in the eastern environs of the Roman Empire. For that service one prefers more prosaic writings, such as Dio Chrysostom's orations or the geography of Strabo. Although the seer's work does not provide us with direct information about urban life in first-century Asia Minor, his language is moored in that social order. Whatever the nature of the ecstatic, visionary origins of that language—John says that he was "in the spirit"—he has a message for Christians who are living in that urban society, specifically in the towns of Ephesus, Smyrna, Pergamum, Thyatira, Sardis, Philadelphia, and Laodicea (Rev. 1:11). That is not to say, however, that the significance of John's message is limited to its first-century Asian social context. As suggested above, past situations may disclose new ways of understanding. Two thousand years have not very much changed how humans adapt to their environment.

In order to understand the highly metaphoric, poetic language of the Book of Revelation, we must do more than read it over and over again; that is, more is required than understanding each of the words and each of the sentences in the book. That more involves the generic dimension of the writing. If a person was not acquainted with the genre "business letter," the closing of a formal, business communication with the intimate language *yours truly* or *sincerely yours* would remain a puzzle no matter how many times it was reread. Phrases, sentences, and even paragraphs often will make sense only when we see them as elements in a particular type of writing, that is, a genre. Scholars have rightly worried considerably about the genre "apocalypse"—what constitutes it, how it differs from other writings, what its essential elements are. I take up several of those issues in chapters 1 and 2.

Here I want to consider the language of the Book of Revelation as an example of a type even broader than apocalypse. John writes as a true cosmopolitan. His vision extends beyond local politics to global issues and beyond global issues to cosmology. And he extends his world by means of *nesting language*. (I am thinking here of small tables or measuring cups that stack by being nested into ever-larger sizes.) John's highly symbolic language nests urban Asia Minor into ever-larger contexts—ultimately into a cosmic vision that includes the whole social order, the totality of nature, and suprahuman divinities that invade but transcend both society and nature.

In scholarly circles, this nesting language is usually called mythic language. Mythic language is the language used by humans to orient and adapt to the most encompassing environment that can be conceived, an environment that usually includes divinities and other cosmic forces. Mythic, nesting language "works" by means of metaphor, symbol, and homologue. So in the opening vision of the Book of Revelation the seer presents Jesus Christ as a suprahuman character by drawing metaphors from the human realm and the realm of nature: the suprahuman character is "like a son of man, clothed with a long robe and with a golden girdle round his breast." Through a simile involving the word *white* wool, snow, and his hair are encompassed (Rev. 1:14). Later it is written that those living in the city of Pergamum in the province of Asia in the Roman Empire live "where Satan dwells" (2:13). Their environs are extended beyond urban social intercourse, beyond the natural terrain and the countryside. Through a homologue between their city and Satan (that ancient *diabolos*) they are oriented to an environment that is cosmic in scope. The same kind of extension occurs at Ephesus: living the proper faithful life at Ephesus is homologous to "eating the tree of life, which is in the paradise of God" (2:7). In the process of nesting or extending the environment through metaphors and homologues nothing is left behind. Minor local issues do not drop out as the seer moves to global and cosmic environments; rather, they are taken up into the cosmic. Everything—local, global, animal, vegetable, mineral, divine—keeps its own place as it is taken into a larger unified system or ordered world. The metaphoric language of myth gathers up the scraps of the world and presents them as parts of a whole.

Anyone acquainted with the mythos of the Hebrew Bible recognizes immediately that the writer of the Book of Revelation is steeped in a specific mythic tradition. When he is transformed psychologically "into the Spirit," he does not leave behind what he knows. The Spirit speaks to him in the language of his own Jewish heritage as it was incorporated into a Christian mythos. At the same time there are more universal elements in this language. For example, we find in the Revelation archetypal themes of the damsel in distress, a dragon-killing hero, a wicked witch (harlot), the rescued bride, and a wonderful city where everyone lives happily ever after (see Frye 1957, 108, 189). Those themes can be traced through folklore or related to the human psyche through depth psychology. Or they and other themes and structures in the Book of Revelation may be paradigmatic in literary criticism. So when Northrop Frye outlines "a few of the grammatical rudiments of literary expression" (1957, 133), he calls upon the Book of Revelation as "our grammar of apocalyptic imagery" (1957, 141). Thus, the language of the Book of Revelation is linked both to a specific tradition and to paradigmatic elements that resonate more universally in religious, psychological, and literary fields.

But I am interested in how this nesting, mythic language of metaphor and homologue relates to the social order of John's time. The whole of the book is moored in the specific social context of first-century, Asian, urban Christianity by the fact that the writer introduces his visions of "that which is to come" by a series of messages to the churches in the province of Asia. Chapters 1–3 take the form of a series of letters prefaced first by a salutation from John (Rev. 1:4–5) and then from the heavenly Jesus Christ (1:11). Most of the key words and root metaphors used throughout the book are introduced in the letters to the seven churches. Whatever other contexts John's visions may be harnessed to, they clearly address the social situation of the Christians to whom he writes. The visionary, nesting language tells us very little directly about that social situation. The seer does not refer to civic political offices and organizations by name nor to trade associations nor to festivals and celebrations of local deities or of imperial government. Nor does the seer give much direct demographic information about early Christians in those cities. Yet the civic organizations and celebrations, as well as demographic features of his audience, bear directly on how we understand his message to those urban Christians.

As Marjorie Reeves points out, common usage associates "apocalyptic" messages with imminent crisis (1984, 40). A particular portrait of crisis involving early Christians has been painted in both popular and scholarly circles along the following lines. Early Christians are characterized demographically as an oppressed minority who belong to the powerless poor. Christians have no part in non-Christian urban life, for that would require them to go against the faith and practice of Christianity. The civic and imperial machinery in the province of Asia works to persecute those Christians, bringing them into the circuses and games to fight with wild animals and skilled gladiators. According to this portrait, mad Domitian is emperor, a "second Nero," who exiles or kills anyone who does not fall down and worship him as "god and lord." In brief, as a type in the sociology of religion, early Christianity is represented as religion against culture. Practitioners of Christianity cannot participate in urban, imperial life, and those who do cannot be Christians. John proclaims to those Christians the message of a blessed kingdom soon to come that will reverse the social status of both Christians and non-Christians: Christians will no longer live in a world of oppression and scarcity, and non-Christians will be made powerless and brought to judgment by the Christian God.

In Part 3 I examine the social order to which those early Christians belonged: the situation of the empire under Domitian, the local conditions of cities in the province of Asia, and the place of Christians in those cities. That examination yields some surprising results. Robert Grant, Edwin Judge, and Abraham Malherbe, among others, have presented convincing evidence that Christians embraced a wide range of status and place in Roman society. Some were probably slaves and down-and-outers, but many were free artisans and small traders, wealthy enough to travel around the Roman Empire, own slaves, and live in houses large enough to hold congregations of Christians. A typical Christian congregation in one of the seven cities to which John writes probably consisted of people from various classes and statuses, men and women, bond and free, with those of greater affluence serving somewhat as patrons to the others. In that regard they were similar to trade guilds and other private religious associations of the time. Most of the members would for the most part share the attitudes and style of life of their non-Christian neighbors (see R. Grant 1977). That revised demographic description of early Christianity leads to a revision of the popular portrait of crisis sketched above. I present elements of such a revision in part 3.

For analytic purposes I have separated the chapters that focus upon the language of the Book of Revelation (part 2) from those that focus on the social order (part 3). That is an artificial separation. John's writing itself is a social activity that makes up a part of the social order; and other elements of the social order permeate his writing. Since language and society cannot be neatly delineated, neither can literary criticism of a work such as Revelation and social, historical analysis of it. By considering how something is said in the Book of Revelation, a literary critic gains information that has both social and literary dimensions; and by investigating the social location of John and his readers, the social historian illumines the style, form, and genre of the Book of Revelation. The communicative act of writing; the syntactical, semantic, and generic dimensions of the writing; its style and structure; the social institutions that support it; the particular social location of the communication; and the particular historical events that surround it are all part of one dynamic cultural system. That is why we are able to place one style of writing (e.g., a business letter) in a social location different from another style of writing (e.g., a personal, love letter). And that is why inquiry into the life of a person whose correspondence we may possess will illumine the meaning of the letters he or she wrote.

As parts of one dynamic cultural system, the literary religious vision of the seer does not operate in a symbolic universe apart from the world of actual, social relations. Further, as parts of one system, each is dependent on the other, and influence is mutual. Religious visions and literary worlds do not arise and change simply in response to more basic, political and economic realities. Nor are social institutions simply the realization of a system of categories derived, for example, from a religious vision or some other symbolic expression. As a social act, the writing of the Book of Revelation makes an impact on the parties who read it or hear it read, and it affects (in however small a way) the social institutions and historical flow surrounding all parties involved. At the same time, a writer like John is limited and shaped by the social role that he fills (a Christian prophet), by how Christian and non-Christian institutions affect that role, and by the language and genre that he uses for communication. Literary religious visions are neither more ephemeral nor more real than social institutions and political policies. Conversely, emperors and armies are neither more nor less constitutive and shaping than words, visions, and religious beliefs. For example, political power rests "not only in taxes and armies, but also in the perceptions and beliefs of men" (Hopkins 1978, 198). The Book of Revelation is something other than simple response to a social, political situation, than a book of exhortation and comfort to a powerless, persecuted minority huddled together in isolation from urban life around them. To bring in language used earlier—instead of simple lines of cause and effect, we shall look rather to feedback loops and reciprocal interaction to understand the relation of the Book of Revelation to the social order of first-century urban Asia.

Humans require all of these mutually influencing modes and forms—social institutions, different genres of writing, social roles, cosmic visions—as they laboriously adapt to their environment. The seer contributes to that laborious process by bringing to bear a Jewish-influenced Christian mythos upon local, historical situations. His is not the only mythos that can be used to interpret those local conditions—certainly most upwardly mobile provincials looked to a more public mythos incorporated in urban and imperial institutions. His is not even the only Christian mythos that can be used to interpret those local conditions—those whom he calls the followers of Jezebel and Balaam proceeded along other lines. But John's contribution to "the laborious process" is interesting and complex as he unifies the world into an organic whole, stitching earth to heaven, the present age to the coming age, local conditions to suprahuman processes, animals to divinities, and fire to water.

I

Orientation

1

Historical Setting
and Genre

The Local Historical Setting

In part 3 I shall discuss in detail aspects of the local situation in and to which John writes the Book of Revelation. Here we need only some basic orientation to the origins of the book so that it will not appear as a floating specter from the past. As a way of beginning to orient to the book, I shall consider briefly those hoary questions—Where? When? Who? In what situation?

The Place

The writer of the Book of Revelation identifies precisely where both he and his audience reside. He writes from the small island of Patmos, one of the Sporades Islands in the Aegean Sea about thirty-seven miles south and west of Miletus, on the western coast of Asia Minor (roughly, present-day Turkey). He writes to churches in seven major cities in Asia, a Roman province situated along the western coast of Asia Minor.

Asia Minor, specifically the western part of Asia Minor where John and his audience were located, was one of the most significant geographical areas in the development of early Christianity. In the fifties of the first century the apostle Paul carried on missionary work in this area (see 1 Cor. 16:19). Letters and tracts such as Ephesians, Colossians, Philemon, and the Pastoral Epistles indicate the continued influence of Paul in this geographical area. The author of 1 Peter writes to the churches of Asia and other provinces in Asia Minor. New Testament writings associated with John—the Gospel and the three letters as well as the Book of Revelation—also, according to tradition, originated in Asia at the city of Ephesus. At the beginning of the second century Ignatius of Antioch writes letters to five churches in Asia as he travels to die in Rome. Through these writings we know of Christian groups in at least eleven cities of the Asian province by the beginning of the second century CE. Moreover, these churches are associated with major apostolic figures such as Paul, Peter, and John.

The province of Asia was also important in the Roman Empire. To become proconsul of that province was a sign of a successful public career. Asia was rich in natural resources and manufacturing — and therefore in taxes. It was located strategically in the empire with regards to both trade routes and military action on the eastern border. Many of the three hundred or more cities in the province of Asia nurtured cultural activities and became centers for libraries, museums, and spectacular monuments. In brief, the locale of the Book of Revelation is significant in both Christian and Roman history.

The Author

According to the Christian apologist Justin Martyr (d. 165 CE), the apostle John wrote the Book of Revelation (*Dia. Tryph.* 81.4). A few years after Justin, Irenaeus, bishop of Lyons, writes that the apostle John, son of Zebedee, wrote both the Book of Revelation and the Gospel of John (*Haer.* 5.30). Most of the church fathers, though not all, follow Justin and Irenaeus (see Kümmel 1975, 469–72). Dionysius of Alexandria (third century) states, however, that "some indeed of those before our time rejected and altogether impugned the book, examining it chapter by chapter and declaring it to be unintelligible and illogical, and its title false. For they say that it is not John's, no, nor yet an apocalypse [unveiling], since it is veiled by its great thick curtain of unintelligibility" (Eus. *Hist. Eccl.* 7.25). Dionysius is more temperate towards Revelation than these unnamed impugners, but he concludes for stylistic and linguistic reasons that the son of Zebedee could not have written the book. Rather, it was written by someone else named John who was buried at Ephesus. Thus, some of the church fathers who assumed apostolic authorship of the Gospel of John did not give the same status to the writer of Revelation. Modern scholarship tends to side with them. The style, vocabulary, and theology of the Apocalypse are sufficiently different from the Gospel of John as to make one conclude that common authorship is unlikely and that the apostle did not write the Apocalypse (see Schüssler Fiorenza 1985, 85–113).

The personal identity of John will probably never be discovered. The name was common in the early church; perhaps he was one of those by that name affiliated with the Christian community at Ephesus. More can be said about John's social identity. He is probably an early Christian prophet who wandered either randomly or by prescribed circuit among the churches of Asia. From other sources we know of such an "office" in the early church. Paul places it high on his list of church offices, second only to "apostles" (1 Cor. 12:28). Elsewhere it is associated with the office of teaching (see Acts 13:1). The *Didache*, or *Teaching of the Twelve Apostles*, an early second-century "handbook" for Christians, understands prophets to be itinerants who go from one congregation to another (*Did.* 11–13). John claims authority as a leader in the churches and comes into conflict with other "prophetic authorities" (see Jezebel, Balaam, Nicolaitans in Rev. 2–3). Still debated are questions about the organizational elaboration of that office: for example, was he a "head prophet" among a school of prophets? Did he head a conventicle splinter group of Christians in each of these cities? Did he deliver his apocalypse to a community of prophets in the churches rather than to all Christians? However one answers those questions, the author of the Book of Revelation was an early Christian prophet who proclaimed a message of revealed knowledge to the seven churches in the province of Asia (see Yarbro Collins 1984, 34–50).

The Date

The author of the Book of Revelation does not give much of a clue about the particular time in which he is writing. In contrast to several references to specific places, the seer's references to the historical situation are either nonexistent or so veiled as to give no certain information to the reader today. Chapters 13, 17, and 18 in the Book of Revelation refer to emperors and to the city of Rome, so that we may be certain that the book was written in the time of the empire; but even Revelation 17, which elaborates on the seven-headed beast (Rev. 13:1) by specific reference to emperors past, present, and future, gives no certain information about the precise time of the writing.

From chapter 17 we can narrow the time down somewhat. The great harlot, referred to earlier in 14:8, is judged. Upon her forehead is written "a name of mystery: 'Babylon the great, mother of harlots and of earth's abominations'" (17:5). She is seated upon a scarlet beast "which was full of blasphemous names, and it had seven heads and ten horns" (17:3). Later the seven heads are identified as seven hills (17:9), and the city that sits upon the seven hills "has dominion over the kings of the earth" (17:18). These two characteristics of the city/woman—power over the earth and "sitting" upon seven hills—identifies her clearly as Rome, the capital of the empire (see Caird 1966, 216). In other words, Rome and all those under her will be destroyed in the pouring out of the seventh bowl (Rev. 16:17–21, cf. 17:15–18).

In connection with Rome's destruction there is an allusion to the "return" of one of the emperors. There are three versions of this return: (1) "The beast that you saw was, and is not, and is to ascend from the bottomless pit and go to perdition" (17:8); (2) "The dwellers on earth . . . will marvel to behold the beast, because it was and is not and is to come" (17:8); and (3) "As for the beast that was and is not, it is an eighth [king] but it belongs to the seven [kings], and it goes to perdition" (17:11). This "coming again" of one of the kings alludes to the Emperor Nero, around whom developed, after his death (or flight), an expectation that he would come again from the East and fight against some or all of the Roman Empire. In Jewish and Christian literature this "revived Nero" is sometimes portrayed as both anti-Roman and an opponent of the chosen people. For example, in the fourth Sibylline Oracle Nero is referred to as follows: "Then the strife of war being aroused will come to the west, and the fugitive from Rome will also come, brandishing a great spear, having crossed the Euphrates with many myriads" (*Sib. Or.* 4.137–39). In the fifth Sibylline Oracle Nero will be destructive "even when he disappears": "Then he will return declaring himself equal to God. But he will prove that he is not" (*Sib. Or.* 5.33–34, cf. 5.93–110). In the Book of Revelation, the Nero legend is associated with the beast from the abyss and with the "eighth" king who is at the same time "one of the seven." He is one of the evil end-time figures who will make war against the Lamb and his followers (17:14). Given the presence of this legend, the Book of Revelation could not have been written in its present form before 68 CE when Nero died, but the legend could have spread quickly after Nero's death.

The identification of Rome with Babylon also provides some evidence for dating the Book of Revelation. In Jewish literature, the enemy Rome is designated Edom, Kittim, and Egypt, as well as Babylon. For the most part, however, the identity with Babylon occurs after 70 CE, that is, Rome is called Babylon after she destroys Jerusalem and the temple. Yarbro Collins thus concludes, "It is highly unlikely that the name would have been used before the destruction of the temple by Titus. This internal element then points decisively to a date after 70 C.E." (1981, 382).

More evidence for dating Revelation seems to be given in the reference to the seven heads of the beast as seven kings (emperors) (17:9–14). Of those seven kings, "five of whom have fallen, one is, the other has not yet come and when he comes he must remain only a little while. As for the beast that was and is not, it is an eighth but it belongs to the seven, and it goes to perdition." One needs simply to figure out which five emperors have already fallen, and then the sixth emperor is reigning during the time that John writes. The earliest possible of the five past rulers would be Julius Caesar who died in 44 BCE. The complete list following Julius Caesar would then be the five emperors of the Julio-Claudian dynasty, the three emperors during the confusion after Nero's death, the three Flavian emperors, and then, if relevant, Nerva and Trajan. Their reigns occurred as follows:

Julius (d. 44 BCE)
Julio-Claudian dynasty (27 BCE–68 CE)
   Augustus (27 BCE–14 CE)
   Tiberius (14–37)
   Gaius (Caligula) (37–41)
   Claudius (41–54)
   Nero (54–68)
Three short-lived emperors
   Galba (68–69)
   Otho (69)
   Vitellius (69)
The Flavians (69–96)
   Vespasian (69–79)
   Titus (79–81)
   Domitian (81–96)
Nerva (96–98)
Trajan (98–117)

The puzzle is twofold: Where should one begin the count, and which emperors should be included in the count? Rowland argues that the simplest solution begins with Augustus and counts each emperor to the sixth, Galba. He thus supports the dating of Revelation around 68 CE, during the upheaval that came between the death of Nero and the accession of Vespasian: "The great uncertainty which was felt throughout the empire during AD 68 could hardly have failed to stir up the hopes of Jews and Christians that their deliverance was nigh" (1982, 406). Rowland here follows the lead of Bishop Lightfoot, B. F. Westcott, F. J. A. Hort, and more recently John A. T. Robinson and Albert Bell (see Rowland 1982, 403), all of whom argued for the chaotic state of the empire after Nero's death as the setting in which Revelation was written.

John Court, on the other hand, noting that the Antichrist tradition is clearly applied to Rome in this passage and not to Jerusalem, concludes that the fall of Jerusalem (70 CE) must have occurred "in the more distant past" and therefore that the present king must be considerably later than Galba (1979, 125). Rome is the Antichrist because of the conflicting allegiance created among Christians by emperor worship (p. 126). Court then begins the count from Nero, the first emperor to be an Antichrist figure, and concludes that the sixth king who presently reigns is Titus (p. 135). Later the author of Revelation adapts his writings to the time of Domitian, when the pretentions of the imperial cult become ever more extravagant and blasphemous (pp. 137–38).

Yarbro Collins concludes that if the kings are to be considered inclusively, the list must begin with Julius Caesar (see *Sib. Or.* 5.12–51, 4 Ezra 11–12); Nero would thus be the sixth, contemporary emperor, which would be an impossibility since the legend of the return of Nero after his death is presupposed in the king list (Rev. 17:11) (1984, 58–64). Galba, the sixth if one begins with Augustus, is also an unlikely candidate, because he reigned prior to the fall of Jerusalem; and the destruction of Jerusalem by the Romans is a necessary prerequisite for identifying Rome and Babylon. She concludes that probably the author by some principle selected certain emperors from the list beginning with Gaius, who made such a negative impact upon Jewish writers of his time. Omitting the three short reigns of 69 CE, the sixth and present king becomes Domitian (Yarbro Collins 1984, 64).

Revelation 17, thus, does not give conclusive evidence for the date of the book. The identification of Rome with Babylon and the reference to Nero as returning from the dead argue for a post-70 date; the list of kings does not justify any precision beyond that.

The most compelling evidence for dating the book more precisely after 70 CE remains the reference by Irenaeus, who came from Asia Minor and knew Polycarp, bishop of Smyrna (d. c. 155 CE). He states that the visions of Revelation were seen "not long ago" but "close to our generation, towards the end of the reign of Domitian" (Iren. *Haer.* 5.30.3 = Eus. *Hist. Eccl.* 3.18.1).[1] As we have seen, a date in Domitian's reign is also compatible with the kings' list in chapter 17. Some scholars still argue for dating the book shortly after Nero's death when several people were vying to be emperor; but when the weight of internal and external evidence is taken together, we may conclude with most scholars that Revelation was written sometime in the latter years of Domitian's reign, that is, 92–96 CE.

Crisis in the Reign of Domitian

Eusebius of Caesarea, the fourth-century Christian historian, laid the groundwork in Christian history for viewing Domitian's reign as a time of persecution and crisis.[2] He says, in a section devoted to the Emperor Domitian, that "many were the victims of Domitian's appalling cruelty." He refers to distinguished Romans and other eminent men who were executed without trial, banished from the country, or had property confiscated. Then he states that the apostle and evangelist John "was still alive [in Domitian's reign], and because of his testimony to the word of God was sentenced to confinement on the island of Patmos." Under Nerva, Domitian's successor, John was allowed to return from exile on Patmos to his residence at Ephesus. Eusebius notes that even non-Christian historians record the persecutions and martyrdoms that Christians such as Flavia Domitilla suffered under Domitian (Eus. *Hist. Eccl.* 3.17–20).[3]

A more critical reading of Eusebius raises doubts about widespread persecution of Christians under Domitian. So Leon Hardy Canfield concludes, after reviewing carefully both Christian and non-Christian sources, that no great persecution occurred under Domitian and if the Apocalypse "does refer to conditions in Asia Minor under Domitian it is the only source for such a persecution" (1913, 162).[4] Recent commentators on the Apocalypse of John support Canfield's conclusion. J. P. M. Sweet, for example, writes, "The letters to the churches [in the Apocalypse] suggest that persecution was occasional and selective, and that the chief dangers were complacency and compromise" (1979, 26).

Although most modern commentators no longer accept a Domitianic persecution of Christians, they do assume that Domitian's increased demands for worship and the "reign of terror" in the years immediately preceding Domitian's death created a critical situation for Christians in Asia Minor. Adolf Harnack, in a discussion of the developing political consciousness of the early church, writes, "The politics of Jewish apocalyptic viewed the world-state as a diabolic state, and consequently took up a purely negative attitude towards it. This political view is put uncompromisingly in the apocalypse of John, where it was justified by the Neronic persecution, the imperial claim for worship, and the Domitianic reign of terror" (1961, 257). Johannes Weiss, another classical church historian from the modern period, notes that not many actual deaths had occurred when the Apocalypse was written but that Domitian's intensified demands for worship, perhaps not by imperial decree but by the importance that he placed on being called "lord" and "god," created a crisis for Christians. Christians in Asia Minor experienced this crisis especially as they no longer were able to claim the special status and privileges (e.g., exemption from emperor worship) given to Jews. The unpopularity of Christians among the local provincials combined with Domitian's religious demands and his general cruelty to threaten the Christian communities (1959, 806–10). More recently, W. H. C. Frend notes that Domitian's increased demands to be worshipped resulted in "intensified apocalyptic fervour among the Christians in the province [of Asia]" (1981, 194). Schüssler Fiorenza underscores how the imperial cult was promoted under Domitian and how he demanded that "the populace acclaim him as 'Lord and God' and participate in his worship"; living as they were in cities that promoted imperial worship, "Christians were bound to experience increasing conflicts with the Roman civil religion since they acclaimed Jesus Christ and not the emperor as their 'Lord and God'" (1981, 62). She and several other commentators suggest that the reference to "Lord and God" in Revelation 4:11 deliberately reflects "political language of the day" (1981, 76).

On the surface there are good reasons for assuming that Domitian did give greater prominence to imperial worship.[5] According to Roman, as well as early Christian, sources Domitian demanded divine worship during his lifetime, most especially at the end of his reign, and generally strengthened the imperial cult, which included the worship of both Roma and the emperor. Pliny the Younger and Tacitus condemn Domitian's evil claims to divinity and tyranny, and Pliny's younger friend Suetonius makes now-famous statements about Domitian's inordinate claims to titles such as "our Lord and God" (*dominus et deus noster*). Dio Cassius, writing about a century later (in the second decade of the third century), repeats and enhances descriptions of Domitian's evil character.

Since Roman historians characterize especially the latter part of Domitian's reign as a reign of terror by a tyrant and megalomaniac who claimed and demanded imperial worship from his subjects, Domitian's reign provides a plausible social, political setting of the Book of Revelation; for one of the major themes in the Apocalypse is unquestionably the conflict between imperial Rome with its divine claims and the rule of the Christian God. John states that he shares his readers' affliction and he perseveres (Rev. 1:9). He is on the island of Patmos "because of the word of God and the witness of Jesus." Antipas has been martyred at Pergamum (Rev. 2:13). In chapter 13 many commentators identify the beast from the sea with the annual docking of a boat carrying the emperor's representative to the province of Asia and the beast from the earth with the provincial cult responsible for "promoting the imperial cult in Asia Minor" (Rowland 1982, 431–32).[6] Later, in chapters 17–18, images center on the city rather than the emperors of Rome, as her economic and commercial power, so destructive to the church, is overcome in a series of eschatological disasters (e.g., Rowland 1982, 433–34). Schüssler Fiorenza thus concludes that "the major part of the work describes in mythological-symbolic language the threat of the Roman political and religious powers" (1981, 31).[7]

Some commentators also call attention to an economic dimension to the critical times under Domitian. Revelation 13:17 refers specifically to the necessity of using coins with the emperor's image in order to enter into economic transactions. Economic transactions could thus be seen as an arm of the imperial cult and a form of oppression (see Schüssler Fiorenza 1981, 173). More specifically, Court links the prices of wheat and barley (Rev. 6:6) with the periodic famines that raged through Asia Minor in this period. The reference to wine may even allude to the opposition expressed by the people of Asia Minor to an edict of Domitian in 92 CE to cut back on the number of vineyards in Asia. As a result, Domitian revoked his edict and allowed the vines to be unharmed (Court 1979, 59–60).[8] Whether that passage can be linked so precisely to the conditions in Asia during the years 92–93 CE or not, others, borrowing from Rostovtzeff, have noted more generally that the prosperity of Asia during the Flavian period brought conflict between the rich and the poor and between Roman governors and the populus (e.g., Yarbro Collins 1983, 744–46).

The Book of Revelation and its Genre

If the Book of Revelation were a distinctive or peculiar work without comparison in form or content, the task of understanding it would involve reading carefully what is said and considering its contents in the context of its historical setting. The Book of Revelation, however, shares a style of writing and a set of motifs with other works from roughly the same historical period; that is, the seer of the New Testament participates in a mystical tradition — a convention of images, themes, styles, and literary forms — that shapes in part his psychological experiences, social perceptions, religious insights, and literary expressions. In literary terms the Book of Revelation belongs to a genre, and an understanding of that work requires an understanding of the genre.

Modern scholars, taking a cue from the Apocalypse of John (in Greek, *Apocalypsis Ioannou*), refer to works written in this tradition as belonging to the genre "apocalypse": a work may be called an apocalypse if it resembles the Revelation of John, that is — in the words of Klaus Koch — if it presents "secret divine disclosures about the end of the world and the heavenly state" (1972, 18). The designation of Jewish and Christian works as apocalypses began in the early church (see M. Smith 1983, 19); but the attempt at literary classification is a modern one, and much debate accompanies any definition of the genre "apocalypse" as well as the notion of genre itself.[9]

The muddle of scholarly debate should not, however, obscure the importance of genre for understanding a specific writing. Recall that one could read and reread as carefully as possible the terms of endearment in one single business letter ("Dear so and so," "Yours truly") and not understand these elements as distinctive to the genre. Moreover, readers cannot recognize something in a specific writing as *generic* (common to the *genre*) unless they have read other examples in the same genre; that is, elements of a genre are discovered through comparing several examples.

In the Christian Bible there are not many examples of apocalypses. Besides the Book of Revelation, there is only Daniel in the Old Testament.[10] Other examples are to be found outside the Christian canon in Jewish writings associated with the names of Enoch, Ezra, and Baruch. Two Christian books written a little later than the Book of Revelation — the Apocalypse of Peter and the Shepherd of Hermas — were recognized by some members of the early church as having special sanctity, with the former included by some circles as part of the New Testament. Among these Jewish and Christian apocalypses, the Book of Revelation is neither first nor last; it has both predecessors and successors.

Characteristics of the Genre

In isolating characteristics of the genre "apocalypse," scholars focus upon the content and style of a work. Philipp Vielhauer (1965), for example, regards the following as fixed, formal elements in this literary genre: (1) the author writes under a pseudonym, a great name in the tradition, such as Ezra, Enoch, or Isaiah; (2) the writing is presented as an account of a vision — a dream, an ecstatic state, or a heavenly rapture; (3) a portion of past history is presented as though it were in the future; (4) farewell discourses, exhortations, prayers, and hymns may be found. According to Vielhauer apocalypses also contain fixed content: (1) There are two ages, the present age and the age to come, which are qualitatively different; (2) the present age is devalued and viewed with pessimism as under the control of Satan, while the age to come is correspondingly glorified as a wonderful time; (3) apocalypses consider the whole world and all peoples, not just Jews, so that everyone is considered as an individual (not simply as a member of a community) — to be resurrected and judged as an individual; (4) God has foreordained everything according to fixed plan, including the activity that brings the imminent end (*eschaton*).[11]

Is Vielhauer's list of fixed forms and essential content adequate for defining the genre? Defining an apocalypse through lists runs into the predictable problem of what to include and what to exclude as basic. Vielhauer, for example, knows that among the things revealed in apocalypses are secrets about heaven, hell, astronomy, meteorology, geography, and the origin of sin and evil; but he concludes that their main interest "does not lie in problems of cosmology or theodicy, but in eschatology" (1965, 587). Michael Stone, on the other hand, argues that speculative interests in such matters as cosmology, astronomy, and the calendar reflect one of the "core elements" of apocalypses (1980, 42, 113–14). As a way of including both eschatological and cosmological speculations, Christopher Rowland argues that disclosure of knowledge through direct revelation is the fundamental characteristic of apocalypses (1982, 21, 357; see also Stone 1980, 29). Koch enumerates a somewhat different list of form and content that an apocalypse must include (Koch 1972, 24–27). Joshua Bloch (1952) and H. H. Rowley (1980) also suggested their own distinctive list of characteristics.[12]

Representative Jewish Apocalypses

In order to get a better sense of how various elements recur in specific apocalypses and how the Book of Revelation is similar to other apocalypses, let us look briefly at some elements of form and content found in three representative Jewish apocalypses: 1 Enoch, Daniel, and 4 Ezra.

The earliest known example of apocalyptic literature has come down to us under the pseudonym of Enoch, who according to Genesis lived in the fifth generation after Adam, prior to Noah and the flood. Of Enoch it is simply stated that he "walked with God, and he was no longer here, for God took him" (Gen. 5:24). Enoch's piety along with his enigmatic ending made him an apt figure for apocalyptic speculation within Judaism and Christianity. This early apocalypse called 1 Enoch or Ethiopic Enoch (the only complete version of the work has come down in the Ethiopic language because of the Ethiopian church's interest in it) is now generally viewed as a composite work of five separate sources written at different times: (1) Book of the Watchers (chaps. 1–36), pre-Maccabean, perhaps late third century BCE; (2) Similitudes of Enoch (chaps. 37–71), mid–first century CE; (3) Book of Heavenly Luminaries (chaps. 72–82), also pre-Maccabean, perhaps the earliest of the sources; (4) Book of Dreams (chaps. 83–90), early Maccabean, circa 165–161 BCE; and (5) Epistle of Enoch including Apocalypse of Weeks (chaps. 91–108), late Hasmonean, circa 105 BCE (but see Nickelsburg 1981, 149–50).[13]

The Book of the Heavenly Luminaries contains some of the earliest material in the Enoch collection and reflects the speculative interests of apocalyptic underscored by Michael Stone. Enoch gives in detail cosmological secrets regarding the movements of the sun and the moon, the twelve winds, the four directions (East, South, West, North), the seven mountains, the seven rivers, and the astronomical laws that establish a solar year of 364 days—a calendrical point of some importance to the author. In this section, Enoch also makes the point that the world as we know it will come to an end and that a better world will replace it—a "new creation which abides forever." Disorder and confusion will occur before the new creation: the moon will alter its course, and chiefs of the stars will make errors in the orders given to them as evil things multiply and plagues increase. The work concludes with other revelations and visions (typical forms in apocalypses) given to Methuselah, Enoch's son.

Chapters 83–90 contain two Dream Visions. One is a brief vision of cosmic destruction in which the sky falls upon earth and earth is swallowed up in the great abyss. Grandfather Mahalalel makes a telling point: "all the things upon the earth shall take place from heaven," that is, the earthly has archetypes in heaven. A second vision presents a portion of "past" biblical history as though it were in the future; this narrative takes the form of an animal allegory (the so-called Animal Apocalypse) in which Adam, Seth, and his descendants Noah, Abraham, and Isaac are all presented as white bulls. Before the great flood, fallen stars come down and pasture with the cows (see Book of the Watchers); their mixed offspring are pictured as elephants, camels, and donkeys. Later the fallen stars are punished by being cast into an abyss "narrow and deep, empty and dark." With Isaac's son, Jacob, the animal symbolism shifts to sheep; by means of fairly transparent symbols, the biblical story of Israel is told up through the restoration after the Babylonian Exile. Then new animals—eagles, vultures, kites, and ravens (the Greeks and their kingdoms)—oppress the sheep until a great horn sprouts on one of the sheep (Judas Maccabeus). With God's help he successfully fights against the vultures, kites, and ravens. Then the Lord smites the earth, and gives a great sword to the sheep to kill all the beasts and birds. Eschatological judgment follows, during which "sealed books" are opened "in the presence of the Lord of the sheep." A new temple is set up; all peoples come and worship the sheep, "making petition to them and obeying them in every respect." Finally bovine symbolism returns with the birth of a snow-white cow with huge horns; all are transformed into snow-white cows, so that the eschatological finale returns to the Adamic vision. In Enoch's panoramic historical review of the world from Eden to the new Jerusalem, the end time becomes the time of beginning. Then Enoch awakes from his vision.

Elsewhere in 1 Enoch there are other themes common to apocalypses including the Book of Revelation. In the Similitudes of Enoch there are references to judgment, the punishment of the wicked, and the dwelling of the righteous in the presence of the Lord and his angels. Special mention is made in the second parable or similitude (chaps. 45–57) to the judging of kings, oppressors, and the economically powerful; and, in contrast, the prayers of righteous ones ascend into heaven on behalf of the blood of the righteous that has been shed. As Enoch ranges through time as well as space, he sees the Son of Man given a name before the creation, the resurrection of the dead in the latter days, and the final judgment. In the third parable Enoch sees the divine throne; the separation of Leviathan and Behemoth, who will become food on the great day of the Lord; various cosmological secrets; angelic measurings to strengthen the righteous; and the final judgment by the Lord and the Son of Man, which will bring reversals of status between rulers and the righteous. In the Epistle of Enoch (chaps. 91–108) Enoch tells Methuselah and all his brothers "everything that shall happen to you forever." After a kind of Jobian soliloquy on who can ponder the thoughts of God words of comfort, exhortation, and warning bring the work to a close.

In Jewish Scriptures Daniel is the only full-blown apocalyptic work.[14] Chapters 7–12 — if not the whole of Daniel — arose in the situation of political conflict between the Jews and Antiochus Epiphanes after 187 BCE, approximately the same time as the Animal Apocalypse in the Dream Visions of Enoch. Daniel 2 tells the story of Nebuchadnezzar's dream and Daniel's interpretation of it. Daniel is able to know the dream and its interpretation through divine revelation, for such knowledge is a mystery known only through revelatory visions. Through the symbolism of a bright and mighty animal, Daniel "foretells" the sequence of four kingdoms, with Nebuchadnezzer's kingdom as its golden head and the Successors of Alexander the Great making up its toes. In the days of the latter, God will break in pieces the kingdoms and will establish his everlasting reign. As with similar narratives in 1 Enoch, the seer is able to see the future, which is fixed and known by God. In Daniel 7 the seer sees in a night vision "four great beasts" coming out of the sea: one like a lion, a second like a bear, then one like a leopard, and a fourth beast "terrible and dreadful" with ten horns. After a little horn appears on the fourth beast, Daniel sees a throne scene with the Ancient of Days who destroys the beasts. Then he sees one "like a son of man" given dominion, glory, and an eternal kingdom by the Ancient of Days. A heavenly figure interprets the vision for Daniel: four kings shall arise out of the earth, but the saints of the Most High shall receive the kingdom and possess it for ever. Detailed explanation is given of the fourth beast and especially of the eleventh horn on that beast. Later chapters (Dan. 8–12) parallel chapter 7 (see Thompson 1978, 211). In the recitation of the kingdoms in Daniel 10–12 grandfather Mahalalel's point is made once again: things on earth take place from heaven — for all earthly events there are heavenly archetypes. Daniel's visions give eschatological assurance to those who are wise and know their God, for the righteous who die have hope at the time of the end.

Around the end of the first century CE, a few decades after the fall of Jerusalem to Rome and around the same time as the Apocalypse of John, several Jewish apocalypses may be dated: 4 Ezra, 2 Baruch, 3 Baruch, the Apocalypse of Abraham, and 2 Enoch (see Collins 1984a, 155–86). Among them 4 Ezra will serve as our final representation of Jewish apocalyptic.[15] Ezra's apocalypse divides into seven sections — first three dialogues with an angel over problems of divine justice, then four visions of the end. In mood, the visions move from despair to hope, from imponderable questions to resolution and consolation (Nickelsburg 1981, 294).

In the first section Ezra is troubled over the problem of the "evil heart" that infects all descendants of Adam. Given the universality of sin, what nation has kept God's commandments? More pointedly, why should Zion be in desolation? Are

Babylonians less sinners than the Jews? The angel Uriel, Ezra's interlocutor, answers in the tradition of Job by asking Ezra to explain certain cosmic phenomena such as the measure of the wind. Moreover, everything in the cosmos has an assigned place (the sea, the plain, sand), and those assigned to earth cannot understand the things of heaven. True wisdom requires revelation. Ezra reiterates a part of his question: he does not ask about heavenly things but about daily life that all Israel experiences, her plight in the world. Uriel then gives an eschatological solution: this age cannot deliver what has been promised to the righteous; but it is hastening swiftly to its end according to a time table determined by God. Certain signs of the end can be seen — chaos, prodigies, the reversal of nature's order (salt waters become sweet; menstruous women bring forth monsters; beasts roam beyond their haunts, i.e., out of bounds), and unrighteousness — but the end will come when it is destined. Evil has been sown, but its harvest has not yet come nor has the harvest of the righteous. The age has been carefully weighted, measured, and numbered; and God will not move until the measure is fulfilled.

The second section follows much the same structure, but this time the dialogue centers more specifically on the plight of Zion. Section 3 also resolves several hard questions about human existence by pointing to the *eschaton* (end-time). In section 4 the dialogue form is interrupted by a vision of a disconsolate women whose son died at his wedding. While she and Ezra speak, she is transformed into an established city, Zion itself. Her story then becomes an allegory for Zion. Section 5 relates a dream vision that is also a reinterpretation of the fourth kingdom in Daniel 7. The fourth beast, here an eagle with sundry wings and heads, signifies the Roman Empire (not, as in Daniel, Alexander the Great and his Successors). A Lion (the Messiah) reproves the eagle and brings its kingdom to an end; this Messiah, from the posterity of David, will himself judge and destroy the evil ones and deliver in mercy the righteous remnant. As an eschatological agent, this Lion of the tribe of Judah parallels the Son of Man in Daniel and the Elect One in the parables of Enoch. The sixth section elaborates on the eschatological agent who is called "Man," presumably drawing again on Daniel 7. The Man comes out of the sea — no one can know what is in the depths of the sea — and flies with the clouds of heaven, finally settling on a high mountain that he carves out for himself. From the mountain (Zion) he will judge and destroy the ungodly and deliver and protect those who have works and faith in the Almighty. The seventh, final section points out similarities between Ezra and Moses. God calls Ezra out of a bush, draws specific parallels to Moses, informs Ezra that $9^1/_2$ of the 12 parts of this age have passed, and then dictates twenty-four books (Hebrew Scriptures) to be made public and seventy to be kept secret until the last day. Ezra deserves this honor because he has devoted his life to wisdom, to studying the law, and to understanding. Likewise, the books contain the spring of understanding, the fountain of wisdom, and the river of knowledge; they will be received by the wise among the people.

Toward a Definition of the Genre

From this sampling in books that everyone associates with the genre "apocalypse," one can see both a variety and a recurrence of elements. Millenarians will recognize the prominence of the end-times, eschatological speculation about the signs and how to interpret them, and descriptions of the end itself. In all of these works knowledge of the end comes through special revelation in the form of visions or heavenly journeys. There is also, however, knowledge revealed about the present world, knowledge in the fields of geography, climatology, meteorology, astronomy, and angelology. Some recite in veiled terms the course of world history, especially in connection with the history of the chosen people. Eschatological speculation is often linked to problems of theodicy — Why do the just suffer? Why doesn't goodness, especially of the chosen people, receive reward?

Must a specific work contain all of those themes to be classified as an apocalypse? Are certain themes more central than others? Are stylistic features also important in classifying a work? Should one also consider its place and function in the life of the people who create it and read it? The problem of defining a genre remains no simple matter, yet our view of the Book of Revelation changes when we recognize that it is part of a generic tradition rather than an idiosyncratic work. Understanding includes proper classification. In order to be more systematic in considering these issues of genre, scholars have generally agreed to analyze separately three aspects of the genre "apocalypse" and to refer to each of those aspects by a different term: *apocalypse, apocalyptic eschatology*, or *apocalypticism*.[16]

Apocalypse refers to a set of writings, a literature, that includes such works as Daniel, 1 Enoch, 4 Ezra, and the Book of Revelation. One asks, What do books or portions of books called apocalypses have in common? What stylistic and linguistic elements distinguish them from other writings?

Apocalyptic eschatology refers to a "religious perspective" (Hanson 1976, 29) or an "attitude of mind" (Koch 1972, 33) that involves certain beliefs about the world and the place of humans in it (see also Schmithals 1975, 10–11, 73). Among the possible ways of conceiving the human situation, how does apocalyptic eschatology describe the place of humans in the world? Among all possible states of the world, which states or which set of states defines the assertions, beliefs, and propositions of an apocalyptic perspective? Most scholars identify the radical transcendence of God as a key element in apocalyptic eschatology (e.g., Hanson 1979, 432). The religious perspective of apocalyptic eschatology reflects transcendence in at least two ways: knowledge of the perspective comes through means that transcend normal human experience (visions, world journeys); and God's activity in saving the world transcends history, that is, God's activity breaks in upon historical realities and human endeavors rather than working through them. Walter Schmithals asserts an extreme form of transcendence when he states that in apocalyptic eschatology, history "is made thoroughly secular, profane. *What* happened in history has no significance theologically. . . . Apocalyptic pessimism toward history expels God from history. . . . The devil becomes the lord of this eon" (Schmithals 1975, 81).[17] Most scholars would want to temper such a statement so that transcendence does not oppose history and ordinary human experience, but expands them. Apocalyptic eschatology would then open the world and human activity to a larger perspective (see Rowland 1982, 29, 92, 175, 475; Koch 1972, 31).

Apocalypticism refers to social aspects of apocalypses and transcendent eschatology: Do apocalypses or transcendent eschatology arise only in certain kinds of social situations (e.g., times of trials and difficulties)? Does the literature or the perspective serve distinctive social functions among those people sharing in it (e.g., to sustain their faith, to give comfort)? Is it possible to speak of an "apocalyptic movement" that can be located in a specific time and place within a specific group of people clearly distinguished from other groups? Scholars tend to describe the social situation and social group connected with apocalypticism as alienated from the socioreligious structures of the society around it and as participating in "an alternative universe of meaning," constructed from the perspective of apocalyptic eschatology, that denies ultimate significance to the social structures of this world (Hanson 1979, 433–34; 1976, 30).

Scholars differ somewhat in how they define or describe each of those three dimensions and in how they relate the three. The community, the perspective, and the literature may be seen as inevitably locked together. Yet it is at least logically possible for a particular writing or a particular group of people to embody the religious perspective of apocalyptic eschatology without sharing, respectively, in the literary form of apocalypse or the social features of apocalypticism. Perhaps it is possible for a work to have the form of apocalypse without the content of apocalyptic eschatology. In any case, it is possible to inquire separately into any one of these three dimensions.

This naming of subdivisions is symptomatic of the present state of biblical studies, in which literary, religious, and social-historical inquiries tend to be pursued separately. Even among those who hold that genre should include linguistic, religious, and social aspects, each subfield is seen as so complex as to require at least a temporary separation of form and content from social situation (see Collins 1979, 3–5). As a result genre studies tend to bracket out apocalypticism (social situation), which is then pursued as a separate issue (Sanders 1983, 450–51; Koch 1972, 21). Thus, the standard definition of the genre "apocalypse," developed by John J. Collins and other members of the Society of Biblical Literature (SBL) in the Apocalypse Group of the SBL Genres Project, refers only to elements of form and content; those who formulated this definition deliberately left out the issue of apocalypticism. They define apocalypse as "a genre of revelatory literature with a narrative framework, in which a revelation is mediated by an otherworldly being to a human recipient, disclosing a transcendent reality which is both temporal, insofar as it envisages eschatological salvation, and spatial insofar as it involves another, supernatural world" (Collins 1979, 9).

A fundamental issue left open in this definition revolves around the relationship between the language-and-religious perspective of apocalypses, on the one hand, and their location in the social order, on the other. Perhaps this relationship cannot be resolved on the level of genre; perhaps the location in the social-historical order will vary from one apocalypse to another. Nonetheless, there are some social issues common to the genre, which we shall take up in the next chapter.

2

The Social Setting
of Apocalypses

With form and content dominating questions of genre, scholars have taken up the social setting of apocalypses as a separate issue. There are several good reasons for that separation. Classic form criticism – a method for relating literature and society – probably linked too closely a particular form with one particular setting; and as Collins, among others, notes, a single apocalypse may be used in different ways in different social settings (Collins 1979, 4). Lack of knowledge about the social-historical setting of specific apocalypses also contributes to the separation, for little is known about when, where, by whom, and for whom most apocalypses were written (see Stone 1980, 72–73, 85). Information about social setting comes for the most part from the apocalyptic texts themselves. Scholars use a text as "a window into the author's *world*," and, as Nickelsburg has observed, in doing so they "see through a glass darkly" (1983, 641). A conventicle setting may have been the provenance for some apocalypses, but certainly not for all (see Schmithals 1975, 46; Collins 1984b, 20–21). Apocalypses were circulated throughout Palestine in various schools (Bloch 1952, 53; Stone 1980, 69), and they were well received among Diaspora Jews outside Palestine, especially those in hellenistic communities (Bloch 1952, 38, 128). The view of the end in apocalypses is not sufficiently different from rabbinic and other eschatologies to warrant the notion that apocalypses arose outside the mainstream of Jewish life (see Bloch 1952, 133).[1] Apocalypses were used in the sectarian context of Qumran, but they were also "scattered in the ranks of all parties of their day" (p. 136). The apocalyptic sections of Daniel arose in the Maccabean uprising, but it is unlikely that 1 Enoch belongs to the same circles or even to a time of political upheaval. The apparent variation in social situation among apocalypses makes it difficult to talk about a genre-specific social, historical setting.

Apocalypse As a Literature of Crisis

Although apocalypses do not arise in just one "setting in life," there is widespread agreement that an apocalypse arises within a particular kind of situation, namely a situation of crisis. H. H. Rowley comments about the author of Daniel, which

Rowley considers to be the first, great apocalypse. "It is fortitude under persecution that he encourages . . . and in this he is the forerunner of other apocalyptists" (1980, 22, 52). Although more recent analysts reject persecution as the setting of all apocalypses, they do continue to link apocalypses to a setting of crisis. So Hanson refers to the "harsh realities" of the Jews in the period between the Babylonian Exile (587 BCE) and the Maccabean era (c. 170 BCE): "With nationhood lost, prophetic and priestly offices taken away, and social and religious institutions controlled by adversaries, world-weary visionaries began to recognize in a mythologized version of eschatology a more promising way of keeping alive a hope for final vindication" (1979, 432). Wilson also suggests that the "chaos of the postexilic period set the stage for the formation of various types of apocalyptic groups" (1982, 87). Rapid social change, especially if cross-cultural contact occurs (such as in the postexilic period of Israel or the rise of early Christianity), is thought to exacerbate disorder, disorganization, conflict, and a sense of deprivation (e.g., Wilson 1982, 84–85), conditions that set the social context for apocalypses.[2] The notion of crisis may be defined in different ways — Collins includes persecution, culture shock, injustice, and the inevitability of death — but a situation of crisis is seen as fundamental to the rise of an apocalypse (Collins 1984b, 22).

A latent social determinism appears here, as in classical form criticism. Through appeals to the sociology of knowledge or to anthropological research on more recent millenarian movements such as Ghost Dance or cargo cults, apocalypses and apocalyptic eschatology are generally viewed as a *function of* the social setting.[3] So Schüssler Fiorenza writes, "A sociology of knowledge approach points out that any change in theological ideas and literary forms is preceded by a change in social function and perspective" (1983, 311). Thus, the form and content of the language change when the setting changes. Hanson's comments on the origins of apocalypticism in Judaism illustrate: "Bleak conditions call into question traditional socio-religious structures and their supporting myths. Life is situated precariously over the abyss" (1979, 433). In response a group may embrace "apocalyptic eschatology as the perspective from which it constructs an alternative universe of meaning" (p. 434). Apocalyptic eschatology affirms that "God, who guides all reality toward a goal, [is] about to intervene to reverse the fortunes of the prosperous wicked and the suffering righteous" (p. 434). That religious perspective — a development from the eschatology of Hebrew prophets — emerged in the harsh realities of postexilic times. While the religious perspective of apocalyptic eschatology should not be identified with apocalypticism as a social movement, Hanson does state that the latter "is latent" in the former: "At the point where the disappointments of history lead a group to embrace that perspective as an ideology . . . we can speak of the birth of an apocalyptic movement" (p. 432). Apocalypses are "produced by apocalyptic movements" and reflect the alienation experienced by those groups (p. 433). As can be seen, the perspectival shifts from prophetic eschatology to apocalyptic eschatology to apocalyptic ideology are necessarily preceded by changes in social, historical situations.

An analysis of the Similitudes of Enoch might thus develop along the following lines. The work arose in a community similar to Qumran around the middle of the first century of the Common Era. Members of that community "resented the rule of the pagan Romans or the impious Herods" and felt oppressed by them. The apocalypse arises out of that crisis of foreign rule and oppression. In response, the writer of the Similitudes uses pseudonymity, reports of a heavenly tour, eschatological predictions, descriptions of judgment scenes that reverse the status of rulers and righteous ones, and other elements of the genre "apocalypse" in order to help members of the community to keep faith and hope. The apocalypse serves several functions in that social setting, but none incompatible with a setting of crisis.[4]

Walter Schmithals: An Alternative

Walter Schmithals offers a decisive alternative to the above solution, for he severs all genetic connections between apocalypses and their social, historical situation: apocalyptic "primarily has its roots within itself, namely in the apocalyptic experience of existence," which is not caused by social, historical forces (1975, 150, cf. 120). Schmithals recognizes that there were "'apocalyptic' situations, that is, times which were so filled with sorrow and destruction, turmoil and oppression, that eschatologically oriented groups saw no more hope at all for this world and concentrated their hopes entirely on a new, coming eon" (p. 141). Not all apocalypses, however, arose in "exceptionally frightful conditions" and in those conditions apocalyptic was "only *one* reaction among various actual reactions" (p. 149):[5] "Apocalyptic does not . . . understand itself to be a reaction to a particular social reality" (p. 145), as though "certain realities inevitably produce a certain understanding of existence" (p. 148). Put differently, an attitude towards the world does not derive simply from "causal structures in existing reality" (p. 148).[6] The decisive factor is rather a predisposition that cannot be derived "but only affirmed or denied, accepted or rejected" (p. 150). Taking a cue from an earlier work by Rudolf Otto, Schmithals grounds the apocalyptic predisposition in "the idea of the transcendence of the divine."[7] The more fully "divine transcendence" dominates a person's understanding of the world, the more likely he or she will be predisposed to apocalyptic thinking.[8]

Perceived Crisis

Schmithal's approach and the social-historical approach of other scholars mark the extremes in relating apocalypses to a social, historical situation: the latter approach makes apocalyptic derivative of social, historical situations; while the former severs connections between apocalyptic and social settings. Most recently, a kind of compromise between those two positions has developed around the notion of "perceived crisis." This notion has arisen, on the one hand, out of a reluctance to break the connection of apocalypses with "social upheaval and turmoil, . . . alienation and powerlessness" (see Nickelsburg 1983, 646) and on the other, out of a recognition that many apocalypses have obviously not arisen from political upheaval and social crisis. "Perceived crisis" becomes a notion or conceptual tool for retaining models that connect apocalypses to social crises while recognizing that the social crises are not necessarily *evident*; an apocalyptic point of view appears to be tied to a particular type of social-historical situation (i.e., crisis), when in fact it is tied only to the piety of the apocalypticist, to his perceptions and his attitude of mind.[9]

What then does *perceived crisis* signify? It is a way of saying that (1) the author of an apocalypse considers a situation to be a crisis but (2) that the crisis dimensions of the situation are evident only through his angle of vision: "The problem is not viewed simply in terms of the historical factors available to any observer. Rather it is viewed in the light of a transcendent reality disclosed by the apocalypse" (Collins 1984a, 32)—that is, the crisis becomes visible only through the revealed knowledge in an apocalypse; prior to that knowledge there is no crisis. People discover the crisis dimensions of their existence by reading an apocalypse. An apocalypse thus functions in a social situation not only to bring comfort, hope, perseverance, and the like but also to cause people to see their situation as one in which such functions are needed and appropriate. An apocalypse can create the perception that a situation is one of crisis and then offer hope, assurance, and support for faithful behavior in dealing with the crisis. In the process the reader or hearer takes on the viewpoint of the writer and sees the human situation from the vantage point of transcendent reality (see Collins 1984a, 32). Thus, the concept "perceived crisis" contributes to our understanding of how an apocalypse functions *in* a social situation; but it sheds no light on the social occasion *of* an apocalypse, for any social situation can be perceived as one of crisis.[10]

Language as Social Exchange

In light of this power of apocalypses to shape perceptions of a social situation, we need to reformulate the relationship between an apocalypse as a written document and its social setting. Rather than asking about the social setting of an apocalypse, we should consider how an apocalypse as a social force relates to other social forces; there is a social dimension to apocalyptic language, for an apocalypse can shape a reader's perception of what the social situation is like. If, for example, through apocalyptic language a U.S. citizen identifies the Soviet Union with the Anti-Christ, that linguistic identification will shape dramatically how that person views the actions of the Soviet Union and thinks the U.S. should respond politically and militarily. Here—as always for human beings—the perception of reality *is* reality.

Through this example one sees that the language and religious vision of the Book of Revelation are not to be relegated to *poesis*, mythos, and ephemeral presence without impact on actual, social relations in everyday life. Literary, religious constructions do not stay neatly isolated as "symbolic worlds" unrelated to power relations in the social world. Literary, religious visions establish at least minimal social-political distinctions, just as social, political realities carry at least low-level symbolic content. Linguistic activity (speaking, writing, reading, listening) is itself social activity and partakes fully in the social world.

We are touching here upon the social, communicative dimensions of language. There is the obvious point that a person normally uses language to communicate with someone else: to inform, to express, to prescribe, or to evoke. Less obviously, that communication depends upon social, conventional grammatical and syntactical structures of a specific language such as English or Greek. A reader of the Book of Revelation draws upon those social conventions in order to understand the message of the book. In addition, communication requires some common ground—some shared beliefs or mutual knowledge about the world—between speaker and hearer. A speaker may explore that common ground or may make an assertion about the world not shared by the hearer, so that the hearer's understanding of what the world is like will conform more closely to the speaker's (see Stalnaker 1978, 322). In the latter case, a speaker may evoke and actualize dimensions of a world never before conceived in the mind of the other.[11]

Further, language is used most frequently in a specific "conversational" context. So John Austin writes: "For some years we [philosophers of language] have been realizing more and more clearly that the occasion of an utterance matters seriously, and that the words used are to some extent to be 'explained' by the 'context' in which they are designed to be or have actually been spoken in a linguistic interchange" (1962, 100). Within a "linguistic interchange" communication includes at least two different dimensions: (1) a proposition, that is, the thing expressed and (2) an intention located in the way that a proposition is expressed. Both propositions and intentions are involved in the normal use of language. For example, the words *There is a crevasse under the snow* can be seen as a proposition; but in the context of conversation, for example, between two mountain climbers, those words are spoken with a certain intention (e.g., "Step carefully"). That intention or manner of speaking is called the "illocutionary" force of a speech act.[12] The meaning of language involves both propositions and their illocutionary force.

In Daniel 6 the satraps and supervisors say to King Darius: "Know, O king, that it is a law of the Medes and Persians that no interdict or ordinance which the king establishes can be changed" (Dan. 6:15). Their statement includes a proposition about law that is shared with the king. Their statement also includes an illocutionary force or point that in the conversational context is both censorial towards the king and approving of the situation; the illocutionary point of their sentence may be captured by a phrase like "We've got you." In order for the king to recognize the force of their statement, he must share with them certain social and linguistic conventions by means of which their point is communicated.[13] Moreover, those shared conventions are a part of a specific social situation in Daniel 6: a conflict between Daniel and the Persian supervisors who seek to best him before King Darius. The force of the statement cannot be fully grasped without considering it within the dynamics of the specific power situation that occasioned the telling.[14]

So long as an audience shares fully in the conventions of the language, both propositions and illocutionary forces are a part of the public record of recorded speech. The illocutionary dimension of speech moves toward the social occasion in which the speech is uttered, but the "complete" social occasion is not contained in the illocutionary act. With regard to the Book of Revelation (and many other biblical books), a reader may become familiar with the language (and even recognize its illocutionary force) and still not be able to locate the exact occasion in which the seer's speech acts entered the flow of other social action. It is also difficult to trace out the consequences of speech acts, but, as seen by the exchange between Darius and the satraps, acts of speech do have consequences after they enter the flow of social action. In sum, the social dimensions of language may be located (1) in the language itself, which includes both what is said and its illocutionary point; (2) in the situation occasioning that language; and (3) in the consequences or effects of the speech activity on further social intercourse.

Social Dimensions of the Genre "Apocalypse"

The move from the use of words and sentences in a conversation to an apocalypse complicates an analysis of the social dimensions of language, for an apocalypse is a more complex literary unit. A simple sentence transmits meaning by combining sounds, morphemes (the smallest units of meaning in a language), and words; an apocalypse transmits meaning by combining sentences, scenes, and visions.[15] As a result, sentences and assertions within a genre must be considered not so much for what they say, their force or point, and their effects as for how they establish these dimensions *on the level of genre*.

The relationship between form and meaning is crucial, for only units of meaning (not formal units) enter into the flow of social intercourse. For example, as sounds, *yawl* and *y'all* may be virtually the same, but those formal elements of sound must be related to meaning before they can be located in a social context; otherwise, one may confuse a context of sailing with that of being addressed by a southerner. So, analogously, one would be on the wrong track in trying to locate 1 Enoch in a social context by trying to find where sheep were oppressed by eagles, for the sentences and scenes about the colorful animals in the Animal Apocalypse (1 Enoch 85–90) are formal elements that contribute first to the meaning of the apocalypse and only then to its social occasion.

In light of the earlier discussion about "perceived crisis," exhortations to remain faithful and consolations in the face of oppression may also be formal elements in the genre and therefore not contribute to any understanding of the social occasion of an apocalypse. Recall that an apocalypse may both create the perception that a situation is one of crisis and then offer hope, assurance, and support for faithful behavior in dealing with the crisis. Those formal elements — scenes portraying crises, consolations to the faithful, exhortations to remain faithful — contribute to meaning on the level of genre, namely, to the revealed knowledge or "the transcendent reality disclosed by the apocalypse" (Collins 1984a, 32). The meaning transmitted by an apocalypse centers on the viewpoint of the writer that is transmitted as revelation.

Just as social conventions determine the place of phonemes, morphemes, and other formal elements in transmitting meaning through sentences, so social conventions determine elements of form and meaning at the level of the genre "apocalypse." That is one reason why "study of the general conventions and assumptions of the genre" is indispensable to understanding any specific example of it (see Skinner 1974, 125). Genres are not literary structures isolated from a social context, nor are they constituted by purely idiosyncratic linguistic forms; they are a part of conventional social exchange involving speakers, writers, hearers, and readers who recognize their communicative force. So Hartman writes: "The genre belongs to a cultural set up which author and reader have in common" (1983, 340; see also Hellholm 1986, 29–33).

Content of Apocalypses

In the definition of *Semeia 14* (Collins 1979), an apocalypse is defined not simply as "revelatory literature" but more specifically as a revelation "disclosing a transcendent reality which is both temporal, insofar as it envisages eschatological salvation, and spatial insofar as it involves another, supernatural world." The inclusion of that specific content is essential to the definition of the genre, but it needs to be understood that an apocalypse does not reveal another world, it reveals hidden dimensions of the world in which humans live and die; that is, an apocalypse is not world-negating but, rather, world-expanding: it extends or expands the universe to include transcendent realities, and it does this both spatially and temporally. Spatial expansion dominates apocalypses with other-worldly journeys: that subgenre reports heavenly tours and ascents in which are described such transcendent realities as the abodes of the dead, cosmological secrets, judgment scenes, and the divine throne (Collins 1984b, 14–18). Temporal expansion dominates the "historical apocalypses." Through symbolic dream visions, revelatory dialogues, scriptural interpretation, and revelation reports the seer narrates prophecy after the fact and makes eschatological predictions (Collins 1984b, 6–14). Although the spatial or the temporal may dominate in a specific apocalypse, both modes of world expansion are present in all apocalypses.

The presence of both modes guarantees that the revelation of transcendence is integrally related to human earthly existence. In apocalypses, there is no radical discontinuity between God and the world (spatial transcendence) or this age and the age to come (temporal transcendence) (see Rowland 1982, 92, 175, 475). A radical transcendence which could sever heaven from earth is tempered by the future transformation of earthly into heavenly existence; and a radical transcendence which could sever this age completely from the age to come is tempered by the presentness of the age to come in heaven. Thus, the presence and interplay of spatial and temporal dimensions in transcendence prevent a thoroughgoing dualism in which the revelation of transcendence would become a separate set of forces without present effect on everyday human activity. The interplay assures that the powers of heaven and of the age to come operate decisively in present, earthly social interaction.[16] Seen from the angle of language as social exchange, a seer is making assertions about the world in such a way as to bring the hearer's understanding of what the world is like into greater conformity with his. His use of the genre "apocalypse" may, at least to some extent, also evoke in the mind of his audience dimensions of a world never before conceived. Or, perhaps more correctly, the genre "apocalypse," through certain widely understood linguistic and social conventions, communicates a certain knowledge and understanding of what the world is like that is genre-specific.

Metaphoric and symbolic language is integral to communicating that knowledge and understanding effectively. Such language operates in the genre to link up correctly the various dimensions of the world that have been expanded through spatial and temporal transcendence. Through metaphor and symbol a seer may report his transformations in space, time, or psychological state. Through the same language here-and-now earthly institutions, powers, and social relations are located properly in the larger, expanded world—for example, by being linked through metaphor and homologue to appropriate suprahuman worldly powers that are both presently locatable somewhere in the expanded universe and eschatologically im-pinging on the here and now. Through that network of language humans are situated—given a place on which to stand—in the expanded world. Apocalyptic language, thus, not only discloses an expanded universe but also orients humans in that larger world.

The Function of Apocalypses

As indicated earlier, the definition of *Semeia 14* deliberately omits any reference to function and social setting. Collins writes, "While a complete study of a genre must consider function and social setting, neither of these factors can determine the definition. At least in the case of ancient literature our knowledge of·function and setting is often extremely hypothetical and cannot provide a firm basis for generic classification. The only firm basis which can be found is the identification of recurring elements which are explicitly present in the texts" (pp. 1–2). That omission has evoked more objections than any other aspect of the definition of *Semeia 14*. At the 1982 Seminar on Early Christian Apocalypticism, David Hellholm argued for the necessary inclusion of function in any definition of the genre "apocalypse." He proposed adding to the *Semeia 14* definition the qualification "intended for a group in crisis with the purpose of exhortation and/or consolation by means of divine authority" (1986, 27). This addition brings into the definition the widespread notion that apocalypses arise from a situation of crisis.[17] David E. Aune also includes "function" in his modification of the *Semeia 14* definition, but, following the lead of John J. Collins, he distinguishes between social and literary functions, with a literary function "concerned only with the implicit and explicit indications within the text itself of the purpose or use of the composition" (1986a, 89). With that caveat he proposes three complementary literary functions: (1) "the legitima-tion of the transcendent authorization of the message"; (2) "a new actualization of the original revelatory experience"; and (3) the encouragement of cognitive and behavioral modifications (see pp. 89–90). Finally Yarbro Collins, in light of Hellholm and Aune's suggestions, proposes to add to the *Semeia 14* definition the qualification "intended to interpret present, earthly circumstances in light of the supernatural world and of the future, and to influence both the understanding and the behavior of the audience by means of divine authority" (1986, 7).

   If all speaking, writing, hearing, and reading are themselves social acts that constitute a portion of the flow of social exchange, apocalyptic language is by definition a part of a social situation. Its existence depends not only on the social conventions of a natural language (e.g., Greek or Syriac) but also on the social conventions that constitute the genre. Social function is not extraneous to genre definition.[18] Nor is there a clear distinction to be made between literary and social functions, for if something is "literary," it is "social." At the same time, one can appreciate fully the objection of Collins and Aune to much of the hypothetical reconstruction of social settings that has been carried out in the name of the historical, critical method. Both underscore that the purpose and use of an apocalypse can only be based on identifiable elements explicitly present in texts under consideration. The issue then becomes, How does a person locate identifiable elements in texts to get at the function and intention of a genre? This question has no easy answer.

Recognition of the social dimensions of language is essential to any move on the social function of a genre. In keeping with the three social dimensions of language outlined above, *function* can refer to (1) social dimensions within an apocalypse, including its social conventions and illocutionary force (on the level of genre);[19] (2) the typicalities, shared conventions, and specifics in the situation occasioning its production; or (3) its effects on ensuing human activity (see Skinner 1970). That taxonomy can help to keep clear where a particular function is to be located in relation to the stream of social activity in and around a text. Analysis of social dimensions within an apocalypse—especially its illocutionary aspects derived from units of meaning on the level of genre—provides the most solid transition from a text to the occasion of its production.

Even with such a taxonomy, however, the term *function* may confuse the relationship between apocalyptic language and its social dimensions. The term *function* may suggest a text's "placement in a social setting," and falsely imply that the genre or "linguistic construction" exists separate from the social order. In order to avoid thinking of genres apart from society, the term *intention* or *illocutionary force* may be a better way to talk about the language-society relationship (see Tucker 1971, 17). Whichever term is used, it should be made clear that genres are embedded in the social process and that the task is to recover aspects of that embedment.

If the writing and reading of an apocalypse are seen as social acts occurring among other social acts, the various dimensions of apocalyptic language interact in many different ways with other elements in the social process: literary aspects incorporate social conventions that make communication possible; religious dimensions make assertions about the world and provide a definite perspective from which to view all human activity; the activity of writing, speaking, and hearing makes a point within a particular social situation and that activity then becomes part of the ongoing presuppositions of further social exchange. In brief, language, genre, and other social processes are integrally related.

Setting an Agenda

In the chapters that follow, I shall test a broad hypothesis about how literary, religious, and social dimensions of apocalypses relate, specifically in connection with the Book of Revelation. The argument will be made that the seer's language does not form a separate "symbolic universe" apart from social, political realities; nor does his apocalyptic message address conflicts, tensions, and crises in the world of his audience. Rather, the seer offers a particular understanding—disclosed through revelation—of what the *whole* world is like, which includes an understanding of how Christians relate to other Christians, to other groups in the cities of Asia, and, more generally, to public social events. This broad thesis has several components.

The Linguistic Vision of the Seer. Every analyst of the Book of Revelation must engage with the language of the book. The multivalence of that metaphoric language needs to be explored so as to respect and appreciate its range and overtones of meanings. The language of Revelation is more like that of poetry than that of a set of directions (as in a cookbook); the language plays through a range of meanings rather than having only one meaning. The analyst must also respect the intertexture of the seer's words: many different organizational devices weave together his words, his sentences, and his scenes. Through attention to that multivalence and intertexture, a vision of the world emerges that includes provincial life in Asia. There is a social dimension to the world constructed by the seer; and ordinary Asian life is to be found in that world vision, not in references that move the reader "outside" the seer's world. For example, the seer's language of comfort, crisis, and exhortation must be understood within the dynamics of multivalence and intertexture of the seer's writing. Social, historical realities are to be found in the interconnectedness of his language, not in correspondences to some external order of reality.

The Social Order. Most recent scholarship continues to assume that the language of Revelation (and of apocalypses generally) reflects and arises in a social, historical situation of crisis. Specifically, in connection with Revelation, that crisis is bound to the reign of Domitian, his reign of terror and his heightened demands for imperial worship. Those assumptions about Domitian and his reign call for a careful investigation of historiographic issues that include both Christian and Roman sources for reconstructing the political situation at the time of Domitian. In connection with this period of Roman history, provincial life in Asia requires special study, for that is where the seer and his audience live and is the social, historical situation of which the Book of Revelation is one part.

The Linguistic Vision and the Social Order. The goal throughout this inquiry is to discover ways to integrate the linguistic vision of the seer with public, social realities or, better, to recognize that the vision itself is a social reality. Each dimension of social activity has its own distinctive structures, forms, and modes of entering into the larger social process; but, as one of those dimensions, apocalyptic language — its generic conventions and symbolic constructions — does not operate in some realm different from other social activity. Acts of speech, writing, and reading enter into the stream of social activity as much as do military victories, economic oppression, or legal actions. Faithful recipients of an apocalypse gain true knowledge about the cosmos, religion, the political order, local economic transactions, and the nature of social life. I seek a framework for integrating literary, conceptual, and social aspects of apocalyptic so that the language, religious sensibilities, and social political experience of the writer, readers, and hearers of the Book of Revelation can be seen as aspects or dimensions of an order of wholeness; for that reflects how language and symbols operate in human life and, more importantly, reflects how language and symbols operate in the Book of Revelation.

II

The Script:
Wholeness and the Language
of the Book of Revelation

3

The Linguistic Unity
of the Book of Revelation

The language of Revelation transmits a tremendous amount of information; the task is to make that language disclose its secrets—about the message that John is sending, the vision of the world that informs his message, his situation in the church and the Roman Empire, the network of social relationships inside and outside the Christian communities addressed by the seer, and the shape of the language itself—its syntax, genre, and ordinary and metaphoric significance. All of that information is interwoven in the seer's language. For analytic purposes, however, we shall consider first the language itself and the vision created by it.

A Synopsis of the Book of Revelation

No synopsis can remain simple and at the same time reflect the subtle connections among words, phrases, sentences, and scenes of Revelation. The seer's text can, however, be blocked out into sections by following certain organizational principles suggested in Revelation itself. The following outline is both approximate and preliminary.

Revelation 1:1–8

Revelation begins with a statement authenticating the visions to John as truly from God, given in a chain of command through Jesus and an angelic messenger. The one who reads, and those who hear, his words of prophecy will be blessed. Thus the complete chain of transmission begins with God and ends with the reading of Revelation in the churches. Verse 4 introduces an epistolary greeting followed by a doxology (1:5–6), a hymnic quatrain (1:7), and then circles back to God described as at the beginning of the epistolary greeting: "who is and who was and who is to come" (1:4, 8).

Revelation 1:9–3:22

John then reports in first-person narrative an account of a vision in which he is ordered to write to the seven angels of the seven churches in Asia: Ephesus, Smyrna, Pergamum, Thyatira, Sardis, Philadelphia, and Laodicea. After an elaborate description of the one who dictates the words, messages to the seven churches are given, all of which follow the same sequential pattern (though not all of the following are in all of the letters): (1) a command to write, (2) an identification of the speaker, (3) a description of the church's situation, (4) accusations against the churches, (5) a call to repentance, (6) a warning that the speaker will come to them, (7) an admonition to listen, and (8) a promise to the victorious.

Revelation 4:1–11:19

At chapter 4 a voice calls John through an open door into heaven. An elaborate throne scene follows (4:1–11) with special attention given to a "slain Lamb" who shares honor and glory with the one on the throne (5:1–14). The Lamb is honored because he has power to open a scroll sealed with seven seals, each of which discloses either earthly destruction or a heavenly scene (6:1–8:1). Between the opening of the sixth and seventh seals John sees two other visions — one of the sealing of the 144 thousand from the tribes of Israel (7:1–8) and one of an innumerable crowd whose clothing is washed in the blood of the Lamb (7:9–17). Then, after preparation of incense and prayers by an angel standing before the altar (8:3–5), seven angels blow serially seven trumpets, which again bring either earthly destruction or a heavenly scene (8:6–11:19). As with the seven seals, visions occur between the sixth and seventh item (10:1–11:14). In connection with the blowing of the fifth trumpet, an opening in the earth appears, a "shaft" to a bottomless pit (9:1–2, cf. 4:1), from which come destructive forces such as deadly locusts (9:3). Also noteworthy are the vision of the mighty angel with an open scroll, which John eats (10:1–11), a command to measure the temple of God (11:1–2), and a story about the fate of two witnesses for God (11:3–13). Considerable dramatic suspense is built around the blowing of the seventh trumpet (e.g., 10:7) which brings heavenly worship, the opening of the heavenly temple, and a theophany (11:15–19).

Revelation 12:1–14:20

Chapter 12 introduces the first of several colorful characters. A pregnant woman clothed with the sun comes into conflict with a great red dragon (12:1–6, cf. 12:13–17); in heaven the archangel Michael battles victoriously and casts out the same dragon (12:7–12); these visions are presented in the impersonal passive *There was seen*, rather than the first person, *I saw*. John then sees a ten-horned, seven-headed beast with great authority arise from the sea (13:1–10) and a two-horned, dragon-voiced beast with the number 666, coming up from the earth (13:11–18). Those visions of terror and danger contrast with three visions of assurance and victory: 144 thousand with the Lamb on Mount Zion singing a new song (14:1–5); three angels proclaiming the eternal gospel, the fall of Babylon, and the fate of the beast-worshippers, respectively (14:6–13); and two reapers of the earth (14:14–20).

Revelation 15:1–19:10

Chapter 15 introduces another series of seven. After a scene of heavenly worship (15:1–4) seven angels with seven golden bowls come out of the heavenly temple (cf. 5:8); from those bowls are poured seven plagues upon the earth (15:5–8). After the seven plagues of God's wrath (16:1–17), theophanic sky phenomena occur once again (16:18–21, cf. 11:19). The sequence of the seven bowls is followed by a series of visions describing the Great Harlot Babylon and her destruction (17:1–19:5) that concludes with a contrasting feminine image of the Bride of the Lamb (19:6–10).

Revelation 19:11–22:5

At 19:11 the heavens open once again and John sees a figure sitting on a white horse, clad with a robe dipped in blood and inscribed "King of kings and Lord of lords," who with his army is to rule the nations with an iron rod (cf. 12:5). Then birds gather to share in the Great Supper of God at which they gorge themselves on human flesh; while the Word of God, victorious over the beast and his followers, either kills them for the birds or throws them alive into the lake of fire that burns with sulfur (19:17–21). The dragon, Satan, is chained in the bottomless pit for a thousand years, while followers of Christ (who did not worship the beast) reign (20:1–6). After their thousand-year reign (the millennium), Satan is loosed again upon earth. He marshals an army, Gog and Magog, to march against the camp of the saints and the beloved city; but fire from heaven destroys them. The deceiving devil is thrown into a lake of fire and sulfur, where, along with the beast and the false prophet, he is eternally tormented (20:7–15). Then John sees a new heaven and a new earth, as well as the new Jerusalem, coming down from heaven, prepared as a bride (21:1–5). The one on the throne speaks, a rare occurrence in the Book of Revelation, declaring all things new and promising a blessed heritage to those who conquer (cf. 2:7, etc.) but damnation to the rest (21:6–8). Finally, the city is described in detail: its wall, twelve gates, twelve foundations, the river of life flowing from the throne of God through its main street, the tree of life with its twelve fruits, and special notice that it has no temple, for God Almighty and the Lamb are its temple (21:9–22:5).

Revelation 22:6–21

The Book of Revelation then ends as it began, with a series of confirmatory statements that the revelation contained in the book is from God and that those who hear it will be blessed (22:6–20). Interspersed are quasi-liturgical exclamations proclaiming that the end is near: "He who testifies to these things says, 'Surely I am coming soon.' Amen. Come, Lord Jesus!" Finally, to balance the epistolary greeting (1:4–5) the seer closes with an epistolary closing: "The grace of the Lord Jesus be with all the saints. Amen" (22:21).

The organization of the Book of Revelation is more complex than the synopsis suggests. The seer's language does not simply flow in narrative or logical sequence; it plays on formal, thematic, metaphoric, symbolic, and auditory levels of association. Moreover, the various levels overlap in such a way that breaks occur at different points among the different levels. As suggested in the synopsis, series of seven constitute units throughout the work, yet other numbering systems overlap these units; for example, the numbering of the three woes (9:12, 11:14, 12:12) connects the seven trumpets that end at 11:19 with the seven visions that begin at 12:1. Different sections of Revelation are also connected and unified by repeated metaphors, symbols, and motifs.

Forms and Composition

Revelation consists of several different component elements—epistolary, prophetic, proverbial, and liturgical forms. Near the beginning of his work the seer addresses the seven churches in an epistolary fashion: "John to the seven churches that are in Asia: Grace to you and peace from him who . . . " (1:4); and he concludes Revelation with an epistolary grace (22:21). Throughout Revelation the seer employs prophetic forms such as an inaugural vision (1:12–20), announcements of judgment or salvation grounded in descriptions of faithfulness or corruption and exhortations to repent (2:1–11), dirges (18:2–3), and angelic interpretations of things seen (7:13–17). Like the Hebrew prophets writing in exile, the seer also draws heavily on liturgical forms.[1] He structures language in the forms of hymns (5:9–10), acclamations (16:7), doxologies (7:12), declarations of worthiness (4:11), and a thanksgiving (11:17–18). Revelation also contains proverbial forms such as aphorism (13:10), beatitude (22:7), and lists of virtues and vices (22:15). Most of those component forms are integrated into a narrative of visionary or auditory reports stereotypically introduced by phrases such as *And I saw* or *And I heard*.

Among the compositional devices by which those smaller elements are ordered and subordinated to a larger, complex work, the most obvious is the sevenfold series: seven letters (2:1–3:22), seven seals (6:1–8:2), seven trumpets (8:2–11:19), and seven bowls (15:1–16:21). Doublets also appear in the work, for example, the number 144 thousand (7:2–8, 14:1–5), plagues of trumpets and bowls (chapters 8–9, 16), two descriptions of the beast (13:1–8, 17:3, 8), two of Jerusalem (21:1–8, 21:9–22:5), and two announcements of the fall of Babylon (14:8, 18:2–3).[2] Marked contrasts of weal and woe provide another means of organization: the first six seals of destruction (6:1–17) contrast with salvation and hope (7:1–8:4), the beast with horns (13:1–18) contrasts with the Lamb on Mount Zion (14:1–7), laments over the fall of Babylon (18:1–24) contrast with celebration over her destruction (19:1–10). Other organizational devices include equivalences of measure, stereotypic words and phrases (especially introductions), metaphoric associations, and scenes that reverse relationships or accumulate images from previous scenes.

The seer composed this "complex literary type" (Koch 1969, 23–24) from a repertoire of those smaller forms and compositional devices. Individual creativity as well as precedents in the genre contributed to the creation of his linguistic vision.

Through that process, those smaller, separate forms are integrated into a distinctive work that constitute Revelation. Each distinctive form now exists as part of a larger whole or as a "moment" in the flow of the seer's language.

Narrative Unity

The linguistic unity of Revelation is established in part through various narrative or sequential links that proceed along a horizontal axis; in the Book of Revelation narrative is made up of several horizontal "threads" that break at different points so that the bundle of threads creates a seamless, unbroken sequence.

Contrasting Units

Commentators often call attention to contrasting oppositions in Revelation as indication of the tension between Christian faith and harsh social, political realities (see App.). So, for example, the innumerable multitude celebrating in heaven and the sealing of the 144 thousand (7:1–17) contrast with the vengeance God delivers upon earth in the opening of the six seals (6:1–17). The celebration of God's kingdom at the blowing of the seventh trumpet (11:15–19) contrasts with those members of an apostate kingdom punished mercilessly by God in the blowing of the six previous trumpets (8:7–9:20). Similarly, the two beasts in Revelation 13 represent a false kingdom in contrast to the community of the Lamb in Revelation 14. The hope of Christians is assured in the scene on Mount Zion (14:1) in spite of the dangers of the beast from the primordial abyss standing on the seashore (13:1; cf. 12:17). Christians will rest from their labors (14:13) in contrast to those who succumb to the pressure of political servitude (14:11). The faithful will feast at the marriage of the Lamb (19:9) rather than be feasted upon by the birds of the air (19:17–18), and they will receive the seal of the Lamb (14:1) rather than the stamp of the beast (13:16).

Contrasting units can be traced through Revelation to illustrate its binary character and to point to conflictual elements in its vision of the world; that is, if a reader seeks out contrast and conflict, those features can be found in the narrative arrangement of Revelation. But contrasting units are only one of several kinds of relations that unify the flow of the seer's language, and contrasting units must be placed alongside those other unifying devices if the linguistic unity of Revelation is to be fully appreciated.

Equivalence of Measure

A fairly simple unifying device is equivalence in numerical measurement. For most people the mention of numbers in Revelation probably evokes 666, the number of the beast (13:18), which the seer claims can be calculated by anyone having some intelligence. Through gematria, whereby a number signifies a letter, many different identifications of that number have been made, the earliest of which is probably the Emperor Nero. My interest here in equivalences of measure has a different focus:

how recurring numbers of equivalent values contribute towards the unity of Revelation.

As previously noted, apocalyptic disasters tend to be organized around the number seven: seven seals, seven trumpets, seven bowls.[3] Themes in the pouring out of the seven bowls are strikingly similar to those of the blowing of the seven trumpets: in each series the first four describe disasters to earth; seas; rivers and fountains; and the sun — respectively. The fifth refers to the kingdom of the beast (16:10, cf. 9:11), the sixth refers to the Euphrates, and the seventh to a heavenly scene that includes theophanic sky phenomena of lightning, thunder, earthquake, and hail. In each series this repetitive unfolding, which occurs again at the pouring out of the seven bowls, culminates in heavenly worship.[4] Those repetitions create a kind of cumulative affect, a sense that one can step in the same river twice.[5]

Within the series of the seven trumpets (8:2–11:19) two references are made to the same time span: (1) in connection with a command to measure the temple of God and the altar and those who worship there, John is told that the nations will trample the holy city for forty-two months (11:2); (2) two witnesses are granted power to prophesy for 1,260 days (=forty-two months) (Rev. 11:3). In the next series of seven visions (12:1–14:20) the same time span is mentioned again. A story is told about a pregnant woman about to give birth to a male child endangered by a great red dragon. After the birth the child is caught up to God and the throne and the woman flees into the wilderness to be nourished there for 1,260 days (12:6).[6] In the third vision of that same series a beast comes forth from the sea and is allowed to exercise authority for forty-two months (13:5). Thus, through equivalence of measure, scenes in the series of the seven trumpets are brought into synchrony with scenes in a later series of visions.[7]

Visions throughout Revelation are also linked through the number twelve. The number itself appears regularly in connection with the new Jerusalem: twelve gates of twelve pearls (21:21), inscribed with twelve names of twelve tribes, accompanied by twelve angels (21:12); a wall with twelve foundations on which are the twelve names of the apostles (21:14); the city measures twelve thousand stadia (21:16); and in it the tree of life produces twelve kinds of fruit, one for each of the twelve months (22:2). Twice twelve elders sit on their twenty-four thrones surrounding the throne in heaven (4:4). Twelve thousand times twelve from the twelve tribes of Israel are sealed (7:4), and the same number (144 thousand) later appear with the Lamb on Mount Zion (14:1).[8]

Reversed Relationships

Sometimes the elements in one scene are transformed in such a way that relationships and actions are reversed as compared to a previous scene. For example, in 7:1–2 John sees four angels standing upon the four corners of the earth, holding back the four winds of the earth so that they do not blow and harm. Another angel coming up from the rising of the sun cries out to the four angels not to harm the earth until the servants of their God are sealed. The only other reference to four angels occurs at 9:13–15, the blowing of the sixth trumpet. John hears a voice from the four horns of the altar saying, "Loose the four angels bound at the great river

Euphrates." Those angels are loosed so as to kill one-third of the people. The two passages form mirror images: in chapter 7 an angel from the East (the rising of the sun) commands the angels on the four corners of the earth not to harm the earth, and they do not; in chapter 9 a voice from the four corners (horns) of the altar commands that angels in the East (Euphrates) be loosed to harm the earth, and they do.[9] Reversal of subject and object of the command results in reversed consequences.

The scene in chapter 10 involving the little scroll ($\beta\iota\beta\lambda\alpha\rho\acute{\iota}\delta\iota\sigma\nu$) reverses several elements in chapter 5, which describes the scroll ($\beta\iota\beta\lambda\acute{\iota}\sigma\nu$) sealed with seven seals. In chapter 10 a mighty angel, crying out as a lion, holds in his hand an open scroll ($\beta\iota\beta\lambda\alpha\rho\acute{\iota}\delta\iota\sigma\nu$ $\mathring{\eta}\nu\epsilon\dot{\omega}\gamma\mu\acute{\epsilon}\nu\sigma\nu$), the contents of which are presumably recited by the angel and the seven thunders (10:3).[10] John is about to write down the contents of this "open scroll" when he is ordered, "Seal [$\sigma\phi\rho\acute{\alpha}\gamma\iota\sigma\sigma\nu$] what the seven thunders spoke, and do not write these things" (10:4). Later he is ordered to eat his words (10:9). In contrast to chapter 10 the scroll of chapter 5 is sealed ($\beta\iota\beta\lambda\acute{\iota}\sigma\nu$ . . . $\kappa\alpha\tau\epsilon$-$\sigma\phi\rho\alpha\gamma\iota\sigma\mu\epsilon\nu\sigma\nu$ $\sigma\phi\rho\alpha\gamma\hat{\iota}\sigma\iota\nu$ $\dot{\epsilon}\pi\tau\acute{\alpha}$) so that no one can open ($\dot{\alpha}\nu\sigma\hat{\iota}\xi\alpha\iota$) it until it is handed over to the powerful "Lion of the tribe of Judah." The sequence of the action in chapter 10 thus reverses that in chapter 5.

After eating the bittersweet scroll John is ordered to prophesy again to people, nations, tongues, and kings (10:9–11, cf. 5:9). This commission to prophesy (cf. Ezek. 2:8, 3:1–3) harks back to the inaugural vision of 1:19–20, where John is ordered to write down the things which he sees, both things that are and that will be. In both instances the commissioning is associated loosely with "mystery" (10:7, 1:20) and with things to come (1:19, 10:7). Because of those connections with chapter 1, some have understood the little scroll in chapter 10 as an introduction to the second part of the Apocalypse.[11] In any case the commission in chapter 1, the unsealing in chapter 5, and the little scroll in chapter 10 are interconnected through these various elements.

Accumulation of Images

Sometimes images occurring at different places in Revelation reappear in a later scene. This type of repetition gives a cumulative effect, as images used earlier are gathered together. A retrospective review of the images in Revelation 19:11–16 illustrates. At 19:11 John sees heaven open (cf. 4:1) and then a white horse with a rider called Faithful and True who judges justly and makes war. The rider's eyes are as a flame of fire, on his head are many diadems, and he has a name written that no one knows but himself. He wears a garment dipped in blood and he has another name, the Word of God. An army in heaven follows him on white horses; they are clad with white, pure linen. From his mouth issues a sharp sword with which to smite the nations, and he will rule them with an iron rod. He will also tread the wine press of the fury of the wrath of God the Almighty. On his blood-dipped garment and on his thigh he has a name written, "King of kings and Lord of lords."

These six verses make up a clearly defined unit introduced by the stereotyped formula *And I saw*.[12] This unit is linked closely with the two following (19:17–18, 19:19–21) by a returning in verse 21 to the warring of the rider with a sword issuing from his mouth. Revelation 19:11–16 is also the first of seven final visions of judgment, victory, and salvation.[13] Thus Revelation 19:11–16 is a narrative unit that links in several ways to the narrative, visionary units immediately around it.

At the same time, however, Revelation 19:11–16 accumulates images from earlier scenes, especially those describing the ultimate judgment and victorious rule of God and his Christ. The rider of a white horse occurs elsewhere only at the opening of the first seal in 6:2 where the first of the four horsemen of the Apocalypse is introduced. There the rider has a bow, is given a crown, and goes forth to conquer (6:2); the language in both visions (6:2, 19:11–16) thus unfolds a warring, conquering king, the first introducing and the second concluding the apocalyptic disasters described by the seer.[14] The phrase "eyes like a flame of fire" (19:12) relates the rider of the white horse to the one who speaks to the angel of the church of Thyatira (2:18) and to the one like a Son of Man whom John sees in his inaugural vision (1:14). That Son of Man also shares with the rider of the white horse a "sharp sword issuing from his mouth" (1:16, cf. 2:12, 19:15). The diadems (19:12), shared only with the dragon (12:3) and the beast from the sea (13:1), are a sign of royalty, as is of course the title "King of kings and Lord of lords" (19:16, cf. 17:14). So, too, "ruling with a rod of iron" (19:15) is a royal image (Ps. 2:9) that links the rider to the male child endangered by the dragon (12:5). As the Lamb conquers through his blood (5:5, 9), so here the royal garment is "dipped in blood" (19:13), probably alluding to the crucified king, though the garment may simply be bloody from battle.

There are four references to the name of the rider on the white horse: (1) in 19:12 he has an inscribed name that only he knows; (2) the name "King of kings and Lord of lords" is written on his thigh (19:16); (3) in 19:11 he is called "Faithful and True"; and (4) in 19:13 his name is the Word of God. (1) That the name is unknown probably identifies the rider of the white horse as outside the human sphere, for mystery also surrounds the name of the land-beast (13:17) and the woman riding the scarlet beast (17:5);[15] (2) a name written on a person establishes identity (13:16–17, 14:1, cf. 3:12, 22:4); (3) the name "Faithful and True" links the one on the white horse to the one addressing the angel of the church of Laodicea, who calls himself "the Faithful and True [Witness]" (3:14, cf. 1:5); (4) the Faithful and True Word of God riding the white horse also parallels "the faithful and true words of God" given to John and authenticated by God himself (21:5, 22:6). (In fact, the Faithful and True Word of God riding the white horse is transmitted as part of those "faithful and true words of God." The words that transmit and the Word on the white horse that is transmitted are one. Form and content, signifier and signified, syntax and semantics are united in the sign of the Word.)

The rider goes forth to make war. Except for Jesus' threat to war against those at Pergamum (2:16), warring occurs only between divine and evil forces: Michael and the dragon (12:7), the dragon and the woman's seed (12:17), the Lamb and the beast with ten kings (17:14), the two witnesses and the beast (11:7), Armageddon (16:16), and Satan and the saints (20:8). Here the rider on the white horse, along with his army wearing white linen and riding white horses, battle the beast, the kings of the earth, and their army (19:19).

Thus Revelation 19:11–16 consists of transformations of a series of linguistic elements that appear earlier in Revelation. Some of those elements first appear in John's inaugural vision of the Christ (1:12–16). Others appear in introductions to the letters, in the description of the first of the four horsemen of the apocalypse, in the fighting Lamb who conquers the beast and ten kings, in the newborn male child threatened by the dragon, and in the blood of the wine press of the wrath of God. That final transformation also takes into itself elements previously formed around the demonic figures of the dragon, the beast from the sea, and the whore on the scarlet beast. If we were to read Revelation aloud, we would notice structures, motifs, images, and perspectives forming and unforming sequentially and then appearing together, so that the scene in Revelation 19:11–16 brings the work to a climax — not in the narrative line but in the concentration of images.[16]

Circularity

Circularity refers to concentric development of a passage so that the ending reflects the beginning.[17] A simple example occurs in 4:1: *"After this* I looked, and lo, in heaven an open door! And the first voice . . . said, . . . I will show you what must take place *after this*."*[18] On a different level the pregnant woman and dragon ring chapter 12, for that chapter begins with their conflict, shifts to a battle in heaven, and then ends with another version of the woman-dragon conflict. Revelation 17:1–19:10 is enclosed by an antithetical ring consisting of the Great Whore clothed in purple and scarlet at the beginning and the Bride of the Lamb clothed in bright, pure, fine linen at the end. Or an even larger unit, 12:1–19:10, is ringed by another set of feminine images: the pregnant woman clothed with the sun and the Bride of the Lamb clothed with pure linen. Circularity consists of a concentric development of words, syntactical forms, or motifs. As with all analysis, location of circularity becomes the more ingenious the more removed it is from the actual language of Revelation.

The seer tends to develop his material concentrically into ever-widening rings. So, for example, several of the eschatological promises to *those who conquer* (the victorious) — an element in the seven letters of chapters 2–3 — reappear in the vision of the New Jerusalem in chapters 21–22: at the beginning of the vision of renewal God makes the link to the letters by saying, "He who conquers shall have . . . " (21:7, cf. 2:11); the tree of life promised to those victorious at Ephesus (2:7) appears in the city (22:2);[19] The victorious at Smyrna will not be harmed by the second death (2:11), a phrase that is made clear in 21:8 (cf. 20:6, 14); those conquering at Sardis will not have their names wiped from the book of life (3:5, cf. 21:27); and to those conquering at Laodicea Jesus promises a seat with him on his throne (3:21), while at 22:5 his servants reign forever in the city with the enthroned God and the royal Lamb.[20]

The final section of the Apocalypse (22:6–21) circles back to the beginning to create one grand circularity. At Revelation 22:8 John once again identifies himself as the seer who hears and sees what he writes (cf. 1:9). At 22:13 Jesus calls himself the "first and last" (1:17) and creates a cumulative affect by including, as well, "alpha and omega" (cf. 1:8) and "beginning and end" (cf. 3:14, 1:5).[21] The chain of command and blessing in 1:1–3 are repeated in 22:6–7, and the assertion that "the time is near" (1:3) is repeated in 22:10.

In the process of circularity and accumulation, placement in the narrative sequence is a significant factor; for earlier occurrences of a term, image, or motif become a given in the narrative line, to be drawn on in the development of a later scene. Thus, the first vision that focuses upon Christian life in Asia Minor provides the givens that the seer loops back on in the heavenly visions. Such recursive activity grounds the seer's visionary scenes in the social life of the Asian province.[22]

Metaphoric Unity in Revelation

In contrast to the narrative unity of Revelation, which interconnects scenes through horizontal threads that run sequentially throughout the work, metaphoric unity interrelates elements of the seer's language vertically. In narrative sequence time is always involved in moving from one scene to another. Metaphoric unity, on the other hand, occurs all at once; that is, a metaphor consists of simultaneous "vertical" layers of language analogous to a musical chord. Metaphoric phrases such as "crown of life" or "wine press of the wrath of God" must be grasped instantaneously; the metaphor cannot be traced through the letters or the words. The seer unifies his work as much through figurative or metaphoric language as through narrative devices.

Similes and Metaphors

A simile compares two different objects through particles such as ὡς, ὥσπερ, and ὅμοιος, all of which may be translated as "like" or "as." An angel cries with a loud voice as when a lion roars (10:3); unclean spirits like frogs come from the mouth of the dragon (16:13); one of the four living creatures around the throne was like an ox (4:7). At 6:14 the seer sees the heavens split, which he compares to a scroll rolled up.

As though to highlight the importance of figurative speech, the seer often introduces some of the most basic elements in his landscape through a simile. The sea, for example, used throughout the Apocalypse as an element coordinate with land, earth, and heaven, first occurs as a figure in the throne scene: "and before the throne there is as it were a sea of glass, like crystal" (4:6).[23] So, too, fire first appears metaphorically as a simile in a description of eyes in John's inaugural vision (1:14). Later, as an independent element, *fire* describes various catastrophies and punishments.[24] *Trumpet* describes metaphorically the voice in the inaugural vision (1:10) and the voice calling John into heaven (4:1), while later (8:2–11:19) trumpets are blown by angels to bring different kinds of eschatological revelations. Several animals are also introduced first through metaphoric speech. The first living creature around the throne is "like a lion" (4:7). In the next chapter, the seer declares that the "Lion of the tribe of Judah . . . has conquered" (5:5).[25] *Serpent* — a significant animal in the action of the Apocalypse — first appears as a simile describing the tails of the horses released at the blowing of the sixth trumpet (9:19). *Serpent* then appears in its own right as an independent object in conflict with the woman (12:9, 14, 15) and as the one bound for a thousand years (20:2). Before an eagle is seen flying in midheaven crying woes (8:13), it has been introduced in a simile describing the fourth living creature (4:7). So, too, the activity of warring first occurs in a simile (9:9). The two witnesses wear sackcloth (11:3), but earlier "the sun became black as sackcloth" (6:12). These transformations from similes to independent entities unify elements in the seer's world. By using key terms as similes, he weaves together elements in nature, animal life, human activity, and forces from both heaven and the underworld.

A metaphor identifies, rather than compares, two elements: in 12:1 the clothing of the woman is not *like* the sun; her clothing *is* the sun. Throughout Revelation, similes and metaphors appear together. In the seer's inaugural vision a figure of awe is created through similes and metaphors combining human and inanimate spheres (1:13–16): the awesome manifestation becomes present visually as "one like a son of man, clothed with a long robe and with a golden girdle round his breast" (1:13); the whiteness of his head and hair is as wool or snow (1:14); his eyes are as a flame of fire (1:14); his feet are like bronze fired in a furnace, and his voice sounds like many waters (1:15); his face is like the sun shining in full strength; from his mouth issues a sharp two-edged sword, and — more striking — he holds in his right hand seven stars (1:16). This portrait is drawn from traditional images in the Old Testament, but it combines metaphorically spheres normally kept separate in everyday experience to produce a creature of awesome, divine proportions.

Later, in the first throne scene, the creator God who sits upon the throne is described by means of images of precious stones — jasper and carnelian — and by a rainbow that looks like an emerald (4:3). Once again the inorganic becomes the means of revealing the divine. In contrast, animal similes become the means of envisioning the heavenly figures around the throne. One is like a lion, the others are like an eagle, an ox, and a man — all with six wings and full of eyes (4:6–8). There are, as well, twenty-four elders with white garments and golden crowns (4:4). Atmospherics, sea, and fire combine to fill out the rest of the presentation (4:5–6). Celestial objects frequently combine to characterize divine forces: a mighty angel comes down from heaven wrapped in a cloud, with a rainbow over his head, with a face like the sun and legs like pillars of fire (10:1). Or a woman appears in heaven clothed with the sun, with the moon under her feet and twelve stars on her head (12:1).

The seer also envisions suprahuman evil forces by breaking the categories of everyday experience through metaphor and simile. The monstrous beast from the sea has ten horns and seven heads. It is like a leopard, its feet are like a bear's, and its mouth is like a lion's. The locusts that come up through the opening from the bottomless abyss combine a stinger like a scorpion, an appearance like horses, though with human faces, women's hair, and lions' teeth (9:7–10).[26]

Insofar as myth typically portrays its characters as such hybrids, Revelation fits the mythic genre. Myth metaphorically transgresses boundaries that normally divide aspects and dimensions of the everyday world: dwellers on earth become drunk with the "wine of fornication" (17:2) or Great Babylon is drunk "with the blood of the saints" (17:6).[27] The New Jerusalem forms a complex boundary with sacred space on earth, eschatological time, and heaven above—a boundary that cannot be charted in an ordinary space-time grid. Finally, the metamorphoses that occur at the opening of a door in heaven in chapter 4 occur simultaneously in several dimensions. The spatial movement "up" (4:1) becomes synonymous with the psychological state of "being in the spirit" (4:2), and the spatial-psychological is in turn linked closely to the temporal future: "Come up here and I will show you what must take place after this" (4:1). Through the identification of metaphor, to go up is to go forward in time, which is to change psychological states.

Irony

Irony is another figure of speech that occurs in the Apocalypse. That is, dissembling and concealing occur, so that true meaning reverses what appears.[28] Ironic language sometimes appears with words that play on different meanings: the Smyrnians are poor but are rich, while the Laodiceans are rich but are wretched and poor (2:9, 3:17). Structural irony—where dissembling occurs in the use of a literary form—appears in the dirges over Babylon in chapter 18, especially the dirge of the angel (18:2–3), who uses that form to rejoice.[29] Kerygmatic irony occurs frequently throughout Revelation, whereby the Christian proclamation and imitation of the "crucified king" are expressed in various ways.[30] The irony of Christian redemption comes out in the language of 5:5–10. John is assured there that the sealed scroll would be opened, for "the Lion of the tribe of Judah, the Root of David, has conquered, so that he can open the scroll and its seven seals." At that point John sees in the midst of the throne "a Lamb standing, as though it had been slain," and heavenly creatures sing to the Lamb about its worthiness to open the book "because thou wast slain." Here the seer, albeit in his own distinctive language, alludes to the irony in the Christian proclamation that the one on the cross reigns as king.[31] Because of the powerless condition of being slain, the Lamb has power and is worthy of receiving "power, strength, honor, and glory" (5:6–9). Life, victory, and power come through crucifixion.

As with the Lamb, so with the Christians who follow the Lamb. Those at Smyrna are exhorted to "be faithful unto death, and I will give you the crown of life" (2:10). For the Christian, life comes through death, power through being powerless. To the Philadelphians it is said, "I have placed before you an open door which no one has the power to close because you have little power and you kept my word and you did not deny my name" (3:8). Here there is a word play on *power* (δύναται/δύναμιν) and an ironic display of powerlessness, for their "little power" is celebrated as a partial reason why no one has power to close the door.[32] In a similar ironic vein Jesus urges the Laodiceans, who think that they are rich but who are actually poor, to buy from him who purchased them by his blood fired gold, white garments, and salve for their eyes (3:18).[33]

The irony of Christian proclamation and imitation occurs in a more subtle form in the message to the Ephesians where those conquering are promised to eat from the "tree of life" (ἐκ τοῦ ξύλου τῆς ζωῆς) (2:7). Reference to the tree of life occurs again at 22:2–3, where it is linked clearly to the cross: "And the leaves of the tree were for the healing of the nations. There shall no more be anything accursed."[34]

Puns and Word Plays

The seer often creates puns and plays on different meanings of a word. To the Ephesians the speaker says, "I know your works . . . how you cannot *bear* [βαστά-σαι] evil men . . . [and are] *bearing up* [ἐβάστασας] for my name's sake." Also, the speaker knows their "toil" (κόπον) and that they "have not grown weary" (κεκο-πίακες) (2:2–3). In the Sardis letter *name* at one point represents external, superficial reality ("You have the *name* of being alive" [3:1]), whereas later he declares that he will not blot out the *name* of the one who conquers (3:5) from the book of life, that is, here *name* represents the deepest reality. Assurance is given the Philadelphians: "Because you have *kept* [ἐτήρησας] my word . . . I will *keep* [τηρήσω] you from the hour of trial" (3:10). The *after this* that rings 4:1 serves as a "superficial connector" at the beginning but as a temporal referent at the end.[35] Earth and serpent create a word play with their *mouths*, as "the earth opened its mouth and swallowed the river which the dragon had poured from his mouth" (12:16). At 1:17 John falls down down "as a corpse" before the one who was "alive, became a corpse, and again lived" (1:18). At 19:20 it is stated that "the beast is caught." Through a play on the action of the verb, the beast is "caught" like an animal (e.g., John 21:3), but the verb also carries here the meaning of "arrest" or "take into custody" (e.g., Acts 12:4). Prepositions sometimes create verbal plays: the difficult proverb in 13:10 probably means that if a person goes away *for the purpose of* [εἰς] making captive, he goes *into* [εἰς] captivity.[36] In the second part there is a play on another preposition: "If someone goes forth *with* [ἐν] a sword to be killed, he will be killed *by* [ἐν] a sword." Prepositional playfulness also occurs at 16:11 where those suffering under the fifth bowl cursed God *because of* [ἐκ] their pain and did not turn *from* [ἐκ] their deeds. On occasion a phrase is used to refer blatantly to different referents. So in the throne scene of chapters 4 and 5 the seer casually identifies the "seven spirits of God" with the "seven torches of fire" (4:5) and later with the "seven eyes" of the Lamb (5:6); similarly, "seven heads" of the beast are both "seven hills" and "seven kings" (17:9–10).

Other verbal playfulness can be seen in the reversals of the Ephesians and those at Thyatira: the Ephesians should do their *first works* (2:5), whereas the latter *works* of those at Thyatira exceed their *first* (2:19). Those at Laodicea must open the door (3:20), but the Philadelphians have an open door set before them (3:8). Those at Smyrna are poor but rich (2:9), while those at Laodicea are rich but poor (3:17). While the scroll is being unsealed (6:12), the 144 thousand are being sealed (7:3).

When considering the meaning of a word at any specific instance, its range of meanings may also be present. Consider, for example, the meaning of *blood* (αἷμα) in the Book of Revelation. A primary meaning is fixed early on in the doxology following the epistolary greeting at 1:5: "to the one loving us and loosing us from our sins by his blood" (cf. 5:9, 19:13). Christian martyrs who follow the pattern of Jesus offer their blood in witness of the gospel (6:10) and are saved by blood. So those who come through the great tribulation wash and make white their garments in the blood of the lamb (7:14), and the brethren endangered by the dragon fallen from heaven are victorious through that same blood (12:11). Blood can also refer to disasters and destruction, especially in association with bodies of water: at the second trumpet, a third of the sea became blood (8:9);[37] the two witnesses had power to turn water into blood (11:6); in a scene where earth is "harvested" blood from the wine press came up to the bridles of the horses (14:20). In certain passages those different meanings of *blood* play together. Babylon the Great Whore is drunk with the blood of the saints and of the witnesses of Jesus (17:6, cf. 18:24), and that blood is exacted from her hand by judgment (19:2) and destruction (18:21-24). In connection with the pouring out of the third bowl of wrath, rivers and streams turn into blood (16:4). Then the angel of the waters declares God just and explains with a *lex talionis* playing on different meanings of *blood*, "You are just . . . because they poured out the blood of the saints and prophets, and you have given to them blood to drink" (16:5-6). Saving blood involves destruction and death; martyrs follow Jesus in pouring out their blood, but that blood brings with it judgment—salvation or destruction. Perhaps in every instance *blood* carries both its positive and negative meanings.

In sum, figures of speech abound in Revelation as comparisons, similes, metaphors, and word plays weave into ever-changing patterns. Syntactical shifts and grammatical changes make novel connections and disclose new associations. The seer writes a language of metaphor, symbol, brief narration, and cultic cries. It illustrates what some call the primary language of religious experience.[38] At times, that "writing style" seems to be in tension with the subject matter of commitment, judgment, and hope; it borders on a facile lightness reminiscent of Ovid's compilation of myth, in which—in contrast to tragedy—the most horrifying content is presented in a light style (see Massey 1976, 24-25). Whatever else may be observed about that language, it spins another set of threads that unify Revelation.

The Old Testament and the Linguistic Unity of Revelation

Revelation shares with all apocalypses the importance of Scripture as a source for its language. The seer, however, rarely quotes the Old Testament directly. Rather, he alludes to it, paraphrases it, and combines various passages from it in order to accommodate the meaning of the Old Testament to his own vision.[39] So, for example, John's inaugural vision (1:12-16) draws from the language of Exodus 25, Zechariah 4, Daniel 7, and Daniel 10, but elements from those separate passages are combined to create a novel figure whom John encounters in his first vision. The beast in Revelation 13 is a composite of the four beasts in Daniel 7, and the visions of Revelation 20-21 are inspired in part by the eschatological descriptions of Ezekiel 37-48. Throne scenes, typical elements in apocalypses, are distinctively portrayed in Revelation (e.g., 4:3) by means of elements drawn from the breastplate of the priest (Exod. 28:18, 39:11) and the garden of Eden (Ezek. 28:13). In the seer's work terms for precious stones in the throne visions of chapter 4 are repeated in the descriptions of the New Jerusalem in chapter 21. Adaptations from the Old Testament thus contribute to both a distinctive and a unified vision in Revelation.

In the preparation and pouring out of the seven bowls of wrath (Rev. 15–16) the Exodus tradition is drawn on in a creative manner. The latter series of seven opens with a Song of Moses (Exod. 15), but in Revelation it is the Song of Moses and the Song of the Lamb. The Lamb alludes to the Passover, which is, of course, a cultic reenactment of the Exodus, and the song may reiterate the singing of Exodus 15 at the evening Sabbath sacrifice. The language of disaster that follows has been shaped in part by the plagues in Egypt. The sores of the first bowl reiterate the sixth plague (Exod. 9:8–12); the second and third bowls, the first plague (Exod. 7:14–24); the fifth bowl, the ninth plague (Exod. 10:21–29); the hail and thunder of the seventh bowl, the seventh plague (Exod. 9:13–35). The drying up of the Euphrates at the pouring out of the sixth bowl is not paralleled in the plagues on Egypt, but it rings the changes upon the crossing of the Red Sea (cf. Isa. 51:10).

In Revelation, however, these disasters are taken into a suprahuman, liturgical context. Frogs, which allude to the plague of unclean animals (Exod. 8:3, Lev. 11:10, 41), are assimilated into the language of Revelation as a way of describing the unclean spirits that come from the mouths of the dragon, the beast, and the false prophet (16:13). So, too, the drying up of the river—an allusion to the Red Sea tradition—becomes associated with a suprahuman conflict between forces at Armageddon (16:12, 16).[40]

In accommodating the Old Testament, the seer consistently adapts it to the Christian message that shapes his work. The Lamb, for example, with precedents in the Jewish tradition, always alludes to the Christian Savior in Revelation. His identify is firmly fixed in chapter 5 as the Lamb that was slain (cf. 7:14). Moreover, the slain Lamb has a place in the suprahuman community of those surrounding the throne of God, a place of great prestige; for the Lamb and God regularly receive worship together (5:13, 7:9, 14:4, 21:22). So at 15:3 the placement of the Lamb alongside Moses is not incidental; for that Lamb overshadows Moses, since the Lamb is "Lord of lords and King of kings"; and those with the Lamb, that is, Christians (not Jews), are the faithful chosen (17:14).

The story of the two witnesses in chapter 11 also illustrates how John modifies the Old Testament. This story depends upon the messianic vision of Zechariah 4, which describes the two olive trees and two lamps standing before the Lord of the earth (see esp. Zech. 4:14). Fire from their mouths expresses the divine presence (e.g., Ps. 18:8). As prophets they have the power of Moses and Elijah, for they can withhold rain (Elijah) and can turn water into blood (Moses). But after these allusions to the Old Testament (11:3–6) those figures built upon Moses, Elijah, and the two Messiahs of Zechariah become reiterations of the pattern of Jesus, their Lord (11:8). They are killed, their bodies are publicly displayed, after 3 1/2 days they are revitalized, and they ascend into heaven on a cloud as a great earthquake shatters the earth (11:7–13).[41] The Christian pattern dominates, and the Old Testament (including Moses) is subordinated to the Christian proclamation, which is implicated in the seer's language. The allusive use of the Old Testament, thus, also contributes to the linguistic unity of Revelation.

The Unity of the Language of Revelation

By taking the shape of metaphor, simile, word play, cultic cry, and symbol, the language of Revelation transgresses fundamental categories of normal language and creates hybrid creatures of awesome and monstrous dimensions. Further, the language constructs through metaphor complex boundaries in space and time that cannot be charted in an ordinary space-time grid. It also makes novel connections and discloses new associations through syntactical reiterations and recursions that appear throughout the sequential movement in the reading of Revelation. Through such narrative and metaphoric devices the language of Revelation is so intertwined that it cannot be easily dissected. The threads crisscross in different ways, sometimes tracing a path forward, sometimes backward, sometimes intertwining through overlays of metaphoric simultaneity.

As the reader reads the last chapters in the Book of Revelation he or she notices an intensification and a more fully developed "end" than was present in earlier chapters. Yet it is difficult to trace out a plot line in the Book of Revelation with a climax at the end. Even narrative progression as a conic spiral moving from the present to the eschatological future (e.g., Schüssler Fiorenza 1981, 26) can be traced only at the expense of other crisscrossing threads of fulfillment that occur throughout Revelation. For example, at the blowing of the seventh trumpet, "the mystery of God, as he announced to his servants the prophets" is fulfilled (10:7, cf. Schüssler Fiorenza 1981, 114). The Apocalypse could end with chapter 11 in the heavenly celebration of God's just judgment of the dead and his eternal reign (11:17-18). Or after the battle of Armageddon the heavenly theophany brings eschatological fulfillment. Terms such as *climax*, *interruption*, and *interlude* help to simplify the complex web by subordinating certain elements, but they inevitably distort the wholeness of the Apocalypse. Even an understanding of the Apocalypse through the metaphor of a web distorts its wholeness, for that metaphor suggests that there are strands that exist throughout the work. The language of the seer is probably more pliable and fluid than that.

The fluid quality of the language that unfolds and than enfolds narrative and metaphoric connections calls into question any structual analysis of Revelation that easily divides the text into sets of oppositions. Such analysis is well-known through the work of the anthropologist Lévi-Strauss and his followers. A text is divided into segments called "mythemes," whose true meanings emerge when they are bundled into opposing groups. So John Gager bundles the "mythemes" of Revelation into two groups, victory/hope and oppression/despair (see App. A). That kind of grouping is problematic, for it ignores the variety of narrative and metaphoric connections that unifies the book.[42] If "web" and "conical spiral" are inadequate images for describing the structure of Revelation, how much less adequate are sets of oppositions! When viewed strictly as a construction of language, Revelation can best be envisioned as a stream: the seer's language flows into and out of images, figures, reiterations, recursions, contrasts, and cumulations as whorls, vortices, and eddies in a stream. That image captures the linguistic unity of Revelation.

4

Unity through the
Language of Worship

Among the various forms and shapes that the seer's language takes, the language of worship stands out. Even a cursory reading of the Book of Revelation shows the presence of liturgical language set in worship. Moreover, as we shall see, the language of worship plays an important role in unifying the book, that is, in making it a coherent apocalypse in both form and content. The scenes of worship are not just "interludes" or "interruptions" in the dramatic narration of eschatology; they take their place alongside these narrations of eschatology to make of the book something more than visions of "things to come."

The language of worship can usually be distinguished from the language around it, yet the elements of that language cannot be described with absolute precision. If liturgical language is introduced by a phrase such as *They worshipped, saying*, it is easily identified. Sometimes, however, the language of worship (liturgical language) appears in the Book of Revelation without such an introduction. Then it is identifiable only by analyzing the form of the language.

Analysis of liturgical language is not an exact science. Writers and worshippers freely adapt traditional forms of worship so that it is impossible to state precisely what the language of worship "looks like." Some observations with regard to style and form can, however, be made about liturgical language in the Book of Revelation, for it follows certain conventions found elsewhere. Worship focuses upon divinity, and the language of worship praises, gives thanks to, or makes requests of the divinity. In Revelation most worship praises God and/or Jesus Christ by affirming their worthiness to receive praise and by recounting their deeds or their qualities. The last type can be simply a short acclamation, such as "Salvation to our God, the one seated upon the throne/And to the Lamb" (Rev. 7:10).[1] Throughout Revelation, liturgical language concentrates upon the object of worship, usually from the perspective of the worshipper who addresses the divine in second grammatical person (*you*) or refers to the divine in third grammatical person (*he*). More rarely, the language reflects the divine perspective, with the deity speaking acclamations in first person (*I*)—(e.g., Rev. 1:8). As Eduard Norden has noted, this pattern is characteristic of liturgical language in general. Liturgical sentences take the form "Thou hast/he has done such and such" or "Thou art/he is such and such." Sometimes the deity speaks saying, "I am such and such" (see Norden 1956, 143–66, 177–201).

Longer liturgical pieces are elaborated in different ways. Norden notes that elaboration often takes the form of participial phrases or relative clauses (Norden 1956, 167–76, 203–7). Thus, a doxology at the beginning of the Book of Revelation develops through participles. "To the one loving us and loosing us from our sins" (Rev. 1:5). Liturgical elements are sometimes elaborated by giving the reason for the praise in a *for* clause. "Worthy are you, O lord and our God, to receive glory . . . for ['ότι] you created all things" (Rev. 4:11). Longer pieces are written in a style called prose hymnody, in which a poetic line consists of a sense unit or one meaningful phrase; often the poetic lines then develop as couplets, with the second line of the couplet completing the first: "Great and marvelous are your works, Lord God Almighty/Just and true are your ways, King of the nations" (Rev. 15:3). In these longer pieces repetition of words, ideas, and images can create interesting sets of relationships among the attributes of the deity or the mighty deeds that God has done. Finally, as we have already noted, the seer's language is filled with metaphor and other figures of speech; that is especially true of the poetic language of liturgy.

Liturgical Language in the Prologue and Epilogue

We saw earlier that images, motifs, and terms loop back in the Book of Revelation to the seven Letters in chapters 2 and 3. That recursion grounds the seer's visionary scenes (Rev. 4:1–22:5) in the Christian communities of the province of Asia (Rev. 2:1–3:22). In the prologue (Rev. 1:1–8) the combination of liturgical and epistolary elements effects similar results. Here the liturgical pieces tend to reflect the apocalyptic, visionary elements, whereas the epistolary elements reflect the specific life in Asia.

The seer introduces his work as an apocalypse (Rev. 1:1). Those three verses (1:1–3) conclude with a liturgical blessing upon the audience:

> Blessed is the one reading out loud
> And those listening to the words of the prophecy
> And keeping the things written in it
> For the time is near.

Those verses are followed by a greeting in the form of a letter: "John to the seven churches that are in Asia: Grace to you and peace" (Rev. 1:4). Here the identities of writer and audience are specified. Within that epistolary greeting John includes liturgical material. He refers to God as "him who is and who was and who is to come" (Rev. 1:4); and he praises Jesus, the only other legitimate object of worship in Revelation, in the form of a doxology. This doxology, like all the others in the Book of Revelation, consists of three elements (not always in the same order): (1) designation of the one receiving the praise, (2) the doxological ascriptions (e.g., *glory* and *power*), and (3) the temporal designate "for ever and ever" (see Deichgrä-ber 1967, 53):

> To him who loves us
> And has freed us from our sins by his blood
> (And made us a kingdom, priests to his God and Father)
> To him be glory and dominion for ever and ever. Amen. (Rev. 1:5b–7)

That doxology is followed by an eschatological cry that also takes a liturgical form:

> Behold, he is coming with the clouds,
> And every eye will see him,
> Every one who pierced him
> And all tribes of the earth will wail on account of him.
> Even so. Amen. [ναί, ἀμήν.] (Rev. 1:7)

The prologue comes to an end with a hymnic affirmation in the first person:

> "I am the Alpha and the Omega," says the Lord God
> Who is and who was and who is to come, the Almighty. (Rev. 1:8)

The first line is in the form of an *I am* saying (see above);[2] variations occur elsewhere in the Book of Revelation:

> "I am the first [ὁ πρῶτος] and the last [ὁ ἔσχατος]. (Rev. 1:17)

> "I am the Alpha and the Omega, the beginning [ἀρχή] and the end [τέλος]. (Rev. 21:6)

> "I am the Alpha and the Omega, the first and the last the beginning and the end." (Rev. 22:13, cf. Rev. 2:8)

The second line of Revelation 1:8 ("Who is and who was and who is to come") is distinctive to liturgical pieces in the Book of Revelation that refer to God. Variations occur, for example, in the Sanctus at Revelation 4:8 and in the hymnic piece at Revelation 16:5. Except for the pastiche of quotations at 2 Corinthians 6:18, "the Almighty" (*Pantocrator*) occurs within the New Testament only in the Book of Revelation — for the most part in liturgical pieces (see Rev. 4:8, 11:17, 15:3, 16:7, 19:6).

That second line repeats the predication about God in the epistolary greeting (Rev. 1:4) and thus structures the various materials in Revelation 1:4–8 as a ring composition with the first occurrence of the ring appearing in an epistolary form, the second in a hymnic cry. Throughout the prologue epistolary elements combine with liturgical elements to establish the divine authority of what is being said.

The epilogue (Rev. 22:6–21) also combines epistolary and liturgical materials, but it tends to focus more on the audience — their faithfulness and its expectations — than on divine authority. A blessing is pronounced on the reader who keeps the words of prophecy in the book (Rev. 22:7, 14, cf. 1:3). Repetition of the eschatological cries at the end of the book gives emphasis to expectation and hope:

> And behold, I am coming soon
> .   .   .   .
> Behold, I am coming soon
> .   .   .   .

> I am the Alpha and the Omega, the first and the last,
> the beginning and the end.
>
> .   .   .   .
>
> I am the root and the offspring of David,
> the bright morning star.
> The Spirit and the Bride say, "Come,"
> And let him who hears say, "Come,"
>
> .   .   .   .
>
> "Surely I am coming soon."
> Amen. Come, Lord Jesus! (Rev. 22:7–20)

After various warnings to the audience as well as eschatological assurances, the book ends like a letter: "The grace of the Lord Jesus be with all the saints. Amen" (Rev. 22:21).

The epistolary elements in the prologue and epilogue make the author and those receiving the "letter" very visible. And those elements locate the work as a whole in the specific space of the province of Asia, where the seven churches reside. The work as letter is a product of an individual, addressed to specific people in specific places. Those epistolary elements combine with liturgical elements that display opposite tendencies: the liturgical is communal, not individual, in origin; so the individual author is suppressed. Moreover, the liturgical cries, the first-person acclamation, the doxology, and even the beatitudes "occur" in a nonspecific space, perhaps a more universal space. Finally, as with all liturgy, the liturgical elements in the prologue and epilogue of the Book of Revelation are centered on the divine, not the human, situation. The combination of epistolary and liturgical elements creates a distinctive setting for the Book of Revelation in the context of the genre "apocolypse." This work is as visionary and "apocalyptic" as any other, but it is grounded in a specific time and a specific place; there is no pseudonymity, for example, to confuse its location in the human world.

Distribution of Liturgical Language in the Visions

Liturgical language is also distributed throughout the body of the visions of the Book of Revelation (Rev. 4:1–22:5). Here it appears in scenes of worship that alternate with dramatic narration of things to come. Those scenes of worship are set in heaven, which in the Book of Revelation is a metaphor for transcendence. There God dwells and is sometimes called simply "the God of heaven" (Rev. 11:13, 16:11, cf. 13:6, 21:10). Heaven is a complex space. It is above the earth (e.g., Rev. 5:3) and contains such things as stars, the moon, rain, and hail (Rev. 8:10, 9:1, 11:6, 12:4, 16:21). There is also, however, another dimension to heaven that cannot be seen by the naked eye but becomes visible through transformational symbols of "going up," "opening," and "Spirit" (Rev. 4:1–2). After the seer's initial vision involving the messages to the seven churches (1:9–3:22), a door opens in heaven and the voice that originally spoke to him (Rev. 1:10) says, "Come up hither, and I will show you what must take place after this" (Rev. 4:1). John is transformed "into the Spirit," a psychological transformation homologus to "being taken into heaven."

Following the cue of the voice, we expect visions of eschatological events, that is, things to come. Instead, the seer sees worship around the throne of God in heaven. At several later points in Revelation, eschatological expectations will be met in the same way—by scenes of present heavenly worship.

The Throne Scene

The initial throne scene—a scene common to apocalypses—is described in detail. The one seated on the throne is like precious stones such as jasper, carnelian, and emeralds (4:3).[3] The throne is the central object; everything else is positioned in relation to it. Surrounding the throne are twenty-four additional thrones on which sit twenty-four elders dressed in white clothes and crowned with golden crowns. Seven lamps of fire burn before the throne, and extending out beyond the lamps of fire is a transparent sea like crystal. Before and around the throne are four living creatures, full of eyes: one is similar to a lion, the second to an ox, the third to the face of a man, and the fourth to a flying eagle. Each of the four creatures has six wings which are also full of eyes (4:4–8, cf. Ap. Abraham 18, 2 Enoch 22).

Worship is an essential element in this heavenly scene.[4] To the accompaniment of lightning and thunder (typical theophanic symbols in the Bible) the four living creatures never cease giving glory, honor, and thanksgiving to the one seated upon the throne; they do this in a liturgical Sanctus or *Kadosh*, the Latin and Hebrew words, respectively, for "holy":

> Holy, holy, holy
> Is the Lord God Almighty [*Pantocrator*]
> The One who was and is and is to come. (4:8)

John's introduction of heavenly worship by this thrice-holy liturgy illustrates his use of the Old Testament and his connection with the apocalyptic tradition. The origin of the thrice-holy lies in Isaiah 6:3:

> Holy, holy, holy
> Is the Lord of hosts
> The whole earth is full of his glory.

There it is sung by heavenly seraphim with six wings (Isa. 6:2). John, however, has the four creatures surrounding the throne singing the thrice-holy, and *their* origins lie in Ezekiel 1:4–28 (cf. Ezek. 3:12–15, 10:1–22). They have four wings (not six, as an Isaiah). Similar heavenly creatures sing the *Kadosh* elsewhere. According to 1 Enoch 39 those who stand before God without slumbering praise him saying "Holy, Holy, Holy, Lord of the Spirits; the spirits fill the earth" (1 Enoch 39:12). In 2 Enoch the six-winged, many-eyed ones who stand before the throne sing "with gentle voice," "Holy, Holy, Holy, Lord Sabaoth, Heaven and earth are full of his glory" (2 Enoch 21:1). In the later 3 Enoch (3 Enoch 1, 35–40) and perhaps in the Apocalypse of Abraham (Ap. Abraham 16), heavenly creatures also sing the *Kadosh*.[5]

The first expression of heavenly worship heard by the seer of the Book of Revelation is thus the song par excellence of apocalyptic visionaries. He, like the other apocalypticists, modifies it according to his own purpose and integrates it into his own style and vocabulary. The seer's second line, "Is the Lord God Almighty [*Pantocrator*]" is an adaptation of Isaiah's "Is the Lord of hosts."[6] The third line of the seer's *Kadosh*, "The One who was and is and is to come," recurs throughout the Book of Revelation (cf. 1:4, 8; 11:17; 15:3; 16:5).[7]

As a liturgical response to the thrice-holy of the four living beasts, the twenty-four elders prostrate themselves before the one on the throne and worship him (as they cast their crowns before the throne) in an elaborate acclamation of the creator God's worthiness to be praised (Rev. 4:11). This form of acclamation recurs in the throne scene (see Rev. 5:9–10, 12); it consists of the predicate adjective *worthy* ($\H\alpha\xi\iota\varsigma$) plus a form of the verb *to be* (in either second or third person) plus *to receive* ($\lambda\alpha\beta\epsilon\hat\iota\nu$) plus ascriptions. This *worthy* form may also include names for the worthy object of praise (e.g., "Our Lord and God") and an explanatory *for* ($\H\sigma\tau\iota$) clause, typical of hymnic, liturgical language. Ascriptions to God (e.g., *glory, honor, power*) in this elaborate acclamation are the same as those found in the doxology.

The first *worthy* acclamation goes like this:

> Worthy are you, our Lord and God
> to receive glory, honor, and power
> for you created all things,
> and by your will they came into being and were created. (4:11)

In these gestures and in this liturgical form, the twenty-four elders collapse into one the human spheres of politics and religion (see E. Peterson 1964, 6). The heavenly scene is portrayed as a temple throne room, the twenty-four elders are kings (political figures) with crowns, and they do obeisance before their king. They acclaim God as king on his throne, present him with golden crowns after the custom of the Roman imperial cult (see Aune 1983, 12–13), and praise him in the form of an acclamation ("Worthy are you") which probably has its origins in the political arena.[8] The address "Our Lord and God" may also resonate the acclamation in the imperial cult of Domitian (*dominus et deus noster*, see Mart. 7.34). The twenty-four elders acclaim God as worthy to receive glory, honor, and power because of the divine act of creation. As creator, God is ruler, a term that allows the seer to draw on both political and religious terminology.

After that litany the liturgical setting is elaborated further through the theme of "worthiness" introduced in Revelation 4:11. The seer sees a sealed scroll in the right hand of the one seated upon the throne (5:1). That scroll has an overpowering quality: no one—either in heaven, on the earth, or under the earth—is worthy (cf. 4:11) to open its seals. John weeps because no one is found worthy. One of the twenty-four elders then comforts the seer by assuring him that there is one worthy to open the seals of the scroll. The elder describes that "worthy one" in the political, religious language of Israel: he is "the Lion of the tribe of Judah, the Root of David," worthy to open the seals because he has conquered (Rev. 5:5). Then John sees a Lamb standing as if slain with seven horns and seven eyes; that animal takes the scroll from the one seated upon the throne. In response, the four living creatures and the twenty-four elders fall down before the Lamb and worship by singing a "new song" (ᾠδὴ καινὴ) in the form of an acclimation that further explains the worthiness of the Lamb (5:9–10):[9]

> Worthy are you
> to receive the scroll and to open its seals,
> for you were slain and by your blood you redeemed for God
> those from every tribe and tongue and people and nation,
> and made them a kingdom and priests to our God,
> and they will reign upon the earth.

The Lion from Judah (the Messiah) is worthy to open the seals because he conquered (5:5) by being slain. The Lion does not simply lie down with the Lamb, he is transformed into the Lamb who is victorious in death. Royal language of political power combines with the religious language of sacrifice (5:6). Just as the Lord God is worthy of receiving worship because (ὅτι) he created all things (4:11), so the Lion/Lamb is worthy of worship because (ὅτι) he was slain (5:9–10).

The symbolism in chapter 5 becomes unambiguously Christian. The sealed scroll remains closed to all except the slain Lamb and his followers. The closed book that reveals and realizes "the things which are to come" is a Christian book — one could almost say a Christian book of liturgy — disclosed only in the worship of the Christian community (whether in heaven or on earth). The messianic king gains victory through his death, and as a result he redeems and creates from all peoples a new, "Christian" people, an international community of kings and priests (political and religious elements combine here also) who will reign upon earth. The "new song" is a Christian song, known and comprehended only by Christians (cf. Rev. 14:3–4). The presence of the slain Lamb in the heavenly temple is one of the fundamental secrets revealed in the Book of Revelation (cf. 7:17).

A myriad of angels joins the other heavenly beings to hymn in response to the song that was sung (5:12):[10]

> Worthy is the Lamb who was slain
> to receive power and wealth and wisdom and might
> and honor and glory and blessing!

In response all creation — in heaven, upon the earth, under the earth, and upon the sea — offers a doxology to God and the Lamb (5:13):

> To the one seated upon the throne and to the Lamb
> Blessing and honor and glory and power
> for ever and ever.

After that crescendo of voices embracing, finally, the whole cosmos, the four living creatures conclude the liturgy with an *amen* (5:14), and the scene ends with the elders in the act of worship (5:14).

Subsequent Scenes of Heavenly Worship

Heavenly worship continues throughout the rest of the visions in the Book of Revelation. The heavenly setting continues to be a throne/temple; the twenty-four elders, the four living creatures, angels, and members of the redeemed community praise God and/or the Lamb; and the forms of praise recur as hymns, doxologies, and acclamations.

At Revelation 7:9 an innumerable crowd from every tribe and tongue and people and nation (cf. 5:9), wearing white stoles and holding palm branches in their hands, appear before the throne and the Lamb. In a loud victory cry (see Deichgräber 1967, 53) they acclaim God and the Lamb (7:10):

> Salvation to our God, the one seated upon the throne
> and to the Lamb.

In response all angels around the throne and the elders and the four living creatures fall upon their faces and worship God in a doxology with seven ascriptions (7:12):

> Amen! Blessing and glory and wisdom and thanksgiving and honor
> and power and might be to our God for ever and ever! Amen.

This scene (7:9–17) emphasizes religious, cultic aspects of the heavenly throne/ temple.[11] The innumerable crowd before the throne worships God "day and night within his temple," and the one upon the throne dwells with them (7:15).[12] That temple existence provides an idyllic life—no hunger, thirst, or hot sun—under the guidance of the Lamb (7:16–17). The situation in the heavenly temple is thus similar to that in the New Jerusalem where those who thirst drink from the well of living water (21:6)—though of course there is no temple in the New Jerusalem. This idyllic scene contrasts with the burning sun which is given to those who blaspheme the name of God (Rev. 16:8–9).

The blowing of the seventh trumpet brings the next sequence of heavenly, liturgical responses. At the blowing of that last trumpet, loud voices in heaven cry in victory (11:15):[13]

> The kingdom of the world has become the kingdom of our Lord
> and of his Christ,
> and he shall reign for ever and ever.

The twenty-four elders who sit upon their thrones before God then fall upon their faces and worship, saying in a thanksgiving (11:17–18):

> We give thanks to you, Lord God Almighty
> Who is and was,
> for you took your great power and reigned;
> and the nations raged, and your wrath came and the time
> of the dead to be judged,
> to give reward to your servants, the prophets, and to the saints
> and to those who fear your name, to the small and the great,
> and to destroy those destroying the earth.

After that lengthy thanksgiving, the heavenly temple opens and the ark of the covenent is seen in the temple. Lightning, noises, and thunder came forth along with an earthquake and hail (11:19).

Heavenly worship occurs again in the description of the battle between Michael, a good angel in heaven, and the Devil (*diabolos*), who is also in heaven. After the victory of Michael over the *diabolos* (12:7–9), a loud voice cries victory (12:10–12):

> Now has come the salvation and power and kingdom of our God
> and the authority
> of his Christ,
> for the accusor of our brethren was cast down,
> the one accusing them before our God day and night.
> And they conquered him by the blood of the lamb and by the word
> of their witness
> for they did not love their lives, even unto death.
> On account of this, rejoice, heavens and those dwelling in them;
> But woe to the earth and the sea,
> for the diabolos has come down to you in great wrath
> for he knows that little time remains.

This victory song includes the basis for victory in typically hymnic style (ὅτι, "for"), a description of those conquering, a summons to rejoice, and finally a contrasting warning of woe. The victory occurs in heaven; here, however, the scene is not that of the heavenly throne/temple but that of a conflict between heavenly powers.

The pouring out of the seven bowls (Rev. 15:1–16:21) unfolds from a heavenly scene just as did the unsealing of the seals. It begins with heavenly worship, in this case, singing by those who conquered and stand on the sea (cf. 4:6). They sing the song (ᾠδή) of Moses and the song of the Lamb. Here is what they sing with harp accompaniment (15:3–4):

> Great and marvelous are your works,
> Lord God *Pantocrator*;
> Just and true are your ways,
> king of the nations.
> Who would not fear you, Lord,
> and glorify your name?
> because you alone are holy,
> because all nations come and worship before you,
> because your just judgments were revealed.

With its rhetorical question, its addressing God in liturgical sentences, and its *because* clauses, this hymn is one of the most structured in all the Book of Revelation. Throughout this song the seer draws freely on the Old Testament, especially the Song of Moses in Exodus 15: both songs are sung "at the sea," as a victory song celebrating the demise of an evil force. There is also a deliberate parallelism drawn between the Lamb and Moses and between the Christian community and Israel (see Deichgräber 1967, 56).

After the victorious complete their singing, the temple of the tent of witness opens (15:5, cf. 11:19): seven angels come forth from the temple with seven plagues, and one of the four living creatures gives to them seven golden bowls filled with the wrath of God. The smoke of the glory of God fills the temple so that no one can enter until after the seven plagues (seven bowls) are poured out (15:8). The pouring out of the seven bowls follows (16:1–21), and a theophany closes out this series of seven.[14]

When the third angel pours out its bowl of wrath upon the waters and turns them to blood, the seer hears the angel of the waters acclaiming (16:5),

> Just are you
> the one who is and was
> the holy one,
> for you made these judgments;
> for they poured out the blood of the saints and prophets
> and you have given blood for them to drink.
> Worthy are they.

This acclamation of God's just judgment contains familiar hymnic elements and ends with an eschatological talion—the eschatological punishment repeats the offense: blood shed/blood drunk. The concluding "worthy formula" has a completely different meaning from those acclamations declaring divine worthiness. The altar repeats antiphonally after the angel of waters a similar acclamation (16:7):

> Yes, Lord God *Pantocrator*,
> true and just are your judgments.

The last liturgical sequence appears in connection with the judgment of Babylon the Whore, in which vengence and retribution play a prominent role (17:1–19:10). A crowd of voices in heaven begins the litany with a victory ode combined with an acclamation of God's justice (19:1–2):

> Alleluiah,
> salvation and glory and power belong to our God
> because his judgments are true and just;
> for he judged the great whore
> who corrupted the earth in her fornication,
> and he has avenged from her hand the blood of his servants.

A second voice responds (19:3),

> Alleluiah;
> and her smoke rises forever.

The twenty-four elders and the four living creatures then fall down and worship God in an antiphonal response (19:4):

> Amen,
> Allelujah.

Then a voice from the throne exhorts (in hymnic style) praise of God (19:5):

> Praise our God, all you his servants,
> You who fear him, small and great.

This liturgy concludes with a final hymn said by a great crowd of voices (19:6–8):

> Alleluiah,
> for the Lord our God *Pantocrator* reigns.
> Let us rejoice and be glad,
> and let us give him glory,
> for the marriage of the lamb has come,
> and his wife has prepared herself.
> (And it was granted her to wear bright, clean linen;
> for the linen is the righteous deeds of the saints.)[15]

This heavenly celebration moves from victory ode to acclamation of divine justice, to cries of alleluiah, to exhortations to praise God because of the marriage of the Lamb. It is followed by a series of visions that describe in narrative form a final conflict between the forces of God led by one seated on a white horse (19:11), the throwing of Satan into the Lake of Fire (20:10), resurrection and judgment (20:13), and finally the New Jerusalem coming down from heaven (21:2). In this renewed, transformed age heavenly worship does not occur, for spatial transcendence has become eschatological transcendence. The transformed age collapses heaven into earth.

Heavenly Worship and Eschatology

In order to understand the relationship between heavenly worship and eschatology in the Book of Revelation, let us return to the definition of the genre "apocalypse" worked out by John J. Collins and other members of the Society of Biblical Literature in the Apocalypse Group of the Genres Project: "a genre of revelatory literature with a narrative framework, in which a revelation is mediated by an otherwordly being to a human recipient, disclosing a transcendent reality which is both temporal, insofar as it envisages eschatological salvation, and spatial insofar as it involves another, supernatural world" (Collins 1979, 9). The relevant part of the definition for an understanding of how heavenly worship functions in the Book of Revelation is "spatial insofar as it involves another, supernatural world."

In brief, scenes of heavenly worship express the spatial dimension of transcendent reality in the Book of Revelation, just as dramatic narratives of things to come express the temporal dimension of transcendent reality in this apocalypse. In most apocalypses that emphasize the spatial, vertical dimension, heavenly journey is described, and the visionary sees such things as secrets of the seasons, store rooms of the winds, sometimes the throne-chariot (*merkabah*) of God and furnishings associated with it. In that connection, heavenly worship sometimes occurs (e.g., Ap. Abraham 17; 2 Enoch 22), but such worship does not dominate other apocalypses the way it does the Book of Revelation. In the Book of Revelation the spatial dimension of transcendence takes the form of heavenly worship.

While some apocalypses emphasize the spatial (especially those with otherworldly journeys) and others (the so-called historical apocalypses) the temporal, one of the distinctive elements of the genre — that is, what makes a piece of writing identifiable as an apocalypse — is the presence and interplay of spatial and temporal dimensions of transcendence. Each is as important as the other. Thus, it would be incomplete to discuss an apocalypse solely in terms of eschatology (i.e., the temporal); for such notions as the eschatological transformation of this world into a "new heaven and a new earth" represent only the temporal dimension of transcendence. To complete the discussion, it is crucial to look at the way in which an apocalypse expresses the spatial dimension of transcendence.

The presence of both dimensions in an apocalypse guarantees that the revelation is integrally related to human earthly existence, that is, that there is no radical discontinuity between God and the world (spatial transcendence) or this age and the age of come (temporal transcendence) (see Rowland 1982, 92, 175, 475). A radical transcendence that could sever heaven from earth is tempered by the future transformation of earthly into heavenly existence; and a radical transcendence that could sever this age completely from the age to come is tempered by the presentness of the age to come in heaven. Thus, the presence and interplay of spatial and temporal dimensions in transcendence prevent a thoroughgoing dualism in which transcendent realities would become separated from everyday human activity (see chap. 2). Through the temporal, heaven touches earth; through the spatial, the future touches the present (see also E. Peterson 1964, 2; Minear 1962). Within this generic framework we can see more clearly the interplay in the Book of Revelation between scenes of heavenly worship and dramatic narratives of things to come: heavenly worship celebrates eschatological realities in the present; and the *eschaton* is portrayed as the "coming down" of heavenly realities.[16]

Eschatological Themes in Heavenly Liturgies

The Book of Revelation envisions a glorious reign of God when God will be king. So in the dramatic narration of Revelation 19:11–22:5 God and his Christ reign over all else. God sits upon his throne in the New Jerusalem and controls all. This same royal theme recurs throughout the heavenly liturgies in the Book of Revelation. In scenes of heavenly worship, God sits upon a throne, and he is worshipped as king. In liturgical pieces, he is acclaimed as king. For example, the voice of the great crowd cries, "Alleluiah, because the Lord our God, almighty, reigns as king" (19:6). Kingship is also implied in the epithet "Lord God *Pantocrator*" as is made clear in the synonymous parallelism between that phrase and "king of the nations" at 15:3 (cf. also 1:8, 4:8, 11:17, 16:7, 19:6). In the presence of the king, worshippers respond appropriately by bowing down and worshipping. The four living creatures and the twenty-four elders fall down before the one enthroned and do obeisance (5:14, 19:4). All the angels join them in worship, acclaiming God in doxology (7:11–12, 5:11–12). Earthlings are urged to do the same (14:7). Only the Lion/Lamb does not do obeisance before the enthroned God; rather, other creatures worship him alongside God (e.g., 5:13).

Worshippers ascribe the same attributes to both God and the Lamb, with *glory* (δόξα) the most common.[17] *Glory*, like other ascriptions, is never clearly defined in Revelation, but it is something that only God and the Lamb are worthy to *receive* (4:11, 5:12–13). Conversely, glory is *given* to God by those who fear him (11:13), whereas unrepentant blasphemers refuse to give him glory (16:9). Glory belongs to God (1:6), fills the temple when God enters it (15:8), and becomes an attribute of the New Jerusalem (21:11), which needs no sun because God's glory shines in it (21:23).[18] Thus *glory* is a term used in both liturgical and eschatological settings; or, perhaps more accurately, the seer may not be making a sharp distinction between liturgical and eschatological situations.

A variety of other ascriptions surround the term *glory* in acclamations to God and the Lamb. Power (δύναμις), honor (τιμή), and praise (εὐλογία) occur most frequently (4:9, 11; 5:12–13; 7:12; 12:10; 19:1). Others include *wealth* (πλοῦτος), *wisdom* (σοφία), *strength* (ἰσχύς), *might* (κράτος), *thanksgiving* (εὐχαριστία), *salva-*

tion (σωτηρία), *kingdom* (βασιλεία), and *authority* (ἐξουσία) (1:6; 5:12–13; 7:10, 12; 11:15; 12:10–12). Three of those ascriptions – power, wealth, and authority – indicate the greatness of evil forces as well as the grandeur of God and the Lamb. These ascriptions are often piled one on another in doxologies in order to indicate the grandeur and worthiness of the object being worshipped.

Creation relates closely to the kingship of the God, that is, the God is king because he has created all things. For example, at 4:11 God is reckoned worthy of receiving glory, honor, and power because (ὅτι) he is the creator. All things (τὰ πάντα) came into existence because God so willed. Other liturgical phrases in Revelation also affirm that the God being worshipped is the creator of all things (see 10:5–6, 14:7).

The Lamb, the only other legitimate object of royal worship in Revelation, receives ascriptions because (ὅτι) he was slain and he redeemed through his blood people from every tribe and nation (5:9, 12). The paralleling of God and the Lamb in the liturgies of chapters 4 and 5 conveys in a subtle but unmistakable manner that creation and redemption are centerpieces of the whole work. Through death the Lamb conquered and was worthy of opening the seven seals of eschatological destruction (5:5). Images of power and awe – seven horns and seven eyes – mix with sacrificial images of death and blood redemption (5:6). So, later, in the vision of the innumerable crowds, those in the temple are able to worship God day and night because their stoles had been washed in the blood of the Lamb (7:14). In the idyllic scene describing those in the temple, the Lamb who is in the midst of the throne above shepherds "is king over" them so that they neither hunger nor thirst (7:16–17). Similarly, at the casting out of Satan from heaven, the brethren conquered Satan through "the blood of the Lamb" and through their witness unto death in imitation of the Lamb (12:11). By placing the slain Lamb in the throne scenes in different ways, heavenly worship becomes a way of expressing the irony of kingship through crucifixion (see chap. 3).

The kingship of the Christ (11:15) results from his role as redeemer. In dramatic narration of things to come he is named "King of kings and Lord of lords" (19:16), and he will reign as king for one thousand years (20:4–6). But the Lion/Lamb who conquers by his death belongs not solely in the eschatological drama; he belongs as well in the world's present structure as revealed in the scenes of heavenly worship. He not only comes as the "pierced one" (1:7), but his crucifixion occurs before the foundation of the world (13:8). Thus, the kingdom of God and the rule of the Messiah – future, eschatological claims – are acclaimed in heavenly liturgies as present, "eternal" realities.

The eschatological theme of God's just judgment is equal in importance to kingship in the liturgies of Revelation. God's judgment can be associated with a specific incident in the dramatic unfolding of the end (for example, the judgment of Babylon at 18:8, 10, 20; 19:2), but in the liturgies judgment refers primarily to the eschatological judgment of the living and the dead. At the blowing of the seventh trumpet, the twenty-four elders give thanks to the Lord God almighty because he reigns as king and because the time for judging the dead and giving proper rewards has come (11:18). An angel flying in midheaven proclaims the same decree: "Fear God and give him glory for the hour of his judgment has come" (14:6–7). Between the pouring of the third and fourth bowls of wrath an angel proclaims justice, and a voice from the altar responds, "Yes, Lord, God almighty, true and just are your judgments" (16:7). The justness of God is also a major idea in the Song of Moses and the Lamb: "Great and marvelous are your works . . . because you alone are holy, all nations come and worship before you, and your just judgments are revealed" (15:3–4). In this song "just judgments" parallel "divine holiness" as they do in the address of the angel of the waters (16:5).

Vengeance is closely associated with just judgment. The pouring out of the seven bowls of wrath on the inhabitants of earth is just, according to the angel of the waters, "because they poured out the blood of the saints and the prophets" (16:6). The judgment applies a *lex talionis*: because they poured out the blood of the saints, God has given them blood to drink. Vengence is even more explicit in the alleluiah following the judgment of Babylon. God is acclaimed because he judges the Great Whore and exacts the blood of his servants from her hand (19:1–2). In those acts of vengeance the cry of the martyred souls under the altar is answered (6:9–10).

Martyred souls "exist" in heavenly worship although they are not resurrected until the final eschatological events described in Revelation 20–21. Prior to their resurrection (e.g., 20:4–6), they cry out for vengeance from under the altar (6:9–11). There is thus a "communion of the saints" in heavenly worship before they are resurrected in the eschatological drama.

Heavenly Worship as Eschatological Celebration

The seer places heavenly worship strategically in his description of "things to come"; that is, the distribution of heavenly worship is not random in the Book of Revelation; it occurs in relation to the dramatic narratives of things to come. The seer links worship to eschatology in two ways. First, eschatological drama has its setting in heavenly worship. Second, heavenly worship celebrates in the present the dramatic finale of eschatological narrative.

The opening of the seven seals, which narrates eschatological drama, flows from the scenes of heavenly worship in chapters 4 and 5, more specifically, from the litany offered to the Lamb (5:9–14). After the new song is sung, the angels respond, and a doxology is said by all creatures, the first seals are opened (6:1). The seer describes four horsemen, sent forth against the earth and its inhabitants for their respective purposes of conquering with a bow, taking peace from the earth, causing famine for the inhabitants of the earth, and killing one-fourth of the earth (6:2–8). That judgment against earth is the beginning of vengeance on behalf of those who were killed because of their faithful witness to the word of God (6:9–11). Those narratives of disclosures are eschatological events; they describe "the great day of wrath." After this dramatic narration, the writer returns once again to heavenly worship. Innumerable persons dressed in white and waving palm branches acclaim the Lamb and the one seated upon the throne: "Salvation to our God who is seated upon the throne and to the Lamb" (7:10). All the angels around the throne respond by falling down and worshipping in the form of a doxology: "Amen! Blessing and glory and wisdom and thanksgiving and honor and power and might be to our God for ever and ever. Amen." (7:12). This heavenly liturgy forms the climax to the opening of the seals. The opening of the seventh seal, which follows this heavenly worship, brings no further dramatic narration. It brings liturgical silence in heaven (8:1).

The drama of the seven trumpets is also introduced by heavenly worship. On the altar before the throne, an angel mixes incense and the prayers of the saints (8:3–4). That sacrificial activity is the setting for the blowing of the seven trumpets. The seer envisions one-third of the earth burned up by hail and fire mingled with blood; one-third of the seas turned to blood; one-third of the ships perishing; one-third of the spring waters turned to bitter wormwood; and one-third of the sun, moon, and stars darkened (8:6–12). The reader or hearer is led to expect a great climactic finale at the blowing of the seventh trumpet. After the first four angels have blown their trumpets and the consequent horrors and terrors have come upon the earth, an eagle is heard in midheaven saying "Woe, woe, woe, to those who dwell upon the earth, because of the trumpeting which the three angels are about to sound" (8:13). Two of the remaining three trumpets bring more horrendous destruction (9:1–21). Prior to the blowing of the seventh trumpet, an angel standing upon the sea and the earth raises his right hand towards heaven and swears by the living God who made heaven, earth, and sea that "time will be no more; in the days of the voice of the seventh angel, when he is about to blow, the mystery of God will be completed, as he announced to his servants, the prophets" (10:6–7). The climactic significance of the seventh trumpet could not be stated more clearly than it is by the angel in this passage. The *eschaton* will come at the blowing of the seventh trumpet. After all this expectation, the blowing of the seventh trumpet brings a surprise. There is no horrendous destruction. There is no decisive battle between the forces of evil and the forces of God. Rather than more dramatic narration of an eschatological event, *the seventh trumpet discloses heavenly worship*: "Then the seventh angel trumpeted, and loud voices were heard in heaven saying: 'the kingdom of the world has become the kingdom of our lord and his Christ, and he shall reign for ever and ever'" (11:15). After that acclamation, the twenty-four elders fall down and worship God in a liturgical thanksgiving, thanking God because he took up his great power and reigned as king, because the judgment of the dead occurred, and because proper reward was given to those fearing his name (11:17–18). Following that long thanksgiving by the elders, the heavenly temple opens and the ark can be seen. Then a hierophantic display of thunder and lightning brings the blowing of the seven trumpets to an end (11:19).

The salvation and kingdom of God is celebrated again liturgically when Satan is cast from heaven (12:9). A loud voice says, "Just now [ʾαρτι] the salvation, power, and sovereignty of our God has occurred along with the authority of his Christ" (12:10). That liturgical affirmation is followed by narratives about a dragon's pursuit of a pregnant woman who is about to give birth to a man-child (12:13–17), a beast from the sea with features similar to the fiery dragon (13:1–10), and a beast from the earth who sounds like the dragon but has the authority of the water beast (13:11–18). Then the seer sees 144 thousand people with the Lamb on Mount Zion singing a new song (14:1–5, cf. 5:9), three angels proclaiming God's victory over the beasts (14:6–13), and two reapers of the earth (14:14–20). Within the visions of

chapters 12–14, thus, God's reign and his salvation and judgment are affirmed
liturgically, even though that set of visions does not end with heavenly worship.

The next series of visions describing eschatological terror is once again introduced through heavenly worship. A vision appears of those victorious over the
beast standing before God on the transparent sea (cf. 4:6) and singing the Song of
Moses and the Lamb (15:3–4). In that song, as in the liturgy at the blowing of the
seventh trumpet, the reign of God and his just judgment are established. The
heavenly temple opens (15:5, cf. 11:19), and seven angels come forth holding seven
future plagues to be poured out upon the earth. Here, again, the cultic place and
eschatological disclosure are linked: the opening into the holy of holies simultaneously opens into the last plagues (cf. 4:1, 6:1). After the pouring out of those
terrors a sacral display of thunder and lightning closes that series (16:17–21).

The visions then shift to the judgment of Babylon the Whore (17:1–19:10).
John sees the Great Whore seated on a scarlet beast with seven heads and ten horns
(17:3). She is clothed in purple and scarlet (in contrast to the white clothes of the
godly), and on her forehead is written, "Babylon the great, the mother of whores
and abominations of the earth" (17:5). She is drunk on the blood of the saints and
those who witness to Jesus. An angel interprets the meaning of the seven heads and
the ten horns of the beast. Another angel declares that Babylon is fallen (18:1–3); a
voice orders Christians to come out of her so that they will not be destroyed (18:4–
8); kings, merchants, and shipmasters wail over their loss of her (18:9–24). After
these eschatological narratives, scenes shift to heavenly worship. John hears the
voices of a great crowd celebrating — in the form of an alleluiah litany — the kingship
of God over the Whore and the just judgment that he gave to her (19:1–5). Alle-
luiahs continue, but the content shifts to the universal kingship of God (19:6):
"Alleluiah, for the Lord our God Almighty reigned." The motive for this acclamation is "the marriage of the Lamb" (19:7), an allusion to the New Jerusalem in the
new heaven and the new earth described later in Revelation (21:2). Thus, the alle-
luiahs celebrate in worship the victory of God over the evil forces and the establishment of the new age.

Eschatological Homologues to Heaven

Finally, the seer links scenes of heaven with eschatological narration by giving to
them similar attributes and by drawing homologues between them. Similes to precious stones describe both the throne scenes in heaven and the eschatological Jerusalem (Rev. 4, 21).[19] Only the heavenly throne and the New Jerusalem are said to be
"fixed" or "situated," the one in heaven (4:2), the other as "foursquare" or "cubed"
(21:16). Only in the heavenly scene of 7:13–17 and in the New Jerusalem are there
references to "washed stoles" (7:14, 22:14) and to God's dwelling with his people
(7:15, 21:3). Only in those two places will people not thirst, for God will give them
water from running springs (7:16, 21:6); only in those two will people worship
(λατρεύω) God and the Lamb (7:15, 22:3). Finally, only in those two idyllic locations does God wipe away every tear from the eye (7:17, 21:4).

In summary, heavenly worship and eschatological drama in Revelation form
homologies and share similar motifs and attributes. Further, eschatological drama

arises in a liturgical setting, and at several points eschatological drama climaxes in heavenly worship and liturgical acclamations. Both celebrate God's kingship, his just judgments, the resurrection of the dead, and idyllic blessedness. The interplay between the spatial transcendence of heavenly worship and the temporal transcendence of eschatological drama establishes one of the most fundamental relationships in the Book of Revelation.

The Worshipping Community

The heavenly scenes in the Book of Revelation assume a fairly complex worshipping community—different kinds of worshippers who worship in different ways at different distances from the throne. Yet there are certain common characteristics shared by all members of this community.

First, it is made up of worshippers located spatially around the throne of God. The four living creatures and the twenty-four elders seem to have a special place in the community. They are closest to the objects of worship, and they are portrayed in distinctive ways. They are royalty—at least the twenty-four elders are—and they function as priests offering incense to God and the Lamb. They also function as singers and musicians. On occasion many angels and myriads of heavenly voices join in worship with these twenty-eight special figures around the throne. The worshipping community even breaks the bounds of heaven. At Revelation 5:13 the worshippers include all creation—those in heaven but also those on earth, under the earth, and upon the seas; they sing a doxology. Between the opening of the sixth and seventh seal, earthlings appear again in the worshipping community. People from every nation, tribe, people, and tongue stand before the throne and the Lamb (7:9). They are those who have gone through the great tribulation and have made their stoles white in the blood of the Lamb (7:14-15); that is, they are portrayed as Christian martyrs before the throne. They hold palm branches in their hands as signs of victory.[20] Those who sing the Song of Moses and the Lamb are also victorious—over the beast and his image and the number of his name (15:2)—on earth. The community of worship breaks down the boundaries between heaven and earth.

Worship is a radical equalizer that breaks down all boundaries in heaven and earth except that between the worshipping community and the two objects of worship. When the heavenly creature with a mighty voice speaks to John, John falls down to worship him, but the voice says, "I am your fellow servant [σύνδουλος] along with your brethren who hold to the witness of Jesus. Worship God" (19:10). Even more forceful is an exhortation at the end of the book. John falls down to worship the angel who showed him all the visions that he recorded. In response the angel says, "I am your fellow servant along with your brethren the prophets and those who keep the words of this book. Worship God" (22:8-9). Those who read, hear, and observe John's words in Revelation are one with John the seer and the angels. Christians in the churches of Asia Minor, the seer, angels throughout the universe, the twenty-four elders, the four living creatures, and the martyred dead under the altar all form, in the words of the late Victor Turner, an egalitarian *communitas*, one community of worship (see also Gager 1975, 33). All—whatever their station and location in the universe—join together in obeisance and submission to the one seated upon the throne and to the slain Lamb.

Earth also enters into heavenly worship in indirect ways. The twenty-four elders and the four living creatures offer bowls of incense, which are the prayers of the saints (5:8), presumably the saints on earth as well as in heaven. The "new song" sung to the Lamb links earthlings to the heavenly Lamb. The seer invites a connection between heaven and the Jerusalem temple on earth (whether or not it was defunct at the time of his writing). Parallels are fairly obvious. The seer calls the heavenly setting a temple (Rev. 7:15; 11:19; 14:15, 17; 15:5, 6, 8; 16:1, 17). An altar stands before the throne (holy of holies) (8:3). The seven "lamps of fire" (4:5) reflect the seven-lighted lampstand in the Jerusalem temple (see Zech. 4:1-6, 11-14); and the "transparent sea like rock crystal" (4:6) probably alludes to the large bronze laver before the altar in the temple, used for ablutions (Exod. 40:7). The twenty-four elders correspond to the twenty-four classes of priests and Levites who oversaw the services in the postexilic temple. Those priests, like the elders in heaven, wore both priestly and royal insignia.[21] In the Book of Revelations these elders are, of course, transformed into Christians, and their royal and priestly combination reflects that of the Christian community (see Rev. 5:10). As in the temple traditions at Jerusalem, inside the heavenly temple stands the ark of the covenant or the tent of witness (11:19, 15:5), and heavenly sacrifices create smoke so thick that no one can enter the temple (15:8).[22]

The worshipping community is clearly a Christian community—no other community would worship the slain Lamb. As we have seen, the symbolism in chapter 5 of the Book of Revelation is unambiguously Christian. The victory of the Lamb is celebrated because he redeems and creates a new Christian people from every tribe and tongue. That specifically Christian worshipping community, whether envisioned on earth or in heaven, is assumed throughout Revelation. This work is addressed to Christians, and relationships in the work are formed from the viewpoint of a Christian community. From the seer's point of view, there are questionable members of the churches in Asia Minor—those who follow prophets or prophetesses who are John's rivals (e.g., Jezebel, the Nicolaitans). John threatens those people with the eschatological appearance of Jesus, soon to come. In all these conflicts and disagreements, however, all worship God and the Lamb, however misguided some of those Christians are from John's point of view. He refers to those who reject Christian worship and refuse to repent (e.g., 9:20-21), but those are not the people upon whom John focuses. His visions are for the Christian worshipping community.

The worship of the one on the throne and the slain Lamb thus serves two complementary functions: First, it establishes an egalitarian *communitas* among Christians, and it helps to establish a clear boundary between Christians and non-Christians, as it places the followers of the Lamb at the center of the universe. The throne and the Lamb are at the center; if one's face (whether on earth or in heaven) is not pointed in the direction of the throne and the Lamb, one looks to the periphery of the universe. Worship establishes what is truly real and therefore what is true: the presence of the Lamb beside the throne is revealed in the seer's vision—a disclosed mystery — but it is no less real and true because it is known only through revealed knowledge. True worship reveals the way things really are, and true worshippers form an egalitarian *communitas* around the center.

Let us be clear at this point. The boundary formed by worship is not a social boundary expressed in a different sphere, as though the social boundary between Christian and non-Christian determines the form of worship in heaven. Rather, the boundaries formed in worship — like the boundaries formed by heavenly forces, by eschatological events, and by social relations in the province of Asia — reveal dimensions of the seer's world. No one dimension can claim causal priority over the others; each equally reflects true knowlege about the world. The social boundary between Christians and non-Christians expresses in the region of social experience that which a liturgical boundary expresses in the region of worship. The whole grid of boundaries (social, eschatological, heavenly/spatial, liturgical, psychological) involves distinctions between true knowledge and deception, authentic self-expression and false consciousness, service to the true god and idolatry. No one boundary region simply codes or allegorizes another boundary region. As in Einstein's notion of time and space, each boundary region is a coordinate in a multidimensional reality.

Hymns in the Worshipping Community

Just as hymns and liturgical celebrations within the Book of Revelation make present the kingdom of God and his just judgment prior to the dramatic narration of those eschatological events, so hymnody generally in the early Christian church celebrates God and his Christ and brings their presence into the worshipping community. Singing involves a double movement. It is a movement from the human to the divine, a human act of praise; but it is also a movement from the divine to the human, a making present of the God. Hymns can be sung only if the Spirit impells.[23] The evocative power of the hymnic word becomes a means whereby one encounters the power and reality of the God to which one responds. Hymn singing, like all other acts of worship, involves both response and encounter.[24] Tertullian writes about the psalmist David that the singer "sings to us of Christ and through his voice Christ indeed also sang concerning himself" (Tert. Carn. 20). Through hymnic performance Christ becomes present. Hymns function within the Book of Revelation as they functioned in the worship life of the early church. More precisely, a reality apprehended in an eschatological context as future was realized as present in the worship of the early church through the evocative power of the hymnic word (see Thompson 1973; Barr 1984, 47–49).

Because hymns were vehicles of sacral power, they were probably sung as preparation for receiving revelatory visions. In a certain kind of Jewish mysticism called *merkabah* hymns prepare the mystic to see the divine glory (see Rowland 1979, 152). Gruenwald notes that hymns are learned to serve as theurgic protectives on behalf of the visionary who ascends and decends through cosmic spheres.[25] Many of those Jewish hymns do not read, however, as magic formulas; they are lyrical songs of praise "in their tone and form" (Gruenwald 1980, 103). The songs learned are "songs of praise said by the angelic beings, and even by the Throne of Glory to

God" (p. 104). Gruenwald concludes, "Whether these hymns were said as autohypnotic means or whether they were recited in heaven as protective means, their numinous quality establishes them as outstanding specimens of Jewish poetry in Talmudic times" (p. 104). Magical papyri from the Greek world may also have been used in receiving revelation (see Aune 1986a, 82–83). It is possible that some of the heavenly, hymnic liturgies found in the Book of Revelation served the purpose of preparing the seer to receive and handle revelatory visions.

Apocalypses in the Worshipping Community

The Book of Revelation itself engages the worshipping community. Reading and listening to the Book of Revelation are themselves liturgical acts in the worship life of Christians in western Asia Minor. The final link in the chain of revelation takes the form of a beatitude for those who receive the work: "Blessed are the reader and those who listen to the words of the prophecy and keep the things written in it" (1:3). The chain of revelation is complete when the Book of Revelation is read in Christian services of worship; that is, the book itself becomes liturgical material for the churches of western Asia Minor (see Schüssler Fiorenza 1981, 19; Gager 1975, 56).

Moreover, the seer receives his visions "on the Lord's Day" (1:10)—*in sacro tempore*—the day of worship in the early church,[26] just as he expects them to be read in the worshipping community. Prophetic revelation is both received and proclaimed in the context of worship. Those comments by the seer square with Paul's, who states that an "apocalypse" makes up a part of the service when Christians gather for worship (1 Cor. 14:26). At the end of a discussion on spiritual gifts, Paul describes a service of worship: it includes, among other things, the singing of hymns and the proclamation of apocalypses (1 Cor. 14:26). Certain rules govern: if one prophet is speaking and someone else gets an apocalypse, the first should give the floor to the second. A form of contagion may here be suggested; that is, listening to an apocalypse may evoke the spiritual experience of receiving an apocalypse. Aune suggests that one of the liturgical functions of the Book of Revelation was to evoke "a new actualization of the original revelatory experience" of the seer (see Aune 1986a, 89). According to Paul every prophet should be allowed to give his revelation, so that all the people can both learn and be comforted. The prophet can use any one of several forms of worship: a prayer, a hymn, a revelation, or even a teaching. The important thing is that the services be orderly and controlled. The true prophet, even when he is "in the Spirit," has control (1 Cor. 14:32). The close connection between worship and apocalypse in the Book of Revelation thus conforms in several respects to what Paul says in 1 Corinthians.

There is, thus, a reflexive character to the Book of Revelation: the use of the Book of Revelation in the churches reflects the interconnection of heavenly worship and eschatological drama within the book. In both Revelation and the early church, worship serves as the context and setting in which eschatological narratives (such as the Book of Revelation itself) unfold. Furthermore, in both Revelation and the churches of Asia Minor, worship realizes the kingship of God and his just judgment; through liturgical celebration eschatological expectations are experienced presently. Hymns, thanksgivings, doxologies, and acclamations realize in the context of worship the eschatological message. Worship, then, becomes a context that integrally relates the visions in Revelation with John's original revelatory experience and the re-presentation of John's experience in the life of the worshipping community. The Book of Revelation, by functioning in communal worship of Asia Minor as heavenly worship functions in the book itself, links heaven and earth. The work mediates its own message.

5

The Seer's Vision
of an Unbroken World

In considering the linguistic unity of Revelation, I have thus far been concerned with the shape of the language itself: how words, phrases, sentences, and larger forms are related in Revelation through narrative and metaphoric devices or through liturgical language. The language of the seer, however, yields other secrets than its own shape; it also transmits a vision of the world or a construction of reality. The choice of terms is important here. The vision transmitted by the seer is not merely a "literary world" or a "symbolic universe"—a vision separate from the everyday life of John and his audience. The seer is constructing an *encompassing* vision that includes everyday, social realities in Asia Minor.

The seer's vision or his construction of reality becomes accessible to us through the language he uses. Thus, in discussing the seer's vision of an unbroken world, our focus remains on the language of the Book of Revelation; but now our concern is with the semantics of that language (its meaning and reference) rather than its syntax (the structure and relationships of phrases and sentences). In the jargon of some linguists, the shift is from linguistic signs as signifiers to linguistic signs as signified. In saying that, we must be careful. The shift does not involve a turn outward, for example, to the "social, historical situation." That is a further step to be explored in part 3. The seer's vision of the world is discovered through exploring the semantics of his language. That exploration is a step in the direction of understanding the social situation of Christians in Asia Minor, but the compass here remains the linguistic construction of the seer.

By understanding the seer's vision as an unbroken world, I offer an alternative to some recent theories about the Book of Revelation. Several scholars uncover a world of conflict in Revelation, a conflict between the seer's religious faith and his experience of Roman society (see chap. 1 and App. A). According to "conflict" theorists the syntax of Revelation alternates from sections of the text affirming religious victory, salvation, and Christian dominion to sections acknowledging social persecution, Christian defeat, and Roman dominion. As we saw in chapter 3, the interconnectedness of the seer's language raises questions about that kind of analysis. The intertwining of narrative and metaphoric elements makes it impossible to divide the text into sets of oppositions. Rather than alternating between clearly demarcated sections of woe and weal, the seer's language is unified syntactically.

A person may, however, write in a style that is unified syntactically and at the same time construct a world of conflict and opposition. So we may pose the following questions about the Book of Revelation: What vision of reality becomes transparent in the seer's apocalyptic language? Does the seer envision a world in which certain elements are in essential conflict with other elements? Does integration and interconnectedness occur only on the syntactical level in the Book of Revelation and not on the semantic level? On the semantic level—the level of meaning and reference—does the seer's language disclose a world of conflict, tension, and crisis? If it does, the interconnections of the syntactical aspects of the seer's language would disguise and dissemble; they would be hiding a world of conflicts and contradictions. However odd that notion may appear, that is the assumption of certain kinds of structural analyses of the Book of Revelation: the interconnections of the seer's language are intended to hide the serious conflicts and contradictions to which the seer alludes in his writing. In the technical language of structural analysis, the surface structure mediates and blurs the tensions and conflicts in the deep structure of the text.[1]

In order to delineate the shape and contours of the seer's world I shall rely heavily on the term *boundary*. As I unravel the spatial metaphor *boundary* in the following sections, the structure and organization of the seer's world should become clear. It is a complex structure, and that complexity will be clarified by locating the fundamental distinctions and discriminations (i.e., boundaries) the seer makes as he constructs a comprehensive vision of reality. The focus on boundary will also provide opportunity to note the nature of the distinctions the seer makes; that is, distinctions between objects or qualities may be absolute and categorical, or they may be relative, with one object blending into the next. I shall argue that the seer's distinctions (boundaries) are of the latter kind, not absolute, firm, or hard, but, rather, blurred and soft.

Boundary Situations

Since *boundary* will recur in different contexts in what follows, I should comment on the term itself. *Boundary* is a term associated with space and spatial demarcations. In common usage *boundary* refers to the outside perimeter of a space: my property is bounded by a curb on the front and a fence behind, that is, the curb and the fence mark the extent of my property, the limits of my land. This common usage of *boundary* as an "outside" limit or perimeter depends on the perspective of one who is inside the boundaries. From an "inside" position boundaries mark the limits of my property; and if the boundary fence is high enough that I cannot see over it, my dog and I will experience the boundary as the limit or extent of space. If, however, I fly over my land in an airplane, the boundaries will be seen quite differently: from that lofty perspective those boundaries mark out and separate my property from other property. Rather than the boundary being a limit or an outside perimeter, a boundary is seen as a mark between two things. In fact, without the boundary the two things might not be distinguishable. Without the fence there would not be two properties, only one. Thus, one can say that a boundary not only marks differences, it *creates* them. A boundary separates and delineates, thereby making a difference where otherwise there would be no difference.

In considering the boundaries that delineate the contours of the seer's world, I view that world from above, as from an airplane. From that viewpoint his boundaries are not outer limits but dividers that create differences and distinctions among the objects in his world. Just as we can learn about how land is controlled by noting where boundaries are placed, so we can learn about the seer's world—fundamental distinctions, values, commitments—by noting where he places boundaries and thereby creates differences. No sharp distinctions need be made between spatial boundaries in the seer's world (heaven, earth, sun, rivers) and boundaries in other sets of relations. *Boundary* can be used in any analysis that locates a set of relations as a "topographical arrangement in space" (Jaspers 1970, 177).

To put it differently, a boundary is formed when two different qualities, objects, or forces come together.[2] A social boundary divides life inside the Christian community from life outside. By paying attention to that social boundary, one can learn a lot about how the Christian community is defined and differentiated from other social groupings. A literary boundary can be located at Revelation 4:1; two different types of literature come together there: seven messages to the seven churches (1:9–3:22) and the ensuing visions (4:1–11:19). By paying attention to literary boundaries in the Book of Revelation, one learns about the structure of the seer's book. One can map out values and morality in John's world by locating divisions between good and evil, that is, where the boundaries between good and evil occur. Insofar as the distinction between good and evil depends upon those boundaries—without them the distinction could not be made—one can also see from the boundaries how the author of the Book of Revelation creates the categories of good and evil as they relate to his world vision.

In brief, I seek to map out the seer's world comprehensively. By mapping the regions, distinctions, and differentiations in the seer's vision of reality, I shall locate boundaries in the seer's world; and those boundaries, in turn, will disclose some of the fundamental structures and networks of relations central to the seer's construction of reality. By proceeding in that manner, I will never abstract the fundamental structures and networks from the specific distinctions and discriminations that the seer makes in his world construction.

Boundaries in Expected Places

The seer recognizes commonplace boundaries. He delineates a three-story universe: earth, heaven above the earth, and the abyss below (see Rev. 5:3). Divine forces come down from heaven, and evil forces come up from the abyss. The seer's celestial realm contains the familiar objects of sun, moon, stars, and sometimes the atmospherics of thunder, lightning, and hail. Birds inhabit the sky (more precisely, "midheaven"; see 6:12–13, 8:13, 12:1, 19:17). With respect to earth there are references to hills and islands, wilderness, and various types of water-bodies such as seas, lakes, and streams (e.g., 1:9, 6:14, 7:17, 8:9, 9:14, 12:6, 12:17, 16:20, 19:20). A few plants are mentioned: trees, grass, and plants in general, all of which are specially protected from evil forces.[3]

The seer mentions animals such as horses, lions, birds, leopards, bears, and frogs (e.g., 6:2, 9:8, 13:2, 16:13, 19:17). Humans are classified as peoples, tribes, tongues, and nations.[4] The seer also mentions social classifications such as kings, great ones, rulers of a thousand, the wealthy, the strong, merchants, and those sitting on horses; most of these are associated with evil. People are also contrasted as small and great, wealthy and poor, free and slave (see 6:15, 13:16, 18:3, 19:18). No male/female contrast is made. When considering godly, faithful people, the seer's categories become more refined. He refers to prophets, servants, fellow-servants, brethren, apostles, saints, "my people," and those fearing God, great and small (see 10:7, 11:10, 18:20, 19:5, 21:3, 22:9).

Sometimes these objects in nature and human characters combine in striking ways to form hybrids. Those hybrids exist in the seer's world as clearly definable objects, but they transgress the normal categories of animal, vegetable, and mineral. By transgressing those ordinary boundary distinctions, the seer creates awesome figures of divinity as well as of monstrous evil. The creator God who sits upon the throne is described by means of images of precious stones, jasper, and carnelian and a rainbow that looks like an emerald (4:3). Heavenly figures around the throne appear like a lion, an eagle, an ox, and a man—all with six wings and full of eyes (4:6–8). A mighty angel comes down from heaven wrapped in a cloud, with a rainbow over his head, with a face like the sun and legs like pillars of fire (10:1). A woman appears in heaven clothed with the sun, with the moon under her feet and twelve stars on her head (12:1).

The red dragon of chapter 12 has ten horns and seven heads as well as a tail that sweeps down one-third of the stars of heaven to earth (12:3–4; only evil forces seem to have tails, cf. 9:10, 19). The beast from the sea has ten horns and seven heads. It is like a leopard, its feet are like a bear's, and its mouth is like a lion's (13:1–2). The locusts that come up through the opening from the bottomless abyss combine a stinger like a scorpion and an appearance like horses, though with human faces, women's hair, and lions' teeth (9:7–10).

Blurred Boundaries among Godly Forces

Although the descriptions of God, heavenly creatures, and demonic forces mix categories that are normally kept separate, each of those divine and demonic beings is distinct and separate. In mapping out the different creatures around the throne or the various beasts from the abyss, one finds that each has a distinctive outline that can be drawn in darker hues, like boundaries on a map. Moreover, each has a distinctive place within the three-story universe: the divine forces belong in heaven; evil forces belong below the earth; and earth becomes a place of conflict between the two. Earth contains creatures (primarily humans) who can be identified with either the godly or the evil forces located elsewhere in the universe.

John creates a world with distinct levels and clearly delineated characters; he also presents those levels and characters in such a manner that they invite comparison. For example, beings in different spheres of the universe may share certain characteristics, so that they are bound together even though they are separated by the stories of the universe. So an effulgence (ἶρις) radiates around the one sitting upon the throne (4:3); later that same radiance is present around the head of one of his emissaries (10:1). The face of that same emissary is "like the sun," a simile which earlier refers to the appearance of the one John saw in his inaugural vision (1:16) and later to the clothing of a woman (12:1). Through that common language, connections and correspondences are made among God on the throne, his Christ, his angels, and other godly beings. Although each may be outlined separately on a map of John's world, each shares common characteristics with the others. Thus, as one becomes familiar with John's creatures, the contours of one evoke the contours of others.

Insofar as Christians are called to imitate their Lord, they come to share characteristics with him. As Christ conquered, so do Christians (3:4–5, 3:21). Moreover, both Christ and Christians conquer through blood. The brethren conquer Satan through the blood of the Lamb and through the word of their witness; "they did not love their souls even unto death" (12:11, cf. 6:9). The innumerable crowd before the throne came through the great tribulation and victoriously stood before God with their clothes made white through blood (7:11–17). Sacrificial language, which underlies most of those references, becomes explicit at 14:4 where the 144 thousand "redeemed from mankind" are designated "first fruits for God and the Lamb."[5] Christ and his followers share not only sacrificial associations but also royal priestly characteristics, that is, characteristics of the sacrificer as well as the sacrificed. Both Christ and his followers are "priests to our God" (5:10).[6] Followers also share the royal status of sonship and they are given "a name which no one knows," a parallel to Jesus (see 2:17, 21:7, 19:12).

John also connects the characters in his three-story universe in more subtle ways. The characteristics of certain humans correspond to the characteristics of divine creatures or heavenly places; that is, godly humans have features *homologous* to divine creatures or divine places or even God himself. Homologous relations are best known from the field of biology, where homologies refer to similar structures with a common origin, for example, the wing of a bat and the foreleg of a mouse. In religious studies, Mircea Eliade has used the term *homology* to show how religious man is a microcosm of larger cosmic structures (see Eliade 1959, 166–70). In the Book of Revelation, one finds homologies other than in the microcosm/macrocosm relationship. Thus, I use the term to refer to any correspondence of structure, position, or character in the different dimensions of John's world. These homologous relations contribute to the blurring of boundaries in the Apocalypse.

John, for example, creates homologies among certain kinds of clothing, holiness, certain colors, and just deeds. God is holy (ὅσιος or ἅγιος), as are his angels, faithful humans, and a city (see 11:2, 11:18, 14:10, 15:4, 16:5). Divine holiness becomes apparent when God reveals his just deeds (τὰ δικαιώματά, 15:4). Just deeds are also attributed to "holy humans" (οἱ ἅγιοι, 19:8). "Holy humans," or saints, thus function on the human plane as God's holiness on the divine plane. Further, the just deeds of the saints are identified with the bright, clean, linen garment worn by the Bride of the Lamb (βύσσινον λαμπρὸν καθαρόν, 19:7-8).[7] A clean, linen garment is also worn by the seven angels who pour out the seven bowls of plagues (λίνον καθαρὸν λαμπρὸν, 15:6) and by the heavenly army support- ing the warring Word of God (βύσσινον λευκὸν καθαρόν, 19.14).[8] Clothing reflects inner qualities and essential characteristics of those who wear them. Jesus urges the Laodiceans to buy from him white clothing (ἱμάτια λευκὰ) to wear, so that the shame of their nakedness not be revealed (3:18). Because of the connection between inner and outer, garments are to be "kept" (τηρέω, 16:15) - a verb used elsewhere in Revelation only in connection with commandments, works, and words (e.g., 1:3, 2:26, 3:10, 12:17, 22:7). Christians at Sardis should not stain (μολύνω) their gar- ments but rather walk in white (ἐν λευκοῖς) with Jesus (3:4).[9]

As seen from this last example regarding the worthy Sardians, whiteness relates homologously to proper garments, righteous deeds, and holiness. The color is first introduced in the inaugural vision, where Jesus' hair is white as white wool, like snow (1:14). In the messages to the churches, those conquering at Pergamum are promised new names written on a white stone (2:17), those not stained at Sardis will walk with Jesus in white (3:4-5), and those at Laodicea are urged to buy white garments to cover their nakedness (3:18).[10] Elsewhere throughout Revelation the twenty-four elders wear white garments (4:4) - as do the slain ones under the altar (6:11); the innumerable crowd before the throne whose garments were made white in the blood of the Lamb (7:9, 14); and the army of the Word of God (19:14). Those people share their colors with the white horse coming forth at the opening of the first seal (6:2, cf. 19:11, 14); the white cloud upon which the one like a Son of Man sat (14:14); and the great white throne of judgment (20:11). The color white thus substitutes in position or structure on the color plane for just judgment, righteous reward, and holiness - themes that span heaven and earth as well as the present and the *eschaton*.

Blurred Boundaries among Evil Forces

With regard to evil forces the following scenes are the most important: the blowing of the fifth trumpet, which reveals a star falling from heaven and opening the shaft of the bottomless pit (9:1-6); the war against the two prophets by the beast ascend- ing from the bottomless pit (11:7); the sign of the great red dragon, his war in heaven and on earth (12:3-17); the beast from the sea to whom the dragon gives power (13:1-10); the beast from the earth who receives power from the beast of the sea (13:11-18); the scarlet beast with the harlot rider (17:1-14); and the Devil bound and loosed in the bottomless pit (20:1-10).

Several common activities connect these demonic forces: ascending from the bottomless pit; making war against the godly; conquering; deceiving; blaspheming; evoking wonder and worship; having authority, power, and kingship. These forces also share common features, that is, seven heads, ten horns, redness, and certain nomenclature. Sometimes they relate in more subtle ways to each other. At 9:11 the angel ruling the abyss is called Apollyon. Revelation 17:8 rings changes on abyss/ Apollyon, for there the scarlet beast goes up from the abyss to apoleian ("destruc- tion").[11] That repeated association blends the angel ruling the abyss (9:11) with the scarlet beast from the abyss (17:8). Through numerical equivalence in the plague sequences—the fifth trumpet and the fifth bowl—a connection is also made between the angel ruling the abyss and the beast from the sea (9:11, 16:10–11).[12] The various beasts described in the Book of Revelation are thus variations on one another. So, for example, the beast from the abyss—introduced abruptly in 11:7— is a transformation or manifestation of the ruler of the abyss in 9:11. At 16:13 a new evil is introduced in the form of a false prophet, but he is not all that new in Revelation, for he shares characteristics with the beast from the earth in 13:11.[13] Through these associations, the identities of the following are blended together, that is, the boundaries separating them are not absolute and hard: Apollyon (9:11); beast from the abyss (11:7); beast from the sea (13:1); beast from the land (13:11); false prophet (16:13); scarlet beast (17:3); the dragon, ancient serpent, Devil, and Satan (20:2, 7); and the opponent of Michael (12:9).

The Great Whore Babylon shares several qualities and functions with those beasts and with other evil forces. Like the beast from the sea, she is incomparable (18:18, cf. 13:4). She shares the color scarlet with the evil beast (17:3–4, 18:16). She causes deception among the nations (18:23), as do the beasts. As sorcerer and one committing abominations, she shares qualities with those evil ones who cannot enter the New Jerusalem (17:5; 18:23; 21:8, 27; 22:15). Finally, she "falls" and becomes a "haunt" ($\phi \upsilon \lambda \alpha \kappa \acute{\eta}$, 18:2), just as Satan is cast into a "prison" ($\phi \upsilon \lambda \alpha \kappa \acute{\eta}$, 20:7).[14]

Like the forces of good, those evil figures have their followers. The kings of the earth are frequently linked to one of these demonic figures (e.g., 9:7, 17:12, 19:19). Others are identified with the demonic by an iconic stamp on the right hand or forehead, which in connection with the beast of the earth (equals false prophet) is said to provide the economic freedom of buying and selling (13:16–17). They are deceived by the false prophet to worship the beast (19:20), but ultimately they shall share with the other demonic forces torment of fire and sulfur (14:9–11).

Finally, there are striking homologies between opponents of the faithful in the seven churches and demonic figures elsewhere in the Apocalypse. The nomenclature of evil is used several times in the letters: those at Smyrna are warned that the Devil is about to throw some in prison (2:10). The Devil is, of course, the dragon of chapter 12, who reappears in chapter 20; the prison ($\phi \upsilon \lambda \alpha \kappa \acute{\eta}$), moreover, in which the Smyrnians are about to be thrown is a horror like the one from which Satan is loosed in chapter 20 and the one into which the Great Whore falls (18:2). Satan's throne at Pergamum (2:13) forms a homologue with the throne of the beast (9:11, 16:10); references to the deep things of Satan (2:24) and to the synagogue of Satan (2:9, 3:9) also link the church's opposition to the demonic. Those of the synagogue of Satan claiming to be Jews blaspheme (2:9), which is otherwise done only by the beast from the sea and the scarlet beast. The prophetess Jezebel, a teacher of whom John does not approve, deceives those at Thyatira—an activity practiced only by Satan, the beast of the earth, and Babylon the Whore. Moreover, her teachings include the practice of fornication; elsewhere in the Apocalypse that term is associated with Babylon the Great Whore (Rev. 17–19), the Balaamites (2:14), and those who at the *eschaton* have to stay outside the New Jerusalem (22:15) and burn in the lake of sulfur (21:8). Thus, prophetic groups in the churches, Babylon the Great

Whore, and those condemned at the *eschaton* share characteristics and function homologously in their respective planes.

Soft Boundaries

Blurred boundaries among forces of evil, on the one hand, and godly forces, on the other, could simply reinforce the notion that there are sharp contrasts in the Book of Revelation between those two camps: Satan and his followers versus God and his faithful. The two groups are set in opposition, and readers/hearers are called on to decide for one or the other. Those with the mark of the beast are thrown into the lake of fire, the second death (19:20, 20:14); and those whose names are in the Lamb's book of life shall enter into the New Jerusalem (21:27). The Laodiceans are called on to display a clear-cut position: be either hot or cold but not lukewarm (3:15–16). Much of the Apocalypse describes conflict between those two groups of forces: those in the churches against their adversaries, two prophets against the beast, woman against dragon, saints against the beasts, witnesses against the Whore Babylon, riders on white horses against false prophet. Their conflicts are described with military language of warring, battles, armies, weapons, victories, and defeats.[15]

Although the seer marks his boundaries well—often as battle lines—those boundaries between good and evil are not hard and impenetrable borders separating the two into separate, limited spheres. Even here distinctions are blurred and boundaries are soft. Evil contrasts with the godly, but evil is not of a fundamentally different order from good. Humans belong to the earthly plane, the divine belong to heaven above, and the demonic belong to the plane below; but those three tiers of the seer's universe are not separated absolutely. Social categories on earth are not impassable, for there are not absolutely bounded divisions among humans. Even divisions of time—past, present, and future—cannot be hardened into "the evil present" and "the blessed future."

Soft Boundaries between Good and Evil

At points in the Book of Revelation, the Lamb and various beasts form dyadic relationships, that is, they become doubles, split images of some more fundamental wholeness. One of the heads of the beast from the sea is described "as slain" (13:3); the same expresson is used of the Lamb in the throne scene (5:6).[16] Further, the beast is healed of his mortal wound so that he lives. By dying and yet living (13:14) he is comparable to Jesus who became a corpse and lived (2:8).[17] A similar pattern is repeated in the description of the scarlet beast who was, and is not, and is to come (17:8). The Lamb and the beast from the sea also share a similar hierarchical position in their respective communities: each is an agent to a higher sovereign (3:21; 12:10; 13:2, 4), yet each is worthy in his own right to receive power and authority from those below (5:12, 17:13); and each wears the royal insignia of diadems (13:1, 19:12, cf. 12:3). Each thus serves as icon of the sovereign above (3:14, 13:14–15) while forming a worshipping community around himself (cf.

Schüssler Fiorenza 1981, 134) with the mark of the beast and the seal of the Lamb functioning homologously in their respective communities.

Soft boundaries also separate feminine images of good and evil. Babylon and Jerusalem — feminine images of cities — embody homologues and similar qualities. Both are clothed with fine linen and bedecked with gold, jewels, and pearls (18:16; 19:8; 21:18-19, 21), and both function as sexual partners in their respective systems (18:3, 19:7-8). *Great city* usually refers to Babylon (Rev. 17-18) but may refer to either Babylon or Jerusalem (16:19); and Jerusalem, the city of God, can even be understood "spiritually" as "Sodom and Egypt" (11:8).[18] In this transformation of Jerusalem Christian interpretation is clearly at work, but the prophets of old also refer to such a metamorphosis of the holy city (cf. Lam. 4:6, Isa. 13:19, Jer. 22:6).[19] The fluidity between godly and demonic cities in the seer's visions points to a common structure within good and evil in the Apocalypse.[20]

Woman in the Apocalypse also moves fluidly between good and evil. At 12:14 the woman who recently gave birth — clearly a figure embodying godly associations — flees into the wilderness from the dragon. In the next reference to wilderness, John is introduced in the Spirit to the woman on the scarlet beast full of blasphemous names (17:3). If the wilderness passages are taken strictly sequentially, the good woman has been transformed in the wilderness into the evil woman on the beast. Wilderness would thus function symbolically as a place similar to chaos with transformational potential for judgment, deliverance, nourishment, punishment, death, and rebirth (12:6, 17:16, cf. Thompson 1978, 95-96, 193).[21]

Even God and Satan, the epitome of good and evil respectively, are not separated by hard, impervious boundaries. Several common aspects blur their boundaries, especially images and symbols related to sovereignty and worship. As God is enthroned, so is Satan (2:13, 4:2). Satanic locusts have golden crowns on their heads like the twenty-four elders around the throne of God (4:4, 9:7), and God and Babylon the Whore are said to rule as king or queen (15:4, 17:18). As supernatural forces with their own spirits (16:14, 22:6), both God and demonic forces are worshipped and glorified (4:11, 13:4, 15:4, 18:7). Through a word play on the Greek word θυμός, God and the Whore offer a similar wine drink to others: all the nations and the kings of the earth drink from the wine of the θυμός of her fornication (14:8, 18:3), and God will give to her and to all those worshipping the beast a drink from the wine of the θυμός of his wrath (14:10, 16:19). The translation (*RSV*) of her θυμός as "impure passion" and of God's as "wrath" reflects the play on the word and one way the seer uses word plays to relate apparently opposing forces.

Soft Boundaries between Spatial Planes

Although humans belong to the earthly plane, the divine belong to heaven above, and the demonic belong to the plane below, those three levels of the seer's universe are not separated absolutely. Creatures descend or ascend through the universe; and as they pass through the different levels, they are transformed in other ways. Movement through spatial planes functions as a transformational experience.

Utilizing the image of an open door, John describes his ascent to heaven (4:1).

As he ascends, he is transformed simultaneously from earth to heaven, from a normal to a "spiritual" psychological state, and from the present to the future. Space, time, and psychological state are assimilated to one another, forming a series of correspondences among different planes (4:2). "Going up" (spatial plane) forms a homology with "in the spirit" (psychological plane) and with an eschatological vision (temporal plane).[22] Once translated into heaven, John is able to see such heavenly visions as the throne of God; the three series of seven seals, seven trumpets, and seven bowls; the fall of Satan and various demonic forces; conflict between divine and demonic armies; and finally, the New Jerusalem.

Elsewhere a cloud becomes the means of moving through the heaven/earth boundary. At 1:7 John declares that Jesus will come on a cloud (cf. 14:14); at 10:1 an angel comes down from heaven clothed in a cloud; and the two witnessing prophets in chapter 11 go up to heaven on a cloud (11:12). Otherwise, movement to and from heaven is signified simply by verbs of motion. John and the two prophets "ascend" ($\dot{\alpha}\nu\alpha\beta\alpha\dot{\iota}\nu\omega$) from earth to heaven (4:1, 11:12). Angels "descend" ($\kappa\alpha\tau\alpha\beta\alpha\dot{\iota}\nu\omega$) from heaven (10:1, 18:1, 20:1) — as do the New Jerusalem (3:12, 21:2), fire (20:9), hail (16:21), and the Devil (12:12). Fire from the altar is thrown (8:5) or poured out (16:2) upon the earth from heaven; in its transformational descent the fire becomes a destructive force.[23]

Passage also occurs between earth and that demonic sphere below the earth. By *opening* that shaft to the abyss below (cf. the "open door" to heaven), locusts from that demonic plane pass onto the earth (9:1), or Satan and his surrogates move back and forth from earth to the realm below.[24] Through images of locks, keys, chains, seals, loosing, and binding John is able to describe controlled movement between earth and Hades (9:1, 20:1-3). Since the realm below represents not only the demonic but also death, movement to and from that realm may also occur in the form of transformation from death to life, or resurrection.[25]

The demonic plane can claim no independent reality, for it derives from the heavenly, divine plane above. Demonic power becomes operative on earth when a "star" fallen from heaven is given a key to open the shaft to the abyss below (9:1). This "star" — in origin from heaven — is apparently later identified as the angel Abaddon or Apollyon, who rules over the bottomless pit (9:11) — and still later as the scarlet beast (17:8).

In chapter 12 a similar transformation occurs when Satan "falls" from heaven. Between two versions of a story about conflict between the evil red dragon and the good pregnant woman, the seer inserts a narrative about the heavenly origin of Satan, the ancient serpent and great dragon called the Devil (12:7-12).[26] It is a striking narrative, for it tells how Satan — the most powerful and concentrated image of evil in Revelation — once served in the divine court. Satan, the ancient serpent, the Devil, emerges here as a transformation of heavenly, divine realities. More specifically, that dragon (hypostatized evil) served in heaven as judge and assessor, that is, as a judicial dimension of God and his heavenly powers (12:10, cf. Job 1:6-12, 1 Enoch 40:7).[27] Through transformational symbols of descent and conflict Satan — whose authority and power lie behind all other evil forces in Revelation — is seen to metamorphose from the divine.[28]

Soft Boundaries in Social Categories

The transformational movement through space has its homologues in the churches of Asia Minor. Social categories distinguishing faithful from unfaithful are not bounded by impassable borders. Furthermore, the social categories are themselves a mixture of different statuses. Those at Laodicea, for example, appear rich, prosperous, and faithful; but they are in reality poor, wretched, pitiable, and naked (3:17); and those at Smyrna appear to be poor but are really rich (2:9).[29] The faithful are admonished lest they fall: first loves may be abandoned (2:4–5); those once alive may become dead (3:1); faithfulness is called for (2:10, 19); garments must be kept (16:15). Only those who persevere to the end will conquer (2:10–11). On the other side, the faithless are urged to be transformed through the alchemy of repentance ($\mu\epsilon\tau\acute{\alpha}\nu\omicron\iota\alpha$, cf. 2:5, 16). A deceiving, self-acclaiming prophetess like Jezebel, who functions and has attributes like the beasts and the dragon, can repent along with her followers (2:21–23). Even the most blasphemous have the possibility of transformation through repentance and can cross the boundary from unfaith to faith (16:9, 11). All can "open the door" (3:20). In the New Jerusalem the probabilities of change diminish, but even there the unfaithful may still have the possibility of repentance. Just as there is ultimately no fundamental dualism between heaven and earth, so there is no final "dualistic division of humanity."[30]

Soft Boundaries in Time

Time takes a curious turn in the Book of Revelation, for past, present, and future are not separated by fixed, absolute boundaries. The seer, rising above time as in an airplane, takes a transcendent view and traces the past, the present, and the future on his temporal map. Boundaries in the future are as visible to him as boundaries of the past, and those future boundaries share characteristics and homologies with present and past. John sees both "what is and what is to take place hereafter" (1:19), and John's God is "the one who is and was and is to come" (1:4, 2:8). The seer's temporal map is analogous to a conductor's score: the conductor "sees" all parts of the score, that which has been played, that which is being played, and that which will be played. Moreover, the spatial arrangement of time in a musical score discloses patterns, motifs, and variations among past, present, and future. The seer, like the conductor, can range freely through time, catching patterns and motifs in his mapping of the aeons. Time can thus be understood as a "topographical arrangement in space," for John portrays the temporal "end" as a detailed mapping of space, that is, as a city let down from heaven.

From the seer's transcendent view of time there is no hard division between the present age and the age to come. No hard boundary separates "the new heavens and new earth" (Rev. 21:1) from the "first heaven and the first earth." The "new" emerges in time as a transformation of the "old." As the first disappears (21:1, $\dot{\alpha}\pi\,\tilde{\eta}\lambda\theta\alpha\nu$), the new comes into being (21:1, 2, 5).[31] The one on the throne declared, "Behold, I make all things new [$\kappa\alpha\iota\nu\grave{\alpha}$]" (21:5). The "new" contrasts with the "old"; but there is a continuity of substance, whether it be a name (2:17), a song (5:9, 14:3), or heaven and earth (21:1).[32] $\kappa\alpha\iota\nu\acute{o}\varsigma$ ("new") could better be translated "renewal" or even "restoration," for the portrayal of the "new" borrows heavily from descriptions of paradise from of old. Then the seer was told, "γέγοναν" (21:6), a term translated variously as "It is done," "All is over," or "These words are already fulfilled." This term, however, also signifies transformation or metamorphosis. Elsewhere in Revelation it is used to indicate transformation into the Spirit (1:10), transformation from life to death (1:18), the sun into blackness (6:12), and water into blood (8:8). The perfect tense of this verb in 21:6 can thus be translated, "All has been transformed." By means of these terms the seer describes "a drastic transformation of existence" typical of eschatological portrayals in the New Testament, but he suggests no sharp dualism between this age and the age to come (see Beardslee 1970, 228).[33]

That eschatological transformation completes what has occurred in the coming of the Christ and is occurring in Christian existence. In a manner distinctive to Revelation the description of the coming of Christ in glory is laced with images of death and the cross. In the Gospels apocalyptic announcements (e.g., Mark 13) are in form and content clearly delineated from the Passion narrative and descriptions of Christ's weakness and humiliation.[34] Paul also keeps the irony of present Christian existence clearly separate from apocalyptic phenomena: humiliation, crucifixion, weakness, and foolishness are not part of Paul's apocalyptic scenarios (1 Cor. 1, 4:8–13; 2 Cor. 4:10; cf. 1 Cor. 15, 1 Thess. 4:16). In Revelation, however, the apocalyptic Christ comes as "the pierced one" (1:7) and the messianic "Lion of the tribe of Judah, the Root of David" is revealed as "the slain Lamb" who is worthy through being slain (Rev. 5:1–14). From the seer's first vision to the final victory of the Word of God seated on a white horse and clothed in a garment soaked in blood (19:13), imagery of sacrifice, blood, and death permeates the visions of the mighty apocalyptic figure. Through that mixing of imagery, Jesus' *parousia* overlays his "first coming" to produce a picture of a mighty warrior as a crucified Lamb. The first and second "coming" of Christ cannot be differentiated as two distinct eras with a clear boundary between them; the second is a radical transformation begun with the first.

There is a permanence to the crucified Lamb that cannot be captured by locating the crucifixion in time, for example "under Pontius Pilate" or "in the first century of the Common Era." To put it differently, the crucifixion is much more than a momentary event in history. That permanence is captured in the Book of Revelation through spatial, not temporal, imagery. The "slain Lamb" appears not only on earth but also in heaven, close to the throne (5:6). The Lamb was not slain at a particular moment in time; rather the Lamb was slain before time. The seer describes that time in spatial language: the Lamb was slain "from the foundation of the world" (13:8, cf. 17:8). The crucifixion is enfolded in the "deep," permanent structures of the seer's vision, and it unfolds in the life of Jesus and those who are his faithful followers. Christian imitation of Jesus through martyrdom and suffering and the homologies formed between Jesus and his followers can thus be seen as temporal unfoldings of a "deeper" order in the seer's world.

Just as John's eschatological visions (time) can be portrayed spatially as a city, so place (topos) is not a fixed, bounded space but a situation with a temporal dimension: in Revelation place is bound up with destiny and contingency. The woman flees into the wilderness to a place prepared for her (12:6, 14); no place is found for the Devil in heaven after the battle with Michael (12:8); if the Ephesians don't repent, their lampstand will be removed from its place (2:5). Items in nature such as islands and hills — even heaven and earth — do not have fixed places (6:14, 20:11).

Homologies connect the heavenly throne (space), the eschatological Jerusalem (time), and the earthly temple at Jerusalem (space). The precious stones in the heavenly throne scene (4:3, 6) derive not from traditional throne descriptions in apocalyptic literature but from descriptions of the eschatological Jerusalem (Rev. 21:11, 18–20, cf. Isa. 54:12). Only the throne (4:2) and the New Jerusalem (21:16) are predicated by the Greek verb κεῖμαι, indicating "situatedness" or "fixed location of place" (in contrast to *topos* above). In the temple at Jerusalem (11:1–2) a spatial boundary divides the temple proper from the court outside. The court can be profaned by the nations, but the temple cannot. So with the New Jerusalem "nothing unclean shall enter it, nor any one who practices abomination or falsehood, but only those who are written in the Lamb's book of life" (21:27, cf. 21:8). Yarbro Collins suggests that the courtyard/temple boundary also parallels the earthly/heavenly boundary in the Apocalypse (1984, 68). The Jerusalem temple and the eschatological city are further linked by the activity of measuring, for only in relation to those two is a measuring rod (κάλαμος) mentioned (11:1, 21:15). Parallels between heavenly throne scenes in the Apocalypse and the Jerusalem temple are fairly obvious (cf. Thompson 1969, 337). Heavenly sacrifices create smoke so thick that no one can enter the temple (15:8). Inside the heavenly temple stands the ark of the covenant or the tent of witness (11:19, 15:5). Through such homologies the sacral space of the Jerusalem temple, the heavenly throne scenes, and the eschatological Jerusalem are overlaid in the mapping of the seer's world. Not only are distinctions blurred among past, present, and future, but also time and space are related as coordinates of a common order.

Transformations, Inner Structures, and Ratios

At the beginning of the discussion on boundary I noted that from a transcendent position, such as in an airplane, a boundary appears not as a limit or outside perimeter but as a divider marking differences. It is from this lofty position that we have mapped boundaries, locating distinctions and categories fundamental in the seer's world. These categorical distinctions in space, time, gender, or moral qualities are created by the seer. For example, he distinguishes between good and evil, heaven and earth, or faithful and nonfaithful in distinctive ways. Thus, as a worldmaker, the seer not only marks differences, he creates differences. He creates a three-story universe rather than a universe with four or seven tiers. He delineates in great detail distinctions among those faithful to God (e.g., prophets, servants, apostles, saints). He blends color, gender, and moral terms to create feminine figures of purity or corruption. In brief, the seer creates a distinctive world by marking out differences that are established at and by boundaries. Boundaries may thus be understood as places where differences touch one another: a boundary is formed when two different qualities, objects, or forces are contiguous. A boundary sets up how these differences are delineated and how "solid" the separation is between them.

In the seer's world we have seen that the boundaries are not very "solid"; they are soft boundaries that blur sharp distinctions. So, for example, the boundary between Christ and Christians is not impervious. Godly objects share characteristics and homologues such as holiness, whiteness, and linen. The various beasts ring changes on each other as they process through the book. More striking are the soft boundaries between good and evil. Beasts and Lamb form dyads; Bride and Whore form counterparts; Jerusalem and Babylon become confused; Satan derives from an aspect of the divine.

As we explored these blurred relations and soft boundaries, the term *transformation* crept into the discussion. *Transformation* became especially prominent when we traced movement across boundaries — whether spatial, social, or temporal. Heaven remains separate from earth, and earth from the abyss, but doors open to allow John and other beings to move between one plane and another. Social boundaries are not hard and fixed. The faithful may "fall," and the faithless may repent. Even the boundary between this aeon and the age to come blurs differences on either side: the new is a transformation of the old, begun with the crucifixion of Jesus which has been "from the foundation of the world."

A boundary separates differences, but it is possible to cross over the boundary. Crossing, however, involves a transformation; characters do not pass through the open door of boundaries and remain untouched. When Satan descends, he is transformed; so is John when he ascends. Passage through a boundary simultaneously transforms the object from what is on one side to what is on the other, that is, earth to heaven, faithfulness to unfaithfulness, good to evil. The phrase *transformational boundary* describes this particular aspect of a boundary situation; that is, a boundary not only locates where differences touch each other; it becomes a place where differences can be transformed into each other.

There is a dynamism to boundaries in the Book of Revelation. Boundaries do not fix limits beyond which it is impossible to pass. Rather, they locate the place where transformations occur, allowing a flow across planes, eras, social categories, or moral values. At the most fundamental level, the seer envisions reality as a world in process, a flow of becoming, a sequence of transformations that unfolds into various planes, eras, qualities, and objects.[35]

From considering the transformational aspect of the seer's boundaries one could conclude that the seer's world is a muddle of confusion — that anything can be changed into anything and that transformations occur randomly and without design or guidance. One wonders, Are there forces or channels guiding the transformations that occur at the seer's boundaries, or are they random and directionless occurrences? The answer is easy: there are guiding forces. Getting at those guiding forces, however, is not so easy. An important clue to guiding forces is given in the earlier discussion of homologies. There we noted that homologies contributed to the blurring of boundaries, for different aspects of John's world are brought into relation with one another, for example, spatial movement, psychological state, and moral character are simultaneously transformed when the seer goes into heaven.

The blurring is not, however, random: homologies point to similarities in the transformations that occur; a definitive relationship is formed between going up into heaven and entering a spiritual state, or moving down from heaven and entering a demonic state. The contour or shape of one boundary is replicated in another. Those replicated contours point the way to fundamental structures and guiding channels in the seer's world. Put differently, the set of relations formed at one boundary is similar to the set of relations formed at another. They are similar because there is an inner structure implicated in every boundary situation that unfolds in all boundary situations.

Measures and Numbers

The seer's use of numbers illustrates how relationships and sets of relationships rather than individual characteristics disclose inner structures of his world. First of all, the seer seems to view numbers and their measurements as an entree into the essential structure of a thing. Measurements of the temple (11:1-2) or the number of the beast (13:18) reveals something fundamental, the essence — at least for those who have understanding and wisdom (13:18, 17:9).

The activity of measuring occurs only two times in the Apocalypse: in connection with the New Jerusalem (21:15-17) and the temple of God (11:1-2).[36] Measuring the New Jerusalem and the temple is not primarily a way of comparing them with some external unit of measure such as a cubit stick. In fact, in chapter 11 no mention is made of an external unit such as cubits or stadia. Measurement is bound up with what is intrinsic to the object being measured. For example, only the temple proper, the altar, and those worshipping — all of which reflect the sanctity of the holy place — are measured. That which is not holy is not appropriately measured in the same measurement. Ezekiel makes the same connection between measurement and sanctity in Ezek. 40-48: there descriptions of measuring alternate with descriptions of the holiness of temple, city, and land. The extent of measurement marks the boundary, in the cases of both Ezekiel and Revelation 11, of that which is essentially holy.[37] The actual numbers in measurements — for example the 144, 12, and 1,000 in the New Jerusalem and the sealed of the tribes — outwardly correspond to an inner, essential measure rather than to some external standard of measurement.[38]

A number such as that of the beast — 666 — may be a numerical code to be translated into some name such as *Nero*, *Domitian*, or *Hitler*, but if so, the code book has been lost. Moreover, a specific number does not necessarily carry a particular significance in the Apocalypse: six does not necessarily signify "evil and incompleteness," nor seven "completeness and fulfillment" (Sweet 1979, 14-15). The creatures around the heavenly throne have six wings, and the dragon, along with the first beast, has seven heads. More often in the Book of Revelation, measures and numbers disclose inner structures. As Sweet suggests, numbers embody "structural elements of the cosmos," manifested in their relations and ratios (1979, 14). If the sixes in the number of the beast are linked to the sixth seal, sixth trumpet, and sixth bowl — all of which allude to the great, prepared day of the wrath of God on which final conflicts with evil occur — the number is associated with penultimacy and evil destruction.

The repetition of the numbers 12, 144, and 1,000 in 7:4–8 and 21:16–17 creates meaning by connecting "people of God" to "the eternal sanctuary provided by God" (Mounce 1977, 381). More specifically, the 144 cubits of the wall of the city connect to the 144 thousand sealed ones redeemed by God (7:4–8); and the incomparably larger cubed city — twelve thousand stadia long, wide, and high — relates to the great multitude that cannot be counted (7:9–12). Or if that specific connection cannot be sustained, the following ratio can be: 144 cubits of wall are to the 144 thousand redeemed as the twelve-thousand-stadia city-cube is to the totality of those redeemed.[39] In that way the individual measurements of wall and city, which make no sense as specific, individual dimensions, take on significance.

Boundaries and Ratios

The term *ratio* is often limited to relationships among numbers, but it need not be. The same is true with the term *rational*. Both derive from the same Latin root and may refer to numbers, but both *ratio* and *rational* belong to the sphere of another cognate, *reason*. All of those cognates refer to fundamental insight a person gains after seeing essential connections and fundamental structures. Newton's insight into gravitation can be expressed as a universal proportion or ratio: as the apple falls, so the moon, and so everything (see Bohm 1983, 21); and that universal proportion discloses something essential in the Newtonian worldview. To grasp essential ratios is to comprehend aspects of an inner structure that unfolds to create a comprehensive vision of the world (see Bohm 1983, 20).

Ratio is a useful term for underscoring the importance of relationships and sets of relationships rather than individuals per se. Just as individual numbers do not disclose insight into the seer's world, so one specific boundary situation does not disclose an understanding of inner structures of the seer's world. Those inner structures are comprehended by comparing boundary situations. As we saw earlier, there are similarities in the contours or shapes of boundaries in the Book of Revelation; thus there are similarities among the transformations that occur at different boundaries in that book. Those similarities in contours and transformations derive from inner, guiding structures that can be formulated as ratios and universal proportions in the seer's world. Thus, by tracing ratios or proportions throughout boundary situations in every dimension of the seer's world, one can disclose fundamental aspects of the structure of that world: God is to Satan as the Lamb is to the beast, as the faithful are to those who deceive, as the Christian minority is to the larger Roman world; heaven is to earth as the eschatological future is to the present, as the temple is to the space around it, as cultic activity of worship is to everyday activity, as being in the Spirit is to normal consciousness. Ratios and proportions can be formed among social, political, religious, theological, and psychological aspects of the seer's vision because all of those aspects unfold an order implicated and replicated throughout the seer's world.

Relationships and transformations at boundary situations tend to be either of two kinds: homologies or contrasts. One of them is usually more prominent. For example, in the letter to those at Smyrna a homology is made between Satan and those who claim to be Jews (2:9), whereas in chapter 12 the "descent of Satan" transforms an aspect of heaven into its contrasting opposite. Both homologies and contrasts are, however, implicit in every boundary situation. Those who claim to be Jews are homologous to Satan because they oppose true Christians who are homologous to God. Satan's "fall" which transforms him into an opponent of God is homologous to the "fall" of Christians such as those at Ephesus (2:5) and contrasts to the reverse movement of "repentance" offered to Jezebel (2:21).[40] Ratios may be formed among any of those various elements: God is to Satan as the faithful (true Jews) are to those who claim to be Jews. Those Jews are to Satan as the faithful are to God. God is to those Jews as Satan is to the faithful. In brief, boundaries channel a discrete segment of "flow" or "becoming" into a homologous or contrasting element along lines consonant with that inner structure reflected at every boundary. Conversely, by tracing the various ratios among homologues and contrarieties, one discloses the fundamental structure implicated in every boundary.

Sexual expressions in Revelation can illustrate that disclosure. Babylon the Great can boast of many sexual exploits. All the kings of the earth fornicate with her, and those dwelling on earth become drunk from the wine of her sexual passion (14:8, 17:2, 18:3). Babylon is therefore called the Great Whore (17:1), full of abominations and uncleannesses from her fornication (17:4). She destroys the earth by her fornicating (19:2). The same term ($\pi o \rho \nu \epsilon \dot{\upsilon} \omega$) is used of Jezebel, the prophetess at Thyatira, who deceived "my servants" into fornicating and who does not seem to want to repent from her fornication (2:20–21, cf. 2:14). Implicit in this homology is a contrasting element to complete the ratio: Babylon is to Jezebel as X is to the faithful Christians. X is, of course, the Bride of the Lamb, who is clothed in fine, bright, pure linen (19:7–8) and who comes down from heaven as the eschatological Jerusalem (21:9–10).[41]

A related homologue occurs in descriptions of eschatological events of judgment and salvation: fornicators experience the "second death," namely, burning in the lake of fire and sulfur (21:8), which is apparently equivalent to dwelling outside the city gates of the heavenly Jerusalem (22:15).[42] Those condemned to the second death contrast with those at Smyrna who conquer (2:11) and with those who share the "first resurrection" and reign as priests with Christ for a thousand years (20:6).[43] They are obviously the ones who will live inside the New Jerusalem, participating in such activities as eating of the tree of life in the restored paradise of God (see 2:7). Fornication is to nonfornication as being outside the New Jerusalem is to being inside the city, as being outside the church proper is to being inside the church proper.[44]

As can be seen from this examination of sexuality, almost any boundary in the seer's world can unwind through quantum leaps into religious, social, political, and psychological realities. No one boundary or one kind of boundary can claim a privileged position in that unwinding. Every boundary reiterates every other as proportions are formed among heavenly worship and Christian celebration, appreciation of Roman culture and demonic excess, political power and insubordination before God, Jewish claims and apostasy to Satan, church boundaries and the boundaries of paradise, or ironic kingship and Christian witness. Any and every object that John encounters gains meaning by taking its place on a boundary in the seer's world and entering into his network of homologies and contrarieties.

Conclusions

In the seer's world boundaries do not seem to reinforce fundamental conflicts and antagonisms between religious promises and social disappointments, bodily mortality and spiritual hopes of immortality, or natural impulses and cultural demands; rather, boundaries provide points of transaction whereby religious promises, social encounters, biological givens, and cultural demands undergo mutual adjustments, form homologous relations, and contribute to the coherence, integrity, and wholeness of Christian existence.

If there was irreconcilable contradiction among religious, social, biological, and cultural dimensions of Christian existence, the seer would be affirming that at the most fundamental level of reality there is an eternal, fixed metaphysical dualism. Such a view is antithetical to John's. Only God is *Pantocrator* (eg., 11:17), and the kingdom of this world has always been, however implicitly, the kingdom of the creator God (see 4:11, 11:15). The syntax and liturgical setting of 11:15 underscore that there is no spatial or temporal dualism between the kingdom of the world and the kingdom of God. God creates and sustains all things. Transformations and changes permeate every boundary and break down every distinction because there is an underlying dynamic system into and out of which all distinctions fold and unfold. God's dynamic power may flow into rebellious vortices and opposing whorls, but ultimately everything and every power derives from and depends upon God. He is the process that binds past, present, and future; heaven, earth, and subterranean demonic forces; faithful followers, apostates, and infidels. For that reason, this monistic flow of divinely ordered being can never quite be compartmentalized into creature and creator, God and Satan, this age and the age to come, or heaven and earth. That is the unbroken world disclosed through the language of the Apocalypse.

This examination of the seer's vision of the world suggests that it does not contain fundamental conflicts. One element or dimension of the vision is not pitted against another; and terms such as *conflict, tension,* and *crisis* do not characterize his vision. Revelation discloses in its depth or innerness a wholeness of vision consonant with the intertexture found at the surface level of his language. At all levels signifiers, signifieds, deep structures, and surface structures form homologies, not contradictory oppositions. The logic of the vision does not progress from oppositions to their resolution. Rather, in all its aspects the language speaks from unbroken wholeness to unbroken wholeness.

III

The Stage:
Roman Society and the
Province of Asia

6

Domitian's Reign: History and Rhetoric

The separation of the seer's language from the social order is an artificial one. John's language—its shape, its genre, the vision it transmits—communicates a message, and that communication is a social act that takes its place in the social order. However visionary John's writing is, it does not operate in a symbolic universe apart from the world of actual, social relations. Moreover, the social order does not exist as a given, simply to be observed with the aid of proper historical tools. The social order is a construction for a person contemporary with it. How much more so for a historian working nineteen centuries later! The historian reconstructs events, imagines social connections, and extrapolates from limited sources so as to construct a plausible order of society. There is room for alternative constructions on issues both small and great. In brief, the "social historical situation" is an imaginative construction built from a critical reading of primarily linguistic evidence—books, speeches, inscriptions, and coins.

In the following chapters I offer a plausible reconstruction of Domitian's reign (chap. 6), social organization in the province of Asia (chap. 9), and the place of Jews and Christians in that province (chaps. 7 and 8). This social order, reconstructed almost entirely from sources other than the Book of Revelation, will provide further evidence that the seer and his audience did not live in a world of conflict, tension, and crisis. Christians lived quiet lives, not much different from other provincials. The economy, as always, had its ups and downs; and the government kept the peace and demanded taxes.

Most scholars today date Revelation to the early 90s CE, the last years of Domitian's principate or rule (see chap. 1). John, of course, was not writing in Domitian's capital; he wrote in the Roman province of Asia, approximately a thousand miles from Rome; nonetheless, the Roman emperor set the direction of political and economic conditions in the provinces. If the Emperor Domitian was a megalomaniacal tyrant, as most commentators on Revelation assume, that perception of Domitian shapes the construction of the social order in the Asian province, which in turn shapes the commentary on the Book of Revelation. If another portrait

95

emerges from a critical examination of the sources, the commentary on Revelation will have to be modified.

First of all, however, a comment needs to be made about the relationship between early Christian and Roman sources. Put simply, commentators on the Book of Revelation (perhaps more generally on early Christianity) tend to approach Roman writings more naively than Christian writings. Frend, for example, writes that historians of the early Christian period have to work on two levels: "that of prophecy and eschatology" when involved with Christian writings and "that of human activity and events" with classical (Roman) authors. Frend concludes, "only seldom . . . do these levels coincide. For the rest, Jerusalem and Babylon move on different planes" (Frend 1981, 183–84). Such a sharp distinction between the two types of sources cannot be made; for Christian sources tell us about human activity and events, and classical authors do not simply report human events. A quotation from Pliny or Suetonius or Tacitus raises thorny historiographic problems and does not directly reflect social and historical realities. In short, the following inquiry into Domitian and his reign is also an exercise in reading Roman historical sources. There is considerable detail on Domitian not related directly to the Book of Revelation, but that detail is necessary in order to present convincingly an alternative view of Domitian and his reign. I shall first present aspects of the standard portrait of Domitian, then an assessment of that portrait along with an alternative.

The Portrait of Domitian in Standard Sources

An official portrait of Domitian was drawn a few years after his death by a circle of writers around Pliny the Younger that included Tacitus and Suetonius. This portrait, created in the early years of Trajan's reign, became the standard for most portrayals of Domitian from the second century to the present. Pliny, a Roman author and senator born around 60, refers to Domitian in several of his letters, which were published from about 105–9.[1] Domitian plays a greater role in Pliny's *Panegyricus* to Trajan, delivered in 100. The Roman historian Tacitus, slightly older than Pliny, refers to Domitian in the *Agricola* (written around 98), the *Germania* (98–99), and book 4 of the *Histories* (100–110). Unfortunately, the last books of the *Histories*, which included Domitian's reign, are not extant. Pliny and Tacitus were good friends who corresponded about hunting, writing, and oratory (see Plin. in *Ep.* 1.6, 1.20, 7.20, 8.7, 9.10). Pliny took great pride in being a friend of Tacitus and contributing to the *Histories*. He sent Tacitus information about the eruption of Mount Vesuvius and the death of his uncle, Pliny the Elder, in 79 (*Ep.* 6.16, 20) and also about the trial of Baebius Massa in 93 (7.33). Pliny writes to Tacitus, "I am delighted to think that if posterity takes any interest in us the tale will everywhere be told of the harmony, frankness, and loyalty of our lifelong relationship" (7.20).[2]

Suetonius was a younger member of Pliny's circle, born circa 70. Pliny helped him at the bar (*Ep.* 1.18); in purchasing property (1.24); and in gaining a military tribunate, though Suetonius turned it down (3.8). Suetonius probably served as an assistant to Pliny during his special appointment under Trajan to Bithynia; in a letter from there Pliny asks for special privileges to be granted to Suetonius who "is not only a very fine scholar but also a man of the highest integrity and distinction. I have long admired his character and literary abilities, and since he became my close friend, and I now have an opportunity to know him intimately, I have learned to value him the more" (10.94).[3] Suetonius wrote *Lives of the Caesars* (from Julius to Domitian) around 120. Those three authors and rhetors set the terms for the understanding and remembering of Domitian's reign.[4]

About a century later (215–20) Dio Cassius wrote his histories. Dio, a senator from Bithynia, claims to have read many books on Roman history, no doubt including the works of Suetonius, if not Tacitus and Pliny (see Millar, 1964, 34–38, 85–86). Dio's account of Domitian's reign is no longer extant, but we do have Zonaras' twelfth-century epitome, which supplements Dio with other works (see Millar 1964, 2–3).[5]

These sources paint Domitian as evil, almost without qualification. Suetonius, the most generous among the sources, says that Domitian began his reign with some "leniency and self-restraint" but that those qualities "were not destined to continue long, although he turned to cruelty [*saevitia*] somewhat more speedily than to avarice" (*Dom.* 10.11).[6] Dio Cassius, writing about a century later, tends to make even more extreme comments about Domitian. He says that Domitian was from the start "not only bold and quick to anger but also treacherous and secretive," traits that resulted in impulsive and crafty behavior (67.1).

The "standard" sources for Domitian and his reign are as follows:

Tacitus
  Agricola (98)
  Germania (98–99)
  Histories (100–110)
Pliny the Younger
  Panegyric (100 CE)
  Letters (105–9)
Dio Chrysostom (c. 40–112)
  Discourses
Juvenal
  Satires (115–27)
Suetonius
  Lives of the Caesars (c. 120)
Dio Cassius
  Roman History (c. 215)
Philostratus (170–245)
  Lives of the Sophists
  Life of Apollonius of Tyana

The most relevant emperors and their reigns are listed in the Chronology, p. xi.

Domitian's Character

Among those writers *saevitia* — savageness, cruelty, barbarity, fury — is a favorite term for describing Domitian and his reign. He loved to be flattered; but he himself was fond of no one, although he liked to appear fond of someone when he was about to kill him (Dio Cass. 67.1.1, 67.4.2).[7] Thus, through devious cunning he sometimes appeared to do good and to express love. So Tacitus writes that when Domitian recalled Agricola (Tacitus's father-in-law) from Britain, Domitian greeted him "as his manner was, with affected pleasure and secret disquiet," for Domitian had "decided that it was best for the present to put his hatred in cold storage" (Tac. *Agr.* 39). Later, when Agricola was on his death bed, Domitian sent physicians down from the palace with regularity, but Tacitus suggests that he did this in order to keep an eye on him (*Agr.* 43).[8] When Agricola died, Domitian expressed sorrow, as even Tacitus has to admit: Domitian "paraded the semblance of a sorrowing heart; his hate was now no longer anxious, and it was his temperament to hide joy more easily than fear" (*Agr.* 43).

The standard historical sources characterize Domitian as mad, too. Sometimes the madness is reported in a light vein. For example, it is said that Domitian used to impale flies on a stylus in solitude, hence the joke, "Where is Domitian? . . . He is living in retirement, without even a fly to keep him company" (Dio Cass. 65.9.4–5, Suet. *Dom.* 3.1). But his madness is also given as the cause of his tyranny and self-importance. As Suetonius says, "From his youth he was far from being of an affable disposition, but was on the contrary presumptuous and unbridled both in act and in word" (*Dom.* 12.3). His tyrannical nature, writes Suetonius, became apparent in 69 when he briefly "ruled" in Rome before his father, Vespasian, returned from Egypt: "He exercised all the tyranny of his high position so lawlessly, that it was even then apparent what sort of a man he was going to be" (*Dom.* 1.3). When in 81 Domitian claimed the throne (after the death of his brother Titus), his tyrannical nature dominated. Pliny describes the palace under Domitian as the "place where . . . that fearful monster built his defences with untold terrors, where lurking in his den he licked up the blood of his murdered relatives or emerged to plot the massacre and destruction of his most distinguished subjects. Menaces and horror were the sentinels at his doors . . . always he sought darkness and mystery, and only emerged from the desert of his solitude to create another" (*Pan.* 48.3–5). He claimed the office of censorship perpetually and he dominated the consulship (see *Pan.* 58.3); he employed twenty-four, not twelve, lictors (heralds who announced the approach of the emperor) and wore "triumphal garb whenever he entered the senate house" (Dio Cass. 67.4.3). In brief, as Pliny says in one of his letters, Domitian displayed "a tyrant's cruelty and a despot's licence" (*Ep.* 4.11.6).

According to the same sources Domitian also engaged in excessive sexual activity. His were unbridled passions (Tac. *Hist.* 4.68). Suetonius comments that Domitian "was excessively lustful. His constant sexual intercourse he called bed-wrestling, as if it were a kind of exercise. It was reported that he depilated his concubines with his own hand and swam with common prostitutes" (Suet. *Dom.* 22.1). Apparently Domitian was bisexual, for Suetonius mentioned that Claudius Pollio used to exhibit a letter that he had preserved from the hand of Domitian as a youth, promising Claudius an assignation; he even reports that Domitian was debauched by his successor, Nerva (*Dom.* 1.1). Dio Cassius simply comments that Domitian was "profligate and lewd towards women and boys alike" (67.6.3). In that regard, authors play down Domitian's statesmanship in 69 and say that he used his temporary authority and power merely for "debauchery and adulteries" (Tac. *Hist*. 4.2, cf. *Agr*. 7). He carried on "with the wives of many men," including the wife of Aelius Lamia, a woman he later married (Suet. *Dom*. 1.3, Dio Cass. 65.3.4). Writers especially enjoy gossiping about Domitian's relations with Julia, the daughter of his brother, Titus. Suetonius says that while Julia was still a young virgin, she was offered to Domitian in marriage; but he refused. Later, when she became the wife of another, he seduced her; and after the death of her husband and her father, he "loved her ardently and without disguise, and even became the cause of her death by compelling her to get rid of a child of his by abortion" (*Dom*. 22.1).[9]

Early Career and Family Relations

When Vespasian was declared emperor in 69, he was still in Egypt, and his elder son, Titus, was busy in Israel bringing an end to the Jewish rebellion. Domitian, Vespasian's younger son, was thus given a brief stint at governing in Rome until Vespasian could return. As we have seen, Roman authors do not speak favorably about that brief period of power. They also use it to show that Domitian was alienated from his father and brother. Domitian is described as power-hungry and as one whom Vespasian had to reprimand. In response to all the offices that Domitian had appointed while Vespasian was in Egypt, Vespasian thanked his son (ironically) for permitting him to hold office and not dethroning him (Dio Cass. 65.2.3).[10] Dio Cassius writes that when Domitian met his father—not at Rome but in the southern Italian town of Beneventum—he was ill at ease, since he knew what he had done and what he planned to do, and sometimes he even feigned madness (65.9.3).[11] After that, Vespasian required Domitian to live at home and to follow in a litter the emperor's chair and that of his brother Titus (Suet. *Dom*. 2.1, cf. Tac. *Hist*. 4.5, Dio Cass. 65.10.1). In response to these pressures Domitian avoided all public service. Tacitus writes, "When Domitian realized that his youth was treated contemptuously by his elders, he abandoned the exercise of all imperial duties, even those of a trifling character and duties which he had exercised before; then, under the cloak of simplicity and moderation, he gave himself up to profound dissimulation, pretending a devotion to literature and a love of poetry to conceal his real character and to withdraw before the rivalry of his brother, on whose milder nature, wholly unlike his own, he put a bad construction" (*Hist*. 4.86, cf. Suet. *Dom*. 2.2).

The sources suggest that Titus, the elder brother, was also the favored son. Titus had a mild nature and was well disposed towards Domitian, but after Vespasian died and Titus became emperor (in 79), Domitian continually plotted against Titus (Tac. *Hist*. 4.52; Suet. *Tit*. 9.3, *Dom*. 2.3). Suetonius gives a heartrending account of the brothers Flavian: "Although his brother [Domitian] never ceased plotting against him, but almost openly stirred up the armies to revolt and meditated flight to them, he [Titus] had not the heart to put him to death or banish him from the court, or even to hold him in less honour than before. On the contrary, as he had done from the very first day of his rule, he continued to declare that he was his partner and successor, and sometimes he privately begged him with tears and prayers to be willing at least to return his affection" (*Tit*. 9.3). At Titus's unexpected, early death (in 81), Dio Cassius reports that Domitian mourned, but it was only pretense: "He delivered the eulogy over him with tears in his eyes and urged that he be enrolled among the demi-gods—pretending just the opposite of what he really desired" (67.2.6, cf. 67.2.4). Earlier Dio Cassius says that according to some when Titus was deathly ill, "Domitian, in order to hasten his end, placed him in a chest packed with a quantity of snow, pretending that the disease required, perhaps, that a chill be administered" (66.26.2, cf. Philostr. *VA* 6.31–32).[12]

Through this rendition of Domitian's relations with his father and brother, Domitian's reign is isolated from that of Titus and Vespasian. He is presented as different from them, a threat to them, and a brother unworthy of them.[13] A friend of Titus could not be a friend of Domitian (see Plin. *Ep*. 4.9.2); Domitian "outdid himself in visiting disgrace and ruin upon the friends of his father and of his brother. It is true, he issued a proclamation confirming all the gifts made to any persons by them and by other emperors; but this was mere vain show; . . . for he regarded as his enemy anyone who had enjoyed his father's or his brother's affection beyond the ordinary or had been particularly influential" (Dio Cass. 67.2.1–2). After Domitian gained power in 81, Dio Cassius says that praising Titus was equivalent to reviling the emperor (67.2.5). These sources can thus praise the Flavian house and still censure Domitian: "This house was, it is true, obscure and without family portraits, yet it was one of which our country had no reason whatever to be ashamed, even though it is the general opinion that the penalty [i.e., death] which Domitian paid for his avarice and cruelty was fully merited" (Suet. *Vesp*. 1.1).

Domitian's Public Rule

During his fifteen-year reign (81–96), Domitian made campaigns against the Chatti, the Dacians, and the Sarmatians—all of whom lived along the ill-defined northern borders of the empire along the Rhine or Danube. As a consequence of the campaign against the Chatti, Domitian took the name Germanicus. The standard literary sources, in predictable fashion, dismiss Domitian's successes among the Germans. Domitian's return, comments Dio Cassius, "filled him with conceit as if he had achieved some great success" (67.3.5). He goes on to say that Domitian did not so much as see any hostilities in his German campaign (67.4.1). Tacitus says that Domitian gained more triumphal awards than victories (*Germ*. 37, cf. Plin. *Pan*. 11.4). In contrast to Agricola, who had real success in Britain, Domitian's triumphs were "counterfeit," "a laughingstock: he had in fact purchased, in the way of trade, persons whose clothes and coiffure could be adapted to the guise of prisoners" (Tac. *Agr*. 39). Pliny comments that even rivers played a part in Domitian's shame—"the Danube and Rhine were delighted for their waters to play their part in our disgrace" (*Pan*. 82.4).

During the latter years of his rule Domitian became—according to these sources—an "object of terror and hatred to all" (Suet. *Dom*. 14.1). His reign is described as a period of confusion, slaughter, and disorder (Latin *strages*, Tac. *Agr*. 45). Tacitus is grateful that his father-in-law did not live to see that political disorder, "to see the Senate-house besieged, the Senate surrounded by armed men, . . . [and]

consulars butchered" (*Agr.* 45, cf. Plin. *Pan.* 76.5). Everywhere were informers (*delatores*) who would indiscriminately make charges of treason against those who had committed no crime.[14] The latter years of Domitian's reign are also said to be in economic disarray: Domitian overspent funds on "grand and costly entertainments," for he loved to appear "in half-boots, clad in a purple toga . . . wearing upon his head a golden crown with figures of Jupiter, Juno, and Minerva" (Suet. *Dom.* 4.1, 4). His massive building program contributed to the empire's reduction "to financial straits" (*Dom.* 5, 12.1). Because of his financial need, no property was safe from seizure (see *Dom.* 12.1). Dio Cassius says that Domitian even had people drugged secretly in order to get their estates (67.4.5), and Pliny refers to forged wills (*Pan.* 43.1). Domitian is said to have plundered the provinces for the same reason; thus the provincials lived in terror of disaster whenever he traveled through (Plin. *Pan.* 20.4). Military affairs were also in disorder. Pliny refers to his experience in the military at Syria in the Domitianic period, when "merit was under suspicion and apathy an asset, when officers lacked influence and soldiers respect, when there was neither authority nor obedience and the whole system was slack, disorganized and chaotic, better forgotten than remembered" (*Ep.* 8.14.7, cf. *Pan.* 18.1, Dio Cass. 67.3.5).

This chaos in the social and cultural realms became manifest in the larger, cosmic world. Domitian, who at the end was anxious and worried, "disquieted beyond measure by even the slightest suspicions," was given several omens (Suet. *Dom.* 14.2). Lightning struck the temple of Jupiter and the Flavians, the palace, and the emperor's own bedroom (*Dom.* 15.2). Inscriptions on one of his triumphal statues were torn off; Fortuna of Praeneste sent back a most dire omen (*Dom.* 15.2). Soothsayers announced the day of his death.[15] Rusticus, whom Domitian had killed earlier, appeared to him in a dream with a sword (Dio Cass. 67.16.1). Then, according to Suetonius, in the midst of all this chaos and disorder, two portents brought assurance: a raven perched on the capitol cried out, "All will be well." And Domitian dreamed "that a golden hump grew out on his back" (Suet. *Dom.* 23.2). A new, golden age was coming, but not during the reign of Domitian. His, according to our standard sources, was a time of *strages*—confusion, disorder, and chaos.

Assessing the Standard Sources

This standard portrait of Domitian is clearly not drawn by neutral observers. At every opportunity the writers defame Domitian by emphasizing his evil actions, by attributing malicious motivation to good deeds, or by omitting favorable aspects of his reign. They present private information and psychological motivation about Domitian to which they could not possibly have access. Moreover, their maligning of Domitian is contradicted in almost every instance by epigraphic and numismatic evidence as well as by prosopography, the study of biographies and public careers of senators during Domitian's reign. The standard sources distort virtually every area of Domitian's public and state activity during the time of his emperorship.

LINCOLN CHRISTIAN COLLEGE AND SEMINARY

Early Career and Family Relations

As we have seen, Domitian's early career, especially his temporary rule (praetorship) in 69, showed his colors, indicating "what sort of a man he was going to be" (Suet. *Dom*. 1.3). Because of Domitian's forced withdrawal from public life at such an early age (about eighteen years old), the standard sources can claim that when Titus unexpectedly died in 81, Domitian came to the principate or emperorship ill-qualified. Furthermore, Domitian's antipathy towards the friends of either Vespasian or Titus meant that he could not have any trusted advisers for his principate.

Several different kinds of evidence do not square with that rendition of Domitian's early career and that account of his relationships with his father and brother. Tacitus himself indicates that in 69 Domitian performed well many acts of public politics. When Domitian entered the senate in 69, Tacitus writes that he spoke briefly and in moderate terms of his father's and brother's absence and of his own youth; his bearing was becoming; and the confusion that covered his face was regarded as a mark of modesty (*Hist*. 4.40).[16] The senate passed a motion made by Domitian to restore former emperor Galba's honors. When asked to locate and punish informers, Domitian judiciously replied that "on a matter of such importance he must consult the emperor" (*Hist*. 4.40). Moreover, Domitian took the lead in the senate, when one group began accusing another of being informers and supporters of recently past regimes "in recommending that the wrongs, the resentments, and the unavoidable necessities of the past be forgotten" (*Hist*. 4.44); later he made a conciliatory motion that the "consulships which Vitellius had conferred" be canceled (*Hist*. 4.47). Also, Domitian reassured the army at the time when cohorts and legionaries were being restored and reintegrated (*Hist*. 4.46). Such activity belies the portrait of Domitian as a royal playboy.[17]

From 70 to 81 Domitian did write poetry and probably philandered.[18] But he continued to have a role in the political life of the empire. Neither Vespasian nor Titus sought to exclude Domitian from official duties. Vespasian was committed to the establishment of a dynasty; he had said to the senate "that either his sons would succeed him or he would have no successor" (Suet. *Vesp*. 25). Vespasian emphasized this dynastic policy through coinage. Kenneth Waters points out that "a lengthy series of Domitianic coins begins early in the reign" of Vespasian and another series continues until 79, when Vespasian died. Similar coinage continues through Titus's short reign.[19] Inscriptions around the empire in the 70s indicate that Domitian held public offices, including the consulate, regularly.[20] Titus and Domitian were jointly chosen to be highest municipal officers around Aquinum in Latium sometime in the early 70s.[21] From Cures in the Sabine region there is a dedicatory inscription to Domitian from sometime between 73 and 76, with the offices of consul . . . priest of all colleges, and *princeps juventutis* (Dessau, *ILS* 267). The last title, which marks a person as a future emperor, is attributed to Domitian in other inscriptions as well (see Newton 1901, 28, 222). In sum, during the reigns of his father and brother, Domitian received the training of a future emperor.[22]

When Domitian acceded to the principate in 81, he not only sanctioned his brother's status of deity, he "seems to have done more for the cult of Titus, than Titus had done for that of Divus Vespasianus" (K. Scott 1975, 62).[23] Domitian completed the temple begun for Vespasian and dedicated it to the cult of Vespasian and Titus; he also built the Porticus Divorum and dedicated the triumphal arch commemorating the Jewish War (K. Scott 1975, 62–63). Most important, he built the Flavian Temple on the site of his birthplace. Statius praises Domitian for consecrating "to his father's line lights that will truly endure, a Flavian heaven" (*Silv.* 4.3.18–19, cf. K. Scott 1975, 64–65). Statius and Martial, poets during Domitian's reign, wrote freely in praise of Titus and Vespasian.[24] They would not likely have done this if Domitian viewed praising Titus as equivalent to reviling himself, as Dio Cassius asserts (67.2.5). Nor does Domitian dismiss all who received honor from his father and brother, as Dio Cassius claims (67.2.1–2). Quintilian had been given special favor under Vespasian as a publicly paid teacher, and Domitian chose him to train the sons of Flavius Clemens, who at that time were heirs designate.

Finally, Domitian's relations with Titus's daughter, Julia, were not as clear-cut as suggested by the standard sources. If indeed she died because Domitian forced her to have an abortion, Martial's epigram at the birth of Domitian's son is puzzling: "Julia [already dead] with her own snow-white finger shall draw thy golden threads, and spin for them all the fleece of Phryxus' ewe" (Mart. 6.3, cf. 6.13, 9.1). It is difficult to see how such a destiny would be viewed as propitious had Julia died in forced abortion. Moreover, inscriptions honor Julia and Domitia jointly.[25] There is point to K. Scott's remark, "I cannot help feeling that normal affection between uncle and niece might have been distorted by the scandal-mongers and enemies of the ruler into the story [of incest] which has come down to us" (1975, 76).[26]

Domitian's Public Rule

Earlier in this chapter I reviewed what Pliny, Tacitus, Suetonius, and Dio Cassius had to say about Domitian's public rule. They speak with a unified, negative voice. In contrast to those post-Domitian condemnations Quintilian, Frontinus, Statius, Martial, and Silius Italicus — writers from the time of Domitian — praise Domitian's military exploits and successes. "Who could sing of war better," writes Quintilian, "than he who wages it with such skill?" (*Inst.* 10.1.91); and Silius Italicus refers to Domitian's successes in Germany as outdoing the exploits of Vespasian and Titus (*Pun.* 3.607). Martial also compares Domitian's victories against the Chatti favorably with Vespasian and Titus's "Idumaean triumph" (Mart. 2.2, cf. 4.3). Frontinus gives more substance to Domitian's abilities, as he describes the various strategies through which Domitian succeeded against the Chatti: concealing the true reason for his departure from Rome (*Str.* 1.1.8), fighting on foot (2.3.23), advancing the frontier (1.3.10), and compensating for land taken — "The renown of his justice won the allegiance of all" (2.11.7).[27] Both Martial and Statius remark on Domitian's modesty for refusing a triumph after the Sarmatian campaign (Mart. 8.15, 78; Stat. *Silv.* 3.3.171, 4.1.34–39, 4.3.159). The writers contemporary with Domitian praise their emperor richly, as Pliny and Dio Chrysostom do their Emperor Trajan.[28]

More important for early Christian history is the general characterization of Domitian's reign as repressive, cruel, and savage (*saevissima dominatio*) and Domitian himself as a suspicious, insecure tyrant, feared and hated by all, who aggrandized himself with new forms of imperial worship.[29] That characterization allows for historians to make a "mark" in early Christian history for dating Christian documents and for seeing them as in some way a response to Domitian's "reign of terror."[30] With regards to the Book of Revelation, Domitian's supposed demand that all call him "Our Lord and God" (*dominus et deus noster*) is seen as influencing the language of the seer at Revelation 4:11 (see chap. 1). This characterization of Domitian and his reign, thus, deserves careful assessment.

Domitian and the Imperial Cult

The imperial cult preceded Domitian by many reigns; it came in with the empire itself.[31] Peoples of the eastern Mediterranean, where worship and deification of rulers had a long history, integrated the worship of Augustus into their public cults.[32] In Rome, too, worship of the emperor developed early, albeit out of a sacral tradition different from the East's. In Rome it evolved from the private worship of household gods and the heads of families as well as from Eastern influence.[33] Even the Roman elite participated in the cult of the emperor, recognizing both Augustus and Julius as divine. Hopkins puts it succinctly: "Several modern historians of Rome have dismissed the evidence of our sources as glib flattery or an insincere exaggeration. Perhaps much was, but that does not explain it away. The idea of the emperor's divinity and close association with the divine persisted and was fostered by Roman notables" (1978, 213).

The imperial cult also continued long after Domitian. As Waters says, "All the evidence undoubtedly points to an increasing emphasis on the divinity of the emperor; this was inevitable, a process which had begun long before Domitian and would continue almost uninterruptedly for two centuries" (1964, 74). That process included Trajan, the modest emperor of Pliny and Dio Chrysostom. Legends and depictions of his coins reflect the imperial cultic tradition;[34] under Trajan the provincials sacrificed to his image, just as they had sacrificed to the images of his predecessors (K. Scott 1932, 164). Like his predecessors, Trajan deifies his relatives, both his natural and adoptive fathers as well as a sister (Plin. *Pan.* 89).[35] Pliny witnesses to the imperial cult under Trajan. In Bithynia wine and incense are offered to Trajan's statue (Plin. *Ep.* 10.96.5), and Pliny himself sets up a statue of Trajan, along with images of former emperors, in a temple built specifically for the imperial cult (*Ep.* 10.8–9). In a letter to Caninius Rufus, who plans to write an epic about Trajan's Dacian wars, Pliny admonishes him to "call the gods to your aid, and among the gods that one [*et inter deos ipso*] whose exploits, achievements and wisdom you are going to celebrate" (*Ep.* 8.4.5, cf. also *Pan.* 14.2). Pliny regularly refers to Trajan as *dominus* in his correspondence with the emperor and often speaks devoutly about him. For example, "The people of Nicaea, *Domine*, have officially charged me by your immortal name and prosperity [*per aeternitatem tuam salutemque*], which I must ever hold most sacred, to forward their petition to you" (*Ep.* 10.83).[36] Dio Chrysostom uses a Greek synonym of *dominus* ("lord") when he boasts, probably about Smyrna, that the god of that city had been the first to proclaim Trajan "master of the world" (*Or.* 45.4).[37]

Those references to Trajan become important when we consider how authors condemn Domitian for receiving the same titles and for deifying his relatives. With regard to the latter, Pliny goes to great lengths to contrast noble Trajan's act to that of evil Domitian (*Pan.* 10.4–6, 11.1–4). Arrogant Domitian required titles such as *deus, dominus, tyrannus*, and *despotes*, whereas modest Trajan requests *civis, parens, pater*, and *homo*.[38]

Authors from Trajan's or later times especially condemn Domitian for demanding the title "Our Lord and God" (*dominus et deus noster*). Suetonius says that Domitian delighted "to hear the people in the amphitheatre shout on his feast day: 'Good Fortune attend our Lord and Mistress [*domino et dominae*]'" (*Dom.* 13.1; cf. Plin. *Pan.* 33.4, 52.6). Suetonius states that Domitian had his procurators send out letters in the name of "Our Lord and God" (*dominus et deus noster*) and that "the custom arose of henceforth addressing him in no other way even in writing or in conversation" (*Dom.* 13.2, cf. Dio Cass. 67.4.7, 67.13.4). Writers also mention the inordinate number of statues of himself that Domitian had erected: "Every approach and step, every inch of the precinct was gleaming with silver and gold" (Plin. *Pan.* 52.3). Dio Cassius, who writes the story a century later, speaks with greater hyperbole: "Almost the whole world . . . was filled with his images and statues constructed of both silver and gold" (67.8.1).

If the statements of these authors writing after Domitian's reign reflect accurately the situation at the time of Domitian, we should expect to find the *dominus et deus noster* title in writings from his time. Fortunately, we have from Domitian's last years works of both Statius and Quintilian commissioned or requested by Domitian himself. If Domitian demanded that he be called "Our Lord and God," as Suetonius and others say, these works should reflect that requirement.

They do not. In a poem celebrating Domitian's seventeenth consulship in 95 — one year before his death — Statius refers to the emperor as *Caesar, Germanicus, parens, Augustus*, and *dux* but never as either *dominus* or *deus* (*Silv.* 4.1). In the same year Statius writes a poem for the opening of the Domitian Road by celebrating Domitian with similar titles — *Caesar, dux, maximus arbiter*, and *parens*.[39] At the beginning of the *Achilleid*, published in 95–96, Statius addresses Domitian simply as *vates* and *dux* (*Achil.* 1.14–19).[40] In the preface to book 4 of the *Institutio Oratoria* Quintilian acknowledges the honor of tutor that is conferred upon him "by such divine appreciation [*judiciorum caelestium*]," by "the most righteous of censors," and by the "prince" (*princeps*) (*Inst.* 4, pref. 2–3). Some form of *dominus* or *deus* would have been natural here if the climate of the times had demanded it. Quintilian does continue by calling upon the aid of "all the gods and himself [Domitian] before them all [*omnes . . . deos ipsumque in primis*] . . . for there is no deity [*numen*] that looks with such favour upon learning" (*Inst.* 4, pref. 5). Such lofty language is, however, much the same as Pliny urges Caninius Rufus to use with Trajan: *Proinde jure vatum invocatis dis, et inter deos ipso, cuius res opera consilia dicturus es*" (*Ep.* 8.4.5). Thus, neither Statius nor Quintilian, each of whom was writing as an official close to the throne at the end of Domitian's reign, uses the titles we would expect, given the statements of Pliny and other standard sources of Domitian's reign. Moreover, among the many inscriptions, coins, and medallions from the Domitianic era there are no references to Domitian as *dominus et deus* (Viscusi 1973, 94).[41] Finally, we must note the counterevidence from Statius, who writes that when Domitian was acclaimed *dominus* at one of his Saturnalia, "this liberty alone did Caesar forbid them" (*Silv*. 1.6.81–84).

In order to reconcile this conflicting evidence, Kenneth Scott proposes a temporal explanation: the longer Domitian reigned the more tyrannical he became. Early in his reign, for example at the time of the banquet mentioned by Statius, "Domitian wished to be considered a *princeps*, a constitutional ruler" (1975, 103). Later in his reign, however, he claimed the powers of a *dominatio* (tyranny) and became a megalomaniac, demanding the honors of a god (p. 109). For evidence, Scott can only turn to the poet and epigrammatist, Martial; for as we have seen, neither Statius nor Quintilian, writers close to the throne, hints at any shift to a *dominatio* or to Domitianic demands for divine address. In Martial's earliest work he refers to *dominus* as a term of reproach, but in his fifth book of *Epigrams* (c. 89) he uses both *dominus* and *deus* as titles for Domitian (K. Scott 1975, 107, 109; Mart *Spect.*; Mart. 5.5, 8; 7.2, 5, 34; 8.2, 82; 9.28, 66). Later, in the second edition of book 10, published in the early reign of Trajan, Martial disavows the flatteries that caused him to speak shamelessly of Domitian as Lord and God (*dominus deusque*). He goes on to contrast Domitian the *dominus* with Trajan the *imperator* and senator (Mart. 10.72). Scott interprets Martial's repudiation of earlier usage of *dominus et deus* as an indication that here "Martial . . . revealed his true sentiments" about Domitian's *dominatio* (1975, 110). Implicit in Scott's remarks is the assumption that Martial uses the *dominatio* terminology in the latter years of Domitian's reign because of official pressure from the crown, a point difficult to maintain in the face of the evidence cited above from Statius and Quintilian.[42]

An alternative explanation better accounts for the use of *dominus* ("Lord") and *deus* ("God") in Martial. As Keith Hopkins has pointed out, "Power is a two-way process; the motive force for the attachment between the king and the gods does not come from the ruler alone" (1978, 198). Martial, a poet who sought, but never gained, entrance into Domitian's inner court, approaches power from below. As a potential beneficiary, Martial probably uses extravagant titles to show his devotion to Domitian, just as he later uses extravagant language of repudiation to show his devotion to Trajan (10.72). Other potential beneficiaries approaching power from below also probably used titles such as *dominus* and *deus* and were eager to display their zeal for Domitian. So, for example, in *Epigrams* 7.34, where Martial is praising the builder of Nero's baths, he himself is concerned with a possible response by the malicious crowd who may say, "What do you set above the many structures erected by [Domitian] our Lord and God [*dominus deusque*]?" Martial here concerns himself not with Domitian but with the crowd who are also the ones calling Domitian *dominus deusque*. They — perhaps including lower-echelon procurators — were the ones who used such titles during Domitian's reign. The danger lies not with imperial policy but with popular opportunism among those seeking benefits from Domitian (see Thompson 1984, 472–73). From a climate of quick accusations made by people approaching power from below one cannot assume imperial repression and tyrannical madness.

A critical examination of the claims made by the standard post-Domitian sources on Domitian's demand to be called *dominus et deus noster* in light of evidence from Domitian's reign suggests that the post-Domitian sources do not reflect political realities from the time of Domitian accurately. Domitian did not encourage divine titles such as *dominus et deus noster*, nor is there evidence that Domitian had become a mad tyrant seeking divinization. The presence of the imperial cult, especially in Asia Minor, is not here being questioned; it had been a significant force in the social life of the Asian province from the time of Augustus. There is no indication, however, that Domitian modified the imperial cult by demanding greater divine honors than either his predecessors or successors (see Prigent 1974, 455–83).

Domitian's *Saevissima Dominatio*

The standard sources written from the time of Trajan or later also assert — as evidence for Domitian's most savage tyranny (*saevissima dominatio*) — that he indiscriminately made charges of disloyalty (*impietas*) and treason (*majestas*) and planted informers (*delatores*) everywhere.[43] No doubt Domitian, like every other emperor, had his informers; perhaps, as McDermott and Orentzel say, informers were necessary to the Roman legal system (1977, 26). It is not evident, however, that Domitian abused his use of them. Josephus praises Domitian in the *Antiquities* for not accepting accusations against him by certain Jews, and he writes that Domitian "gave command that a servant of mine, who was a eunuch, and my accuser, should be punished" (Joseph. *Vit.* 76).[44] In a section where Suetonius still has some good things to say about Domitian, he writes that Domitian "checked false accusations designed for the profit of the privy purse and inflicted severe penalties on offenders; and a saying of his was current that an emperor who does not punish informers hounds them on" (*Dom.* 9.3). Suetonius limits that saying to the first part of Domitian's reign, before he turned cruel and avaricious (*Dom.* 10.1). Dio Cassius refers to a version of the same saying but explains it away on different lines: Domitian would instigate informers to give evidence but then have them punished so "that they alone should appear to have been the authors of the wrongdoing. It was with this same purpose that he once issued a proclamation to the effect that, when an emperor fails to punish informers, he himself makes them informers" (Dio 67.1.4). The saying itself fits with the evidence from Josephus; the explanations by those retrospective writers make the saying compatible with the view of Domitian standardized in Trajan's reign.

Authors writing after Domitian's death in 96 associate Domitian's informers with his arbitrary charges of treason (*majestas*). Pliny says that Domitian brought charges of *majestas* to "incriminate men who had committed no crime" (*Pan.* 42.1). "It was enough," Suetonius writes, "to allege any action or word derogatory to the majesty of the prince [*majestas principis*]" (*Dom.* 12.1). Any offense to the *princeps* was an offense to the state; even a slighting of Domitian's gladiators, according to Pliny, could bring recriminations (*Pan.* 33.3–4).[45] Suetonius says that Flavius Sabinus was killed because a crier inadvertently announced him as emperor elect instead of consul (*Dom.* 10.4).[46] Pliny's friends, Senecio, Junius Arulenus Rusticus, and Helvidius Priscus the Younger were killed by Domitian for, respectively, writing a life of Helvidius Priscus the Elder, publishing eulogies of Paetus Thrasea and Helvidius Priscus the Elder, and writing a play about Paris and Oenone that really censured Domitian's divorce (*Dom.* 10.3–4).[47] Dio Cassius reports that the sophist Maternus was killed because he referred to tyrants in a practice speech (67.12.5).[48]

Quintilian, however, writes (towards the end of Domitian's reign) in unequivocal terms about the evil of tyranny: "Suppose a man to have plotted against a tyrant and to be accused of having done so. Which of the two will the orator, as defined by us, desire to save? And if he undertakes the defence of the accused, will he not employ falsehood with no less readiness than the advocate who is defending a bad case before a jury?" (*Inst.* 12.1.40). Regarding the danger of speaking favorably about Paetus Thrasea, Martial (during Domitian's reign) refers to "great" (*magnus*) Thrasea in book 1 and to Thrasea's "constancy" (*constantia*) in book 4 (Mart. 1.8, 4.54.7).[49] The critic Peter Howell is surprised that Martial could still refer to Thrasea and stoic virtue at such a late date, for by this time (88–89) Domitian should be a mad tyrant (1980, 126). Sherwin-White sums up nicely: "In all these cases [of Pliny's friends] the anecdotal sources relate only isolated elements of the charges against the accused. The elder Helvidius had formed a faction out of his friends and relatives against Vespasian," and Domitian "made a clean sweep of the Helvidian coterie, late in his reign, after trying to placate them with high office or the offer of it" (1966, 243).

Throughout Domitian's reign Martial refers favorably to people like Paetus Thrasea and Caesonius Maximus from the reigns of Claudius and Nero, and to Cato, Pompey the Great, and Brutus — figures from the past who championed free speech and senatorial power.[50] Furthermore, although Martial praises Caesar Domitian for reviving the Lex Julia against adultery, he also pokes fun at it and even criticises it implicitly. To Faustinus he writes, "Since the Julian law, Faustinus, was reenacted for the peoples, and Chastity was commanded to enter our homes, 'tis the thirtieth day — perhaps less, at least no more — and Telesilla is now marrying her tenth husband. She who marries so often does not marry; she is adulteress by form of law; by a more straightforward prostitute I am offended less" (Mart. 6.7).[51] Suetonius says that Domitian was "so sensitive about his baldness, that he regarded it as a personal insult [*contumelia*] if anyone else was twitted with that defect in jest or in earnest" (*Dom.* 18.2). Yet Martial freely ridicules Labienus (5.49), Phoebus (6.57, 12.45), and Marinus (10.83) because they are bald. Szelest shows that those elements in Martial's epigrams help to explain why Martial was never accepted into Domitian's inner circle, but those same elements witness to the freedom in Domitian's reign to laud the opposition and to comment on Domitian and his social programs.[52]

Neither Pliny nor Tacitus limits Domitian's violence (*saevitia*) to the last years of his reign; Tacitus describes the whole fifteen-year reign of Domitian as one of violence (*Agr.* 3, cf. Plin. *Pan.* 52.7); only its intensity shifted. Scholars, however, generally tend to follow Suetonius' schema of a good emperor gone bad; Suetonius gives no precise time when the shift to evil occurred, but the standard portrait usually marks it around 89, after the revolt of Saturninus.[53] Brian Jones (1979) has shown the difficulties with that schema. We need only summarize his prosopographical analyses.

Saturninus's revolt in Upper Germany in 89 resulted in some reorganization of the army and punishment of rebellious officers, but it was a military, not a senatorial, affair (B. Jones 1979, 30–32, 35). No policy shift in appointments occurs after 89 to suggest greater conflict with the senate or a suspicious, fearful attitude on the part of Domitian. A few years later there was conflict between Domitian and the Helvidians: several members of that coterie were killed or exiled, and philosophers were expelled from Rome, probably in 93 (see Sherwin-White 1966, 763–71). Not long before Domitian was murdered, Epaphroditus and Flavius Clemens were killed. Those few unconnected facts, however, offer no basis for assuming an "ever-increasing ferocity" by Domitian. In 95, for example, Lucius Maximus, whom Dio Cassius (67.11) suggests worked against Domitian in the Saturninus revolt by burning incriminating papers, received a second consulship, the only one in Domitian's reign. In contrast to Pliny's comments that Domitian himself dominated the consulship, the records show that Domitian took only three consulships in the last eight years of his reign; in 96, the year of his death, Domitian was not consul.[54] As Waters observes, "The aristocracy, the post-Augustan nobility that is to say, continued to receive the consulships to the very end" (1964, 66). Domitian's assassination was a palace affair and did not involve the senate or the praetorian prefects. Some senators may have rejoiced at his death, but the body as a whole did not. Silius Italicus — member of a wealthy, prominent senatorial family — shows genuine admiration for Domitian long after his death.[55] Domitian had no heir, and with the advent of a new dynasty there was sufficient support to abolish his memory by erasing his name on monuments, inscriptions, and the like. According to Fulvio Grosso, however (as cited by Brian Jones), for the empire as a whole only thirty-seven percent of the inscriptions are effaced, suggesting "that not all senators agreed with the official decision" and "that it was by no means implemented universally."[56] There is no reliable information on the number of senators who opposed Domitian at any time during his reign. Prosopographical analyses show, however, that Domitian "gradually abandoned the Flavian monopoly of the ordinary consulships," that he promoted deserving senators, and that he tried to mollify his opposition by honoring them with suffect consulships (i.e., appointments made during the year) (B. Jones, 1973, 91). Those would not seem to be the policies of a cruel tyrant, suspicious of those around him and fearful of sharing power.

Trajan's Reign as a New Era

Those who establish the official view of Domitian write in the early years of the reign of Trajan. What they say about Domitian's reign does not square with epigraphic, numismatic, and prosopographical evidence from the Domitianic period. Furthermore, their assertions about Domitian are disconfirmed in the writings from Domitian's rule. What they say about Domitian is shaped by their situation under Trajan. To understand why they shape Domitian's portrait the way they do, we must inquire more fully into the period of transition between Domitian and Trajan.

Continuities and Discontinuities

In contrast to the upheaval that came with the change of dynasties in 69, the transition from Domitian, the last of the Flavians, to Trajan and the Antonines went smoothly. After the brief, ineffectual reign of Nerva (96–97), Trajan became emperor in 98. Analyses of prosopography and imperial policy show how smoothly the transition was made. Administrative personnel remained much the same. Careers of talented public officials such as Gnaius Octavius Titinius Capito, Fabricius Veiento, and Corellius Rufus, as well as Pliny and Tacitus, continue uninterrupted in the transition.[57] Syme notes that "apart from persons related by blood or marriage, Trajan's most powerful allies [were sought] among the consuls of the last decade of Domitian's reign. . . . Next after the recent consuls come the men of praetorian rank, commanding legions or governing provinces in the last years of Domitian" (1958, 1:50–51).[58] Trajan also followed Domitian in electing more senators from the East and using more equestrian officials. Imperial policy continued along the same lines: autocracy in relation to the senate, the assumption of imperial titles and use of legends on coins,[59] road building in the East, prohibition of vines, policy toward Christians, use of Domitian's newly established offices for cities, and perhaps even in the basic strategies of military and foreign policies.[60] Moreover, Trajan uses Domitian's official correspondence and imperial decisions as precedents for his dealing with Asian problems (see Plin. *Ep.* 10.58, 60, 65, 66, 72).

Nerva and Trajan were not strangers to the Flavians. Nerva received his first consulship in 71 as colleague with Vespasian and his second in 90 with Domitian. The whole of Trajan's record of public life (*cursus honorum*) occurred under the Flavians: quaestor in 78, praetor in circa 84, then to the Seventh Legion. During his time there he apparently helped to suppress the mutiny of Saturninus in 89, and in 91 he received an ordinary consulship with Marcus Acilius Glabrio.[61] Later writers go to great pains to explain away those connections with Domitian. Dio Cassius acknowledges that Nerva was in Domitian's closest circle but claims that Domitian would have killed him had not a friendly astrologer declared to Domitian that Nerva "would die within a few days" (67.15.5–6, cf. Mart. 12.6). Pliny prays for Trajan's safety in the *Panegyricus*: "This is no new concern we ask of you [Jupiter], for it was you who took him [Trajan] under your protection when you snatched him from the jaws of that monster of rapacity [Domitian]; for at the time when all the peaks were tottering to their fall, no one could have stood high above them all and remained untouched except by your intervention. So he [Trajan] escaped the notice of the worst of emperors [Domitian], though he could not remain unnoticed by the best [Nerva]" (94.3).[62]

There were of course discontinuities. The personalities of the emperors were different: perhaps Domitian lacked tact and enjoyed being alone, perhaps Nerva was diffident, and Trajan diplomatic; we cannot be sure, for our sources are hardly trustworthy in portraying psychological nuances. More importantly, neither Nerva nor Trajan had a backlog of dynastic support—political legitimacy and divine relatives—which Domitian had by virtue of being the third Flavian. The most important discontinuity was marked out during Nerva's brief interlude. Nerva advertises his reign as a "new era." Coin legends proclaim *libertas publica, salvus, aequitas*, and *justitia*. He also proclaims it in edicts, one of which happens to come down to us in Pliny's correspondence: "There are some matters, citizens, which need no edict in happy times like ours, nor should a good ruler have to give evidence of his intentions where they can be clearly understood. Every one of my subjects can rest assured without a remainder that, in sacrificing my retirement to the security of the State, it was my intention to confer new benefits. . . . No one on whom the fortune of the Empire has smiled, shall need to renew his petitions in order to confirm his happiness. Let my subjects then permit me to devote myself to new benefactions, and be assured that they need ask only for what they have not hitherto been granted" (*Ep.* 10.58.7-9). Nerva was probably committed to certain "republican" ideals and was willing to relinquish the power of the "sovereign" (*princeps*). Nerva and his cohorts set out, probably with genuine intention, to change the balance of power between *princeps* and senate and thereby to inaugurate a new era. Tacitus echoes that official dogma in the early work *Agricola* (c. 98): "Now at last heart is coming back to us: from the first, from the very outset of this happy age, Nerva has united things long incompatible, Empire and liberty" (*Agr.* 3). But as Syme says, Nerva's short reign was "an interruption in the development of the Roman government in the Flavio-Antonine period" (1958, 2:631).

Nerva had little effect in the course of imperial development. His successor, Trajan, returned to governing as a strong *princeps* supported by the military. There is a breach in continuity between the *principatus* of Nerva and the *imperium* of Trajan (cf. Syme 1958, 1:12; Waters 1969, 386–87). Although Trajan rejects Nerva's notion of *principatus*, he retains Nerva's propaganda about a break with the past. Trajan claims to continue the Nervan "new age" of *libertas*, senatorial power, and freedom of expression. Trajan's official propaganda can be seen in a rescript to Didius Secundus, preserved in the *Justinian Digest*: "I am aware that the property of persons who have been relegated has been confiscated to the Treasury [*fiscus*] by the avarice of former times,[63] but a different course is agreeable to my clemency, as I wish to give this additional example to show that I have favored innocence during my reign."[64] In brief, Trajan follows Domitian in areas of state personnel and imperial policy; but in propaganda and ideology, he follows Nerva and propagates a "new era" from the death of Domitian.

Trajan is not unique in proclaiming a break with "former times." Politicians from all ages have used the idea of "newness" to establish and to legitimate themselves in the political arena.[65] Nor are Trajan's ideological themes of rejecting tyranny and divinity unique; both Augustus and Vespasian are portrayed as emphasizing the same themes at the beginning of their respective dynasties. Through a combination of social-political realities, senatorial ideals, imperial ideologies, and tendencies in Roman historiography, Roman sources portray the first ruler of a new dynasty as righting the balance between senate and *princeps* that has become imbalanced through the successive rulers of the previous dynasty.

Rhetoricians of a New Era

In order to propagate his ideology effectively Trajan required more than edicts and rescripts. Not a man of letters, he needed proficient writers and orators who could both articulate and transmit his ideology. C. P. Jones observed, "So far from regarding literature with amused tolerance, Trajan ensured that his reign and conquests would be celebrated directly or indirectly by a flourishing of Greek and Latin letters" (1978, 116). Trajan found many of his needs met in Pliny the Younger, Dio Chrysostom, and Tacitus. Each of these figures no doubt had his own complex set of motives for supporting Trajan and beginning to write under him. Dio's motives were the least complicated: exiled under Domitian, he was honored by Trajan. Pliny and Tacitus, however, had successful public careers under Domitian and were no doubt concerned to establish their loyalty to old friends returning from exile and to the new regime (see Plin. *Pan*. 45.5). Moreover, they were probably genuinely excited by the possibility of helping to shape the direction of the new dynasty. Trajan valued their particular form of oratory and writings in ways that would have been, at the most, redundant under a Domitian with the authority and legitimacy of the Flavian dynasty.

Dio Chrysostom (Golden Mouth) from Prusa became, in the words of C. P. Jones, an "eloquent witness of the 'new age' and its vaunted conciliation of the principate with liberty"; "His works exemplify the way a benevolent monarchy can mold opinion. Largely through Dio, Trajan transmitted to posterity a picture of himself as the ideal king" (1978, 53, 123). Dio delivered his first discourse on kingship before Trajan at Rome in 100, the same year as Pliny's panegyric to Trajan. In it he characterizes the ideal king "to whom the son of Saturn gives the sceptre" (1.11); and, as Jones points out, this ideal king "turns out to have a distinct resemblance to Trajan" (1978, 116–17).

Pliny the Younger runs a close second to Dio Chrysostom in his praise of Trajan as benevolent founder of a new era of liberty. Like Trajan, Pliny was well known to Domitian. He moved into senatorial status — probably in the late 80s — as *quaestor principis*, attached to the emperor Domitian as a courier between senate and *princeps*. He became praetor in 93, during which tenure he was involved with public prosecution (see Plin. *Ep*. 3.11, 7.33). At the end of that year several of his friends were either exiled or put to death "so that," he says later, "I stood amidst the flames of thunderbolts dropping all round me, and there were certain clear indications to make me suppose a like end was awaiting me" (*Ep*. 3.11.3–4, *Pan*. 90.5). Elsewhere Pliny says that after 93 his career was checked by Domitian and even that he would have been brought to trial had Domitian lived longer (*Ep*. 4.24.4–5, 7.27.14; *Pan*. 95.3–4). From Pliny's own writings we would assume that his next appointment after praetor occurred under Nerva in 98, when he was made an officer over the main state treasury (*praefectus aerari saturni*) (*Ep*. 10.3A, 5.14.5, 10.8.3; *Pan*. 91.1, 90.6); that fits with his statements about a checked career and danger in the last years of Domitian's reign. According, however, to his public career (*cursus honorum*) on an inscription from Pliny's hometown of Comum, immediately after the praetorship in 93 Domitian appointed Pliny to the important post of treasurer of the military (*praefectus aerari militaris*).[66] Contrary to Pliny's own statements, his career was not checked after the trouble in 93; indeed it was furthered.[67] Pliny claims, however, to come to restored liberty under Trajan "awkward and inexperienced" (*Ep*. 8.14.3, 9.13.4; *Pan*. 48.3, 76.3). Perhaps Trajan was a bit slow in using the young Pliny in his administration, as Syme suggests, but Pliny came to be a strong supporter and an effective proponent of Trajan's ideals. He spent his last years as Trajan's special envoy to Bithynia (see Syme 1958, 1:83, but cf. Plin. *Ep.* 10.2).

Pliny's *Panegyricus* was delivered orally at Rome in 100 in honor of the Emperor Trajan.[68] Pliny announces at the beginning that the new era requires a new response, a new word in a new style. "It is my view," he writes, "that not only the consul but every citizen alike should endeavour to say nothing about our ruler which could have been said of any of his predecessors. . . . The sufferings of the past are over: let us then have done with the words which belong to them. An open tribute to our Emperor demands a new form. . . . Times are different, and our speeches must show this" (2.1-3). The new speech should be frank and sincere, removed from both flattery (*adulatio*) and constraint (1.6). No word under Trajan need be prompted by fear, for Trajan is not one to be flattered as a divinity and a god. He is a fellow-citizen (*civis*), not a tyrant (*tyrannus*); our father (*parens*), not our overlord (*dominus*) (2.3). Never forgetting that he is a man himself, he delights in the liberty, not the servitude, of his subjects (2.4-5). Dio Chrysostom follows the same line. The good king may be called father, but not master (δεσπότης); he rules for the sake of all men, therefore no one need feel terror or fear around him. He values sincerity and truthfulness, not unscrupulousness and deceit (*Or.* 1.22-26). Dio Chrysostom says to Trajan, "You delight in truth and frankness rather than in flattery and guile" (*Or.* 3.2, cf. 3.12-13). In his third discourse Dio Chrysostom spends more than twelve sections making clear that he does not flatter in his praise to Trajan (see Waters 1969, 399)! From these concurrences in Pliny and Dio, C. Jones concludes, "There is no evidence that one influenced the other; rather, they express the ideology of a particular time. . . . By their coincidences, among them their coinciding claims to frankness and spontaneity, they are revealed as the servants of his [Trajan's] wishes" (1978, 118-19). Pliny himself gives us the clue when he advises Vettenius Severus about writing a speech of tribute to the emperor: Pliny says concerning his own panegyric, "I made a point of avoiding anything which looked like flattery [*adulatio*], even if not intended as such, acting *not on any principle of independence but on my knowledge of our Emperor*" (*Ep.* 6.27.2, my emphasis). Pliny and Dio speak in a style pleasing to the emperor. Their adulation of him as *princeps* and bearer of *libertas* reflects the image that Trajan seeks to propagate; their preoccupation with sincerity and frankness of speech without *adulatio* is a correlate to Trajan's image as a modest, human, and "plain" man.

There seems to be a certain fluidity in the forms of praise that Pliny uses. In a letter to Vibius Severus in which Pliny tells about reading his revised *Panegyricus* before some friends, he observes, "This is yet another tribute to our Emperor: a type of speech which used to be hated for its insincerity has become genuine and consequently popular today" (*Ep.* 3.18.7). How new, then, is the form of speech in the *Panegyricus*? Here Pliny suggests that while past and present types of speech remain the same, under Trajan laudation is genuine.

Tacitus managed a successful career under Domitian: praetor in 88, he was appointed to the board *quindecimviri sacris faciundis* that directed the Secular Games (*Ann.* 11.11.1); he was away from Rome, probably as a legionnaire, during the period 89-93; and consul designate in 96, most likely selected by Domitian (see

Syme 1958, 1:70). In retrospect, however, Tacitus characterizes those times of success as a period of submissive silence, deprived "even of the give and take of
conversation." For fifteen years under Domitian those who managed to live, says
Tacitus, did not open their lips (*Agr.* 2–3, cf. 45).[69] In contrast, Tacitus praises not
only the most blessed era of Nerva but also Trajan's daily increase of the happiest of
times (*Agr.* 3). Tacitus laments that his father-in-law did not live to see Trajan's
most blessed era, a reign that Agricola had frequently prophesied in Tacitus's
hearing (*Agr.* 44). After completing a successful public career under Trajan with his
proconsulship of Asia in circa 112, Tacitus retired from public life, presumably to
devote full time to his writings. In the *Agricola* and later in the *Histories* he
promises to record for posterity the happy times of Nerva and Trajan, "an age in
which we may feel what we wish and may say what we feel" (*Hist.* 1.1, cf. *Agr.* 3).
But like Statius who kept putting off his work on Domitian's feats in Germany,
Tacitus never fulfilled his promise to Nerva and Trajan. We do not know the reason.
Perhaps he became disaffected with Trajan, as the expectations of Trajan's early
years were not realized in ways that satisfied Tacitus;[70] perhaps the promise was
nothing more than a literary convention;[71] perhaps he simply did not live long
enough to do all that he meant to do. In any case, the early works of Tacitus reflect
the official propaganda of Nerva and Trajan. Scholars may debate the extent to
which the *Agricola* is a political tract, but in it Tacitus unquestionably joins with
Dio Chrysostom and his close friend Pliny in flattering Trajan according to the style
acceptable to the emperor.

For all those writers, contrast with Domitian serves as a device for praising
Trajan (C. Jones 1978, 118). Speaking before Trajan, Dio Chrysostom tells how
Hermes led Heracles to the two peaks — the "Peak Royal" and the "Peak Tyrannus."
The former represents Trajan; the latter is characterized by all the stock traits of
Domitian in the retrospective writings (*Or.* 1.67–82). More explicitly, in the forty-
fifth discourse, Dio Chrysostom contrasts the most-hated Domitian, under whom
he suffered exile, with "the most noble Nerva" and the "benevolence and an . . . interest" that Trajan has shown him (*Or.* 45.1–2).[72] In the *Agricola* Tacitus sets up the
same contrast: our former servitude versus our present good (*Agr.* 3). Agricola was
not permitted to see "the light of this most happy age," but through his hastened
death he escaped the *strages*, "that last period, in which Domitian drained the state,
no longer at intervals and with respites of time, but with, as it were, one continuous
blow" (*Agr.* 44). The contrast reappears at the beginning of the *Histories*, on which
Waters remarks, "Tacitus' attempt (*Hist.* 1.1) to define *libertas* as the absence of
adulatio seems to fail; flattery continued unabated under the new regime, often
thinly disguised as denigration of Domitian; the change of men produced no
change of manners" (1964, 69). In the *Panegyricus* Pliny underscores the importance of contrast in his praise of Trajan: "Eulogy is best expressed through comparison, and, moreover, the first duty of grateful subjects towards a perfect emperor
[*optimus imperator*] is to attack those who are least like him: for no one can
properly appreciate a good prince who does not sufficiently hate a bad one. . . .
With all the more assurance, Conscript Fathers, can we therefore reveal our griefs
and joys, happy in our present good fortune and sighing over our sufferings of the
past, for both are equally our duty under the rule of a good prince" (*Pan.* 53). Pliny

constructs the *Panegyricus* on that principle as he sets Domitian's evil and weakness against Trajan's goodness and strength.[73] What a pleasure it is for Pliny to be suffect consul during September, a month of triple rejoicing "which saw the removal of the worst emperors [Domitian], the accession of the best [Nerva], and the birth of one even better than the best [Trajan]" (*Pan.* 92.4)!

Those examples of the use of contrast illustrate how a retrospective presentation of Domitian and his reign serves as foil in the present praise of Trajan. Their relationship is constructed to form contrasting pairs in which each member reciprocally shapes the other. Domitian has to be the opposite of Trajan—the more evil Domitian, the better the Optimus. Domitian's evil tyranny displays the life of liberty under Trajan, just as Trajan's humanness requires Domitian's exaggerated divinity. The reigns of the two emperors oppose each other: Trajan's *principatus* requires of Domitian a *dominatio*.[74] As a paired member, Domitian inevitably becomes a flat character, for there is no room to develop him as a complex figure composed of a variety of attitudes and motivations. He is Trajan's foil, so our writers' conceptions of Trajan shape their presentation of Domitian at least as much as any other single force.

The opposing of Trajan and Domitian in a binary set serves overtly in Trajan's ideology of a new age as well as covertly in his praise. Newness requires a beginning and therefore a break with the past; such a break is constructed rhetorically through binary contrast. Propagandists for a new age have to sharpen both edges of their two-edged sword: both the ideal present and the evil past have to be exaggerated. The sharper the contrast, the clearer the break and the more evident the new era. As we have seen, in most respects Trajan carried on the policies and administration of Domitian; therefore adjustments had to be made in order to fabricate a break with the Flavian period. On the other hand, several factors in the Flavio-Antonine era made the task easier. There is a fluidity not only in forms of praise, as we have seen, but also in terms like *libertas* and *principatus*. None of our writers supports political revolution or radical social change. Most are chary of extending power and privilege: Pliny's paternalism towards his slaves is one thing; egalitarianism quite another. Only subtle differences distinguish emperors and dynasties in their shifting mix of liberty and stability, of republican forms and imperial control. Moreover, in an empire where friend and foe "appeared identical down to the smallest detail" (MacMullen 1966, 243), when newness is announced, one can never be sure who, if anyone, has switched sides.[75]

Ideology—a verbal product—need not be embodied in practice, and pundits will differ on whether it was embodied or not. But ideological factors in the Flavio-Antonine era make for considerable leeway when it comes to matching word and deed in Trajan's "new age" and to contrasting that age with Domitian's time. We need not mar a Pliny's commitment to Trajan with cynical response. His support and that of the others were genuine. But present commitments that involve contrasting pairs—combined with fluid, pliable terms—can shape perceptions of the past (and present) decisively.

7

Christians in the Province of Asia

On the basis of the previous chapter it would at best be hazardous to assume an empirewide political crisis during Domitian's reign. Writers such as Pliny, Tacitus, Suetonius, and Dio Chrysostom cannot be appealed to uncritically as evidence for corruption, terror, and intensification of imperial demands for divinity under Domitian. As we have seen, Domitian served those writers as foil for Trajan: Trajan's rule as proper "leader" (*princeps*) was enhanced by portraying Domitian as a tyrant, just as Trajan's proper humanity was enhanced by Domitian's exaggerated divinity. Domitian was no more or less demanding an emperor than those who came before or after him. It would be a mistake to interpret the Book of Revelation as a response to Domitian's supposed excessive claims to divinity or to a reign of terror at the end of Domitian's rule. Domitian provided economic and political stability for the whole empire at least as well as did other emperors at the end of the first century and the beginning of the second.

John of Revelation is concerned, however, with Christian life in a very specific location in the eastern part of the Roman Empire. He writes to seven churches in seven cities in the province of Asia along the western edge of Asia Minor. He experienced his visions on the island of Patmos, off the city of Miletus. Thus, within the Roman Empire, Revelation is to be associated with urban life in the province of Asia. Conditions there are more important to him than life at Rome, and the social and political institutions and arrangements in those cities of Asia are the backdrop for Revelation. For that reason we turn now to local conditions in Asia. The procedure will be to move in ever-larger concentric circles, beginning with Christian life in the province of Asia (the present chapter), then Jewish life in that province (chap. 8), and finally, social, economic, and political life in the cities of Asia where Christians lived and worked (chap. 9). Throughout these chapters special attention will be given to Ephesus, Smyrna, Pergamum, Thyatira, Sardis, Philadelphia, and Laodicea—the cities of Asia mentioned in the Book of Revelation.

Distribution of Christian Communities

Evidence for Christian communities in the cities of Asia comes for the most part from Christian sources. The earliest sources are the letters associated with Paul's name; later are the churches mentioned in the Book of Revelation and Ignatius. Ignatius sends letters to Asian churches at Ephesus, Magnesia, Tralles, Philadelphia, and Smyrna and to Polycarp, bishop at Smyrna. Thus, correspondence with Ephesus, Smyrna, and Philadelphia is found in both the Book of Revelation and Ignatius; Pergamum, Sardis, and Thyatira are referred to only in the Book of Revelation; and Magnesia (on the Meander) and Tralles are mentioned only in Ignatius.

From the late first and early second centuries, there are no inscriptions or coins that mention Christianity and few references to Christians by non-Christian writers. This complicates our attempt to place Christians in the social world of Asia, for the Christian sources refer only incidentally to factors that could contribute to a social profile. Christian writers are concerned primarily about life and thought within the churches rather than about demographic features of early Christians.

Ephesus

Our earliest reference to Christians in Asia comes from a letter of Paul written from Ephesus, probably in the early 50s, about forty years before the seer wrote the Book of Revelation. Writing to the Corinthian church, Paul states that he plans to remain at Ephesus until Pentecost, "for a door for me opened, a door great and powerful [or effective], and those opposing are many" (1 Cor. 16:9). The sentence is highly metaphoric, and it is not altogether clear what is being said about Paul's activity in Ephesus. The dramatic contrast between the opposition and Paul's open door has something to do with missionary opportunities there (see 2 Cor. 2:12), but little more can be concluded from it.[1] From the same letter it is clear that Paul is not alone at Ephesus. There are, or have been, other Christian leaders there; he asks the Corinthians to send along Timothy if he should come to Corinth; and through Paul "the churches of Asia" send greetings, as do Aquila and Prisca, along with the church that meets in their house (1 Cor. 16:19, cf. Rom. 16:3).

Paul stayed at Ephesus for three years — at least according to the writer of the Acts of the Apostles (20:31), written between 80 and 100, thirty to forty years after Paul's letter to Corinth. Although this source cannot be taken uncritically as a statement about social relations in the time of Paul, it does reflect at least one Christian's understanding, towards the end of the first century, of the emergence and presence of Christianity at Ephesus.

According to that writer, Paul first stopped at Ephesus as he was traveling from Corinth to Syria.[2] He was accompanied there by Priscilla (Prisca) and Aquila (Acts 18:18–19), who, like Paul, were Jewish tentmakers. According to Acts 18:2 Aquila was a native of Pontus (northern Asia Minor), but Aquila and Priscilla had more recently come to Corinth from Rome, as a result of the Emperor Claudius' expulsion of the Jews from Rome (Acts 18:1–2, see Smallwood 1981, 210–16). Paul first met these two in Corinth, where the three worked together in common occupation.

Then they travelled together to Ephesus, where, as we have seen, Priscilla and Aquilia eventually established a house church and became a patron of the church at Ephesus (see 2 Tim. 4:19). Still later, according to Paul's letter to the Romans, they have a house church at Rome (Rom. 16:3-5).[3] From their mobility, occupation, and house churches we may conclude that these church leaders and devout Christians were fairly wealthy independent artisans, originally a part of the Jewish community in Asia Minor.[4]

After a brief contact with the synagogue at Ephesus, Paul goes on to Syria. At this point the author of Acts introduces a story about Apollos. Apollos is a learned (*logios*) Alexandrian Jew, well versed in the Scriptures, and a teacher "in the way of the Lord" (Acts 18:24-28). The adjective *logios* implies "rhetorical ability, perhaps also rhetorical training" (Meeks 1983a, 61). He taught accurately concerning Jesus, but he knew only John's baptism. When Priscilla and Aquila (the woman is again mentioned first) hear him in the synagogue, they take him aside to teach him "the way of God more accurately." Apollos then wishes to cross over to Achaea, more specifically to Corinth, "and the brethren encouraged him" to do that (Acts 18:27).[5] Apollos travels independently. Paul says in his Corinthian correspondence that he strongly urged "our brother Apollos to visit you . . . but it was not at all his will to come now. He will come when he has opportunity" (1 Cor. 16:12, cf. Titus 3:13). Apollos is thus another Christian convert from Judaism who has independent means and in addition is probably educated in rhetoric.

Still according to Acts, "While Apollos was at Corinth, Paul passed through the upper country and came to Ephesus. There he found some disciples" (19:1). This account reads like another version of Paul's entrance into Ephesus. No mention is made of Apollos, Priscilla, or Aquila. The disciples Paul found at Ephesus had been baptized only "into John's baptism," which in the earlier account was said of Apollos (Acts 18:25). They are then baptized "in the name of the Lord Jesus"; when Paul places his hands on them, the Holy Spirit comes upon them and they speak in tongues and prophesy. Altogether there are about twelve of them (Acts 19:1-7). For about three months Paul argues in the synagogue about "the kingdom of God." Because of opposition and disbelief among those in the synagogue, Paul withdraws with his disciples to "the hall of Tyrannus." He argued daily there for two years, "so that all the residents of Asia heard the word of the Lord, both Jews and Greeks" (Acts 19:8-11).

A somewhat curious story is told in Acts 19:11-20 about Paul's healing powers, Jewish exorcists, and magic at Ephesus — all of which, according to Acts, contribute toward a church greater in number and purer. After a powerful display of the exclusiveness of Christian exorcism in which several Jewish exorcists are injured by evil spirits, a number of Christian believers come forward confessing that they practice magic. They then burn all their books having to do with magic. "So," writes the author of Acts, "the word of the Lord grew and prevailed mightily" (Acts 19:8-20). There was a magic dimension to religion at Ephesus. Artemis of Ephesus was associated with the *Ephesia grammata*, originally six powerful, magical words (see Kraabel 1968, 55). Moreover, Artemis' magical powers are referred to in later Jewish and Christian traditions.[6] At or near Ephesus a few magical Jewish amulets have also been discovered (see Kraabel 1968, 56-59). In the story in Acts the "many who were now believers" who confessed that they had been practicing magic arts secretly could have come from either Jewish or Gentile backgrounds.

After Paul makes plans to travel on to Macedonia but before he leaves Ephesus another conflict arises, this time from silversmiths who make "silver shrines of Artemis" (on Artemis see Oster 1976). The writer of Acts makes Paul so successful not only among Ephesians but more generally among Asians that the occupation of the silversmiths is threatened as well as the temple of their great goddess Artemis. The craftsmen stir up a crowd who drag "Gaius and Aristarchus, Macedonians who were Paul's companions in travel," into the theater. Paul tries to go into the theater but is held back by disciples and asiarchs who were friends of his (see Meeks 1983a, 61). In the midst of the confused assembly Alexander, a Jewish spokesman, tries without success to establish calm. Finally the town clerk establishes calm, assuring the crowd that Artemis is not in danger. He warns the crowd that they have brought men who are "neither sacrilegious nor blasphemers of our goddess." If Demetrius and the craftsmen have a complaint, there is an orderly procedure to follow, for the city has courts and "there are proconsuls." Things must be settled in regular assembly, for "we are in danger of being charged with rioting today, there being no cause that we can give to justify this commotion." The clerk dismisses the assembly. Then after the uproar ceases, Paul exhorts the disciples and departs from them for Macedonia (Acts 19:21–41).

Later, when Paul is on his way to Jerusalem, he calls the elders from Ephesus to Miletus for a final exhortation. This speech in Acts (20:18–35) also contains some information about the writer's views of the Ephesian church. Paul refers to "Jewish plots" against him while at Ephesus.[7] Paul says in this speech that he taught in Ephesus in public and "house to house." Paul recognizes the elders as "bishops" (*episkopoi*) over the flock of Ephesus (cf. Phil. 1:1; 1 Tim. 3:1; Titus 1:7; 1 Pet. 2:25). He says that after he leaves, "savage wolves" will come among the flock and even from among the "bishops" men will speak "perverse things" to draw away the disciples. Paul then concludes by making clear that he labored with his hands while at Ephesus and thereby met his own needs and the needs of those with him; by such toil one helps the weak and gives rather than receives.

The theme of Paul's "laboring with his own hands," found also in his correspondence to the Thessalonians and Corinthians, may also give some clues about the social location of some Christians at Ephesus.[8] According to Acts Paul makes contact with Aquila and Priscilla "because he was of the same trade" (18:3); that is, when Paul went into a town, he made initial contacts not only through the synagogue but also through his trade. Further, Malherbe argues that the shift in Ephesus from synagogue to the *schole* of Tyrannos (Acts 19:9) was a shift not to a "lecture hall" for wandering philosophers but to a guild hall "which was frequently named for a guild's patron" (1983, 90). Paul would thus have preached for two years in a union hall—a *collegium* or club of those who practiced the same craft. In this case the club would have included tentmakers, tentmaking being the craft of Paul, Aquila, and Priscilla. Such clubs attracted people of varied social rank and status; they could be headed by wealthy people—even of the senatorial class—and include ordinary plebian members. In that regard, such trade unions reflected the cross-section of the population, just as early Christian churches did (see Malherbe 1983,

88–91). In such a context Paul would be preaching to artisans and tradespeople of modest, and sometimes of relatively affluent, means (see Malherbe 1983, 86). Onesiphorus and his household, an Ephesian mentioned in an early-second-century Christian work, may have been a convert of that sort; for he is said to have "refreshed" Paul often and to have been a patron of the church at Ephesus. Moreover, he had the means to travel to Rome (2 Tim. 1:16–17, 4:19). Paul's trade also comes into play in the miracles he performed at Ephesus (Acts 19:11–12). The "handkerchiefs" and "aprons" carried to the sick were taken from Paul's workshop. That activity suggests that Paul carried on missionary preaching while at work (see Hock 1980, 37–42). Also, the conflict with the silversmiths can on one level be seen as a conflict between artisan guilds.

At two points the writer of Acts suggests that he knows of different forms of Christianity at Ephesus: first, there were Christians at Ephesus who had different kinds of relations to followers of John the Baptist and Judaism. Secondly, Paul warns the bishops at Ephesus that "savage wolves" will come, even from among the "bishops," and will pervert Christian disciples. Köster (1971, 155) has suggested the following strands of Christianity at Ephesus: "the originally Pauline church," represented by the writer of Ephesians and the author of Luke and Acts; "a Jewish-Christian 'school' engaging in a daring interpretation of the Old Testament" (e.g., Cerinthus);[9] a Jewish-Christian conventicle led by the writer of the Book of Revelation; and the Nicolaitans referred to in Revelation 2:6.[10] Central to these variations of Christians were their use and assessment of the Old Testament and their views of the nature of Christ (see Johnson 1972, 187–88).

Generally, among the variations of Christianity at Ephesus Paul and the writer of Acts are open to social intercourse between Christians and non-Christians in their urban social setting. So, too, is the author(s) of the Pastoral Letters, associated with the Pauline Ephesus connection. The Nicolaitans, of whom the seer of Revelation is so critical, are also open to cultural accommodation to urban society (see Aune 1981, 26–29). Among the Christian groups at Ephesus only the writer of the Book of Revelation seems to be hostile towards urban culture and opposed to any Christian accommodation toward it (see Rev. 2:6, 14, 20, Yarbro Collins 1983, 740–41, see also Johnson 1972, 192–93).

Churches in the Lycus Valley

The only other Christian communities in Asia to which there is more than passing reference from the time of Paul are the three cities nestled in the Lycus Valley that joins the Meander River about a hundred miles east of Ephesus.[11] In that area were Laodicea (one of the seven churches of the Apocalypse), a prosperous cultural center located at a major crossroad on the southern route from Ephesus through Anatolia; Colossae, a smaller town about ten miles further on the main road east; and Hierapolis, a textile town five or six miles north on the road to Tripolis and Philadelphia (see Col. 4:13, 16). The letters of Paul include two to Christians at Colossae — Colossians and Philemon.[12] From these two we learn a few things about the constituency of the Colossian church. In the Colossian letter the writer refers to Epaphras, who is from Colossae (4:12) and may have founded the church there (the wording of Col. 1:7 is not definite on that point.[13] According to that letter, Epaphras is with Paul working hard for the three churches in the Lycus Valley (Col. 4:13).[14] Tychicus, whom Paul is sending to Colossae so that they may know Paul's situation and receive comfort, may have been from Colossae.[15] In contrast to Epaphras and Tychicus, Archippus is mentioned as a leader living at Colossae; Paul urges him to fulfill his ministry (*diakonia*) from the Lord (Col. 4:17). Nothing more is known about him.[16]

Along with Tychicus Paul is also sending Onesimus, who is definitely from Colossae (Col. 4:9) and is most likely the runaway slave Onesimus of the letter to Philemon. If the letter to Philemon was written earlier, so that the episode with Philemon lies in the past, Onesimus has become a fellow worker with Paul.[17] In any case one learns something about the social position of Onesimus' master (or former master) in Paul's letter to Philemon. He has a house in Colossae large enough to have a house church—and a guest room for Paul (Philem. 22)—and he owns at least one slave (Philem. 2). Along with Philemon, Paul greets Apphia "our sister"—thus, if she is Philemon's wife, she is greeted in her own right—and Archippus "our fellow soldier" (Philem. 1–2). As we have seen, nothing is known of Archippus' wealth or social status in the larger community, though he is a leader in the church at Colossae.[18]

The Colossian letter also refers to correspondence with Laodicea. Paul writes to the Colossians, "And when this letter has been read among you, have it read also in the church of the Laodiceans; and see that you read also the letter from Laodicea" (Col. 4:16). That Laodicean letter is not, however, extant. Earlier in the Colossian letter, Paul wants the Colossians to know "how greatly I strive for you, and for those at Laodicea" (2:1). The church at Laodicea, as well as the one at Hierapolis, is mentioned, too, in connection with Epaphras: the author of Colossians writes, "For I bear him witness that he has worked hard for you and for those in Laodicea and in Hierapolis" (Col. 4:13).

In later Christian literature the church at Colossae does not appear, but the churches at Laodicea and Hierapolis do. That Christians shared in the prosperity of Laodicea is attested in the Book of Revelation where, in an ironic tone typical of the author, the material wealth of the Laodicean Christians is contrasted to their poverty in Christ (Rev. 3:17). They are urged to buy gold from Christ (Rev. 3:18, see Calder 1922–23, 326–27). Hierapolis is best known for its famous Stoic philosopher, Epictetus.[19] In Christian history Hierapolis is known for Papias, the milennarian lover of oral tradition who, according to Eusebius, was "of exceedingly small intelligence." He lived in Hierapolis in Trajan and Hadrian's reign and was a friend of Polycarp of Smyrna (Eus. *Hist. Eccl.* 3.39, see also Johnson 1975, 109). Eusebius also locates the apostle Philip and his daughters there (*Hist. Eccl.* 3.39.9).

Thyatira and Pergamum

The author of Revelation refers to the churches at Pergamum and Thyatira as communities infested with false prophets. The church at Thyatira is commended except for the fact that it allows "the woman Jezebel who calls herself a prophetess" to teach and beguile "my servants to practice immorality and to eat food sacrificed to idols" (Rev. 2:20–21). In the letter to Pergamum the teaching of Balaam — probably the seer's derogatory label for the Nicolaitan teaching — is also said to involve eating "food sacrificed to idols" and practicing immorality (Rev. 2:14–15). Finally, as we saw above, the Nicolaitans are also mentioned in the letter to the Ephesians (Rev. 2:6): the Ephesian Christians "hate the works of the Nicolaitans" and thus do not allow them to be a part of the church at Ephesus. There the Nicolaitans seem to be outsiders seeking to establish themselves in the Ephesian church. In contrast, the adherents of Balaam *were* a part of the church at Pergamum (Rev. 2:14).[20] They were among those who "did not deny my faith even in the days of Antipas" (2:13). So also the prophetess Jezebel and her followers were a part of the church at Thyatira, whose "latter works exceed the first."[21] At Pergamum and Thyatira John is in conflict with local church members, not heretical outsiders. One could perhaps argue that they belong to different house churches in those cities, but there is no indication of that (see also Schüssler Fiorenza 1985, 154).

The offense of the Nicolaitans and Jezebel is great in the eyes of the seer — they are called upon to repent — but their evil behavior is not of such magnitude that he refuses to recognize them as members of the churches. That is an important point to remember when trying to ascertain what the "offense" is in those groups. Their offense is stated by means of two closely connected infinitives: "to eat food sacrificed to idols and to practice immorality" (φαγεῖν εἰδωλόθυτα καὶ πορνεῦσαι) (2:14). The order of the activities is reversed in the description of Jezebel (2:20). The offense alludes to the story of Israel's idolatry with Baal of Peor at Shittim. According to the story in Numbers 25, the daughters of Moab "invited the people to the sacrifices of their gods, and the people ate, and bowed down to their gods" (Num. 25:2).[22] In this context, "eating food sacrificed to idols" is an act of idolatry, "a bowing down to their gods." This close link between the two activities is kept in the indictments in the Book of Revelation: to eat food offered to idols is to commit idolatry (πορνεῦσαι). The second infinitive in 2:14 should thus be understood in its metaphoric sense, "to commit idolatry," rather than simply its more literal meaning, "to fornicate."[23] The fault of the Nicolaitans and Jezebel, then, centers on the issue of eating meat offered to idols and the meaning of that activity.

In order to understand the social significance of this activity we may draw upon Theissen's study of the conflict at Corinth regarding the eating of meat offered to idols.[24] In Greek and Roman cities people had opportunity to eat meat in a cultic setting both on public occasions (in extraordinary situations such as victory celebrations or funerals and in regular calendric festivals) and at more private settings (clubs and associations including guild meetings or small, private banquets in a temple) (see Theissen 1982, 127–28). These banquets and parties expressed social connections and common causes, whether of the whole province, the whole city, a specific group (such as tentmakers, purple-clothiers), or private friendship. Thus, the eating of meat in a cultic setting posed fundamental questions about how Christians should relate to social institutions in the professional and civic order. Theissen makes a convincing argument that the issue affected different social strata in different ways. In brief, wealthier Christians involved in economic or civic responsibilities would have to (and want to) engage in the eating of meat in a cultic setting more frequently than poorer members of the church, who would have occasion to eat meat primarily in large public festivals or social clubs that included members from all strata of society (Theissen 1982, 127–32).

Among Christian leaders there was no universally agreed-on position about the eating of meat in a cultic setting and its meaning for a Christian. All agreed that there was only the one true God and that therefore all other cults involved the worship of idols (1 Cor. 8:4–6). But were those idols harmless to Christians who might share the eating of meat in a non-Christian setting? Paul answers that question in terms of the strong (those Christians who know that idols are nothing) and the weak (those whose consciences are troubled by participating in the eating) (e.g., 1 Cor. 8:1–13). Paul argues that the eating of meat does not matter in itself so long as the weak brother is not destroyed (1 Cor. 8:11). Probably Theissen is correct in thinking that Paul thus preserves the "privileges of status enjoyed by the higher strata" of the Corinthian church, for "private meals with consecrated meat continue to be allowed in principle" and "participation in cultic meals" is not excluded in principle (1982, 139). Only in cases where different strata in the Corinthian church would be together — the strong with the weak, as in a public festival — should Christians abstain. It is probably not coincidental that Paul defends his working at a trade in the context of defending eating meat offered to idols (1 Cor. 8:13–9:7); for he probably ate such meat in his trade connections.[25]

John of the Book of Revelation objects to Christians participating in professional and civic life. Writing to Ephesus, he praises their rejection of the traveling Nicolaitans who try to impose themselves and their apostolic authority upon the Ephesian church.[26] He orders the churches at Pergamum and Thyatira to purge themselves of those false teachings propagated by local leaders: the Nicolaitans, teachers in the local community, and Jezebel, a Christian prophet. In contrast to John's teachings they allow Christians to share in the eating of meat in a cultic setting, which would no doubt occur in the same contexts in the Greek cities of Asia as in Corinth. In other words, those whom the seer views as his opponents (we do not know how they view him) hold to a view similar either to the one Paul espouses in the Corinthian correspondence or to that of the "liberal strong" at Corinth (since idols were "nothing," the eating of meat offered to idols need not offend a Christian's conscience). Theissen presents evidence that this is also the view of Christian gnostics, who were open "to the culture of antiquity" (1982, 132–36). The "strong" at Corinth — to some extent Paul himself, Jezebel, the Nicolaitans, and Christian gnostics — thus shared in a similar attitude towards urban life, although they did not share the same philosophical and theological beliefs. Since we learn of the seer's opponents only through his language and perspective, it is impossible to know whether the Nicolaitans and Jezebel would have agreed more with the apostle Paul or with the "strong" at Corinth regarding Christian obligations to the church as a whole, the model of the crucifixion in Christian living, and the "not-yet" character of Christian existence in relation to the coming kingdom of God.[27] Whatever their differences in theology, their similar attitudes towards urban society reflect a similar social position; that is, those groups consisted of, among others, people who were in crafts and trades, who shared in civic responsibilities, and who were on the whole probably of the wealthier stratum of leadership in early Christian churches.[28]

The author of the Acts of the Apostles mentions a woman from Thyatira named Lydia whom Paul met at Philippi in Macedonia at a place of prayer (*proseuche*), presumably a synagogue (16:13–14, see Kraabel 1968, 156–57). She is identified in Acts as a seller of purple cloth and a worshipper of God (*sebomene ton theon*). From the latter phrase we may associate her with the Jewish community: either she was a Jew, a proselyte, or one who was faithful to some, but not all, of the Jewish law (see Meeks 1983a, 207, n. 175). According to Acts she converts to Christianity, with her household, and prevails successfully upon Paul and his companions to stay at her house (16:14–15). Some time later the owners of a soothsaying slave girl whom Paul had healed drag Paul and Silas before the archons of the city and get them imprisoned. After Paul is released from prison, he goes to Lydia's house, sees and exhorts the brethren there and leaves Philippi (Acts 16:40). From the wording of verse 40, one may conclude that Lydia has a house church at Philippi. We thus get several clues about Lydia's status. She is a businesswoman dealing in a luxury item, a sign of wealth. She has a house large enough to house several guests and to serve as a house church. She travels extensively in connection with her trade (see Meeks 1983a, 62). Since Lydia is a Jew from Thyatira who converts to Christianity after leaving Asia, her story tells us little about the Christian church at Thyatira. Nonetheless, it gives further indication that some prosperous Jewish traders from Asia (like Aquila, above) converted to Christianity. Jewish converts of similar social status were also probably a part of the church at Thyatira.

Sardis and Laodicea

Sardis is the other Asian city mentioned only in the Book of Revelation. As we shall see in chapter 8, the city of Sardis supported a large, well-integrated Jewish community throughout the first and second centuries CE. The first Christians at Sardis were probably drawn from the Jewish community there. The city itself was flourishing at the end of the first century CE: it was a judicial (assize) center; the temple of Artemis (of Sardis) had the right of asylum—a status sought by many temples; and after the earthquake of 17 CE private, as well as imperial and sacral, gifts indicate wealth in the community (see Broughton 1938, 723 and chap. 9 below). Sardis's favorable position at the crossing of five roadways (to Philadelphia and Laodicia southeast, to Acmonia east, to Thyatira north, to Magnesia ad Sipylum and Smyrna west-northwest, and to Ephesus south) made it a lively commercial center. The point is that Sardis was a thriving Asian city with a significant Jewish community— the social ingredients for emerging Christianity. Yet Paul makes no reference to a church at Sardis, nor does the writer of Acts, nor does Ignatius. Without the reference in the Book of Revelation, we would not know that a church existed in Sardis at the end of the first century.

Among the seven letters in the Book of Revelation, only those addressed to Sardis and Laodicea do not mention specific adversaries, either Jewish (Smyrna and Philadelphia) or Christian (Ephesus, Pergamum, and Thyatira). Both Sardis and Laodicea are prosperous communities in major commercial centers, and the writer of Revelation speaks of both in an ironic mode: the Laodiceans appear wealthy but are actually in poverty (3:17); those at Sardis have a reputation for being alive but are dead (3:1). Probably both Christian communities had accommodated well to their urban setting, from the seer's point of view *too* well. Whereas poor and powerless churches receive encouragement and assurance from the seer (Smyrna and Philadelphia), these wealthy and relatively powerful ones receive unmitigated censure. There is another correlation that may be significant: Sardis and Laodicea are known to be centers of Judaism, yet the seer makes no mention of oppression from Jews in those prosperous Christian communities. In contrast, the poor and powerless Christian communities at Smyrna and Philadelphia, where Jews have a less-significant presence are, according to the seer, harassed by Jews or, in his language, so-called Jews of the synagogue of Satan (2:9, 3:9); that is, (1) receiving praise from the seer, (2) little accommodation and assimilation to urban Asian society, and (3) conflict with Judaism in the local city seem to go together, as do (1) condemnation from the seer, (2) accommodation and assimilation, and (3) no conflict with Judaism. To put it differently, a Christian community that sets up high boundaries between itself and the rest of the world and that holds to a concomitant "separatist" definition of the church (which of course the seer supports) sees both Judaism and Greco-Roman society as demonic; but a Christian community that is less concerned with sharp boundaries and exclusive self-definitions seems to have little conflict with either Judaism or Greco-Roman urban institutions. Probably Christians at Sardis and Laodicea view their Jewish and Greco-Roman neighbors much as do the Nicolaitans and the prophetess Jezebel. The seer of the Apocalypse cannot allow such trafficking with the world, for it stains Christian character (Rev. 3:4).

Toward the end of the second century Sardis is heard from again through the voice and pen of Melito, who wrote, among other items, an Apology to Marcus Aurelius and a Paschal Homily (see Wilken 1976, 57–58). Melito was a Quartodeciman — one who celebrated Easter on the Jewish Passover (the fourteenth day of the month Nisan) — and for one such occasion he offered "a prolonged, bitter, personal attack on 'Israel'" (Kraabel 1971, 81). In order to account for the vitriolic nature of this sermon, Kraabel proposes that one cannot look to either Quartodeciman theology or other Christian literature attacking Jews. Rather, the homily addresses the situation of Jews and Christians in Sardis toward the end of the second century. Melito takes an antagonistic stance against the large and powerful Jewish community in Sardis. Because of his close connection to Judaism through the Quartodeciman practice, Melito may have felt it necessary to differentiate his Christianity from Judaism. Further, his apology on behalf of Christianity to the Romans may have also put him on a collision path with the Jews (see Kraabel 1971, 84). At any rate, Melito's *Paschal Homily* points to serious conflict between his form of Christianity and Judaism in Sardis at that time.

Philadelphia and Smyrna

Both the seer and Ignatius address Christian communities at Ephesus, Smyrna, and Philadelphia. I discussed Ephesus above; there remain the two churches that the

seer lauds without qualification. In the letter to the Philadelphians, the seer links
their "little power" with their keeping Jesus' word and not denying his name (Rev.
3:8). Alongside the *imitatio Christi* theme, there is probably reflected here a social
attitude of the seer: the Christians at Philadelphia are keeping high boundaries
between church and world and are refraining from much participation in the cultural and social life of the city. The same high boundaries set the church off sharply
from the synagogue (Rev. 3:9). There is no indication of persecution by the Jews.
The seer is the one making the sharp contrast and opposition between Christians
and Jews.

In the letter Ignatius wrote a few years later to the Philadelphians, there is
within the church some form of "Judaizing," but as Schoedel notes, little is given as
to its exact character (Schoedel 1980, 35). Perhaps there were Gentile Christians
impressed with Judaism (I *Philad.* 6.1), or perhaps there were people in the church
simply "fascinated with the idea of Judaism as presented in scripture . . . but not
tempted to become deeply involved in the actual practice of Judaism" (Schoedel
1980, 35). In either case, Schoedel suggests that "the line dividing Judaizers from
true Christians in Philadelphia was not as obvious to others as it was to Ignatius"
(1980, 35).

Smyrna, in contrast to the seer's portrait of Laodicea, appears to be in poverty
but is rich (Rev. 2:9). Here, according to the Book of Revelation, Jews at Smyrna
(who, as at Philadelphia, belong to the synagogue of Satan) are blaspheming
Christians. We may assume that the seer in this letter is enforcing the same high
boundaries between church and world as he did in the Philadelphia letter. Thus, the
"blasphemy by the Jews" may reflect his concern to sharpen the distinction between
Christian and Jew. *Blasphemia* could possibly refer to "charges of anti-social behavior" reported to proper officials against Christians (Sweet 1979, 85), but the
term generally refers to "slander" in connection with humans (e.g., 1 Cor. 10:30);
that is, Jews at Smyrna "slandered" Christians, just as Christians, including John,
"slandered" Jews (though with less clout).

Ignatius does not discuss "Judaizers" in the letter to Smyrna, though he assumes
that Christians there have come from both Jews and Gentiles (I *Smyrn.* 1). He is,
rather, concerned with those who do not take seriously the fleshly incarnation of
Christ.[29] There is nothing about Jews in Ignatius' letter to Polycarp (of Smyrna)
either. According to the later *Martyrdom of Polycarp*, at Polycarp's murder (c. 154
CE) the Jews are a part of the multitude at Smyrna who cry out for Polycarp's death
(*Mart. Polyc.* 12). That account, however, has been shaped along lines reflected in
the Gospel at Matthew's death of Jesus. At Smyrna Jews are zealous in assisting to
kill Polycarp (*Mart. Polyc.* 13); and they are concerned with Polycarp's body
(*Mart. Polyc.* 17), just as they were with Jesus' body in Matthew's Gospel.

Finally, Ignatius provides some evidence that the Christian community at Smyr-
na was made up of different social classes. He urges Polycarp not to despise "slaves,
whether men or women" and he urges slaves not to desire freedom at public cost (I
Polyc. 4). The wording suggests that slaves did not make up a large percentage of
Christians at Smyrna. Among specific people mentioned by Ignatius, the woman
Alce stands out. She is "a name very dear to" Ignatius (I *Smyrn* 13; I *Polyc.* 8).
According to the *Martyrdom of Polycarp* she was an aunt of Herod, the officer

of the peace who arrested Polycarp, and sister of Nicetes, who tried to convince Polycarp to observe imperial cult (*Mart. Polyc.* 6, 8) and later appealed to the magistrate at the prompting of Jews not to give up Polycarp's body (*Mart. Polyc.* 17). Holding a city office such as "officer of the peace" required "the possession of wealth and a readiness to spend" (Magie 1950, 649). Alce belonged to a wealthy family (see also Schoedel 1980, 47). Schoedel aptly compares the social mix at the church at Smyrna — which is typical of the other churches in Asia — to a private religious association in Philadelphia attested to in the first century BCE (see Schoedel 1980, 47). There one Dionysius, by order of a divine dream, opened his house as a shrine for sacrifices and purification rites to several gods. Both men and women were welcome, "both bond and free," so long as they upheld the command-ments of the cult.[30]

Magnesia and Tralles

Ignatius also writes to churches at Magnesia and Tralles in the Meander River Valley southeast of Ephesus, which sent embassies to Smyrna while Ignatius re-mained there. Writing to those at Magnesia, Ignatius underscores the separation of Christianity from Judaism: "for if even unto this day we live after the manner of Judaism, we avow that we have not received grace" (I *Magn.* 8). Apparently there was an issue of some Magnesians observing the Jewish Sabbath rather than the Lord's Day (I *Magn.* 9). Ignatius, of course, argues for the Lord's Day and con-cludes his exhortation with the words, "It is monstrous to talk of Jesus Christ and to practice Judaism. For Christianity did not believe in Judaism, but Juda-ism in Christianity" (I *Magn.* 10).[31] In the letter to those at Tralles, about fifteen miles east of Magnesia on the road to Laodicea, Ignatius attacks docetism (the doctrine that Jesus Christ only *seemed* to have a human form), as he had in the letter to Smyrna.

William Schoedel examines some aspects of "social experience" in the Ignatian corpus. Ignatius' understanding of church-world relations "cannot be shown to be determined by class conflict" (Schoedel 1980, 46). Drawing on the work of Ramsay MacMullen, Schoedel demonstrates that the social bonds in the church, as in the empire generally, "were forged not horizontally (embracing whole classes of people) but vertically (with the lower classes looking to the upper classes for patronage)" (1980, 46–47). So in Ignatius' churches one finds a range of social classes. As we have seen from the church at Smyrna, the churches embraced both slaves and high-ranking citizens. Thus the churches "were like the *collegia* and private religious associations in appealing to people across a broad social spectrum" (Schoedel 1980, 47). Ignatius himself was "a person of some standing" with "a significant openness to aspects of pagan society and culture." Schoedel sees this openness reflected (1) in Ignatius' use of popular rhetorical methods and conventions of the Hellenistic letter, (2) in common social values related to marriage and family life, and (3) in his modeling the Christian community after elements in urban organization: "The presence of these elements in the letters of Ignatius indicates that 'early catholic' Christianity maintained a relatively positive attitude towards the things of this world" (Schoedel 1980, 51).

Social Status of Christians

As Wayne Meeks points out, "status" is a complex, multidimensional phenomenon (1983a, 54). Variables such as power, occupational prestige, income or wealth, education and knowledge, religious and ritual purity, family and ethnic group position, and local community status are involved in deciding a person's status. Since no one is likely to give the same weight to each of those variables, status will be determined in part by which variables are used, what weight is given to each, and from whose perspective the ranking is given (for example, how others see a group or how a group sees itself).[32] In other words, we probably lack the data to develop very subtle models of social status and social stratification in early Christianity.

A few passages give direct information about the social status of early Christians. One is 1 Corinthians 1:26. There Paul writes that "not many of you were wise according to worldly standards, not many were powerful, not many were of noble birth [*eugeneis*]." Theissen has argued for a clearly sociological import to those terms: there were some at Corinth who were influential, educated, and well born. Moreover, he has shown that though a minority, "representatives of the upper classes" were "apparently a dominant minority" (1982, 73).[33] References to slaves in the early church also indicate a variety of classes. The many references to slaves may be one reason why early Christians have often been identified only with the poorest and least powerful.[34] R. M. Grant notes, however, that "slaves were not significant in the church" (1980, 24). Moreover, directives to slaves usually imply that there were also slavemasters in the church, and Christian writers tend to take the point of view of the owners rather than the slaves (see Meeks 1983a, 64). Pliny the Younger, who writes to the Emperor Trajan about Christians in Pontus at about 112 CE (see chap. 6) states that people being examined as Christians or former Christians were "of every age and class, both men and women" (*Ep.* 10.96.9).[35] Pliny also refers to two slave women whom Christians called deaconesses (*Ep.* 10.96.8).

For the most part, however, clues about status must be drawn indirectly. For example, the Christian activity of reading and writing may tell us something about their status. R. M. Grant, in an essay on second-century Christianity, points out that Christians "belonged to a sector of society in which literary education was highly valued and a disproportionate number of persons wrote books." This sets them off from both the lowest and highest classes of Roman society (1980, 21).

Grant also suggests that something can be learned about class status from the kinds of punishments Christians received from the state. Drawing on an observation of Peter Garnsey that "punishments did not so much suit the crime as suit the criminal," Grant notes that Paul appeals confidently to Caesar; whereas Ignatius, "obviously lower in status, fearfully lists the dire punishments he anticipates at Rome" (I *Rom.* 5.3); the proconsul threatened Polycarp (also of lower status) with fire and wild beasts (R. M. Grant 1980, 23–24). So in Pliny's dealing with Christians in Bithynia Roman citizens are treated differently from those who are not citizens, and free people are treated differently from slaves. He tortures two female slaves without hesitation.

As we have seen in analyses of individual Christian communities in the province of Asia, status is also reflected in having a house church and in traveling. The owners of house churches probably served as "patrons" to the church much as patrons supported private clubs and guilds (see Malherbe 1983, 69–75). Social analyses of patterns of travel are readily available. Christians traveled a lot, sometimes in connection with their trade, sometimes with the support of a particular church or specific churches. Travel cost money, but individual Christians and churches could afford to do it.[36] In sum, Meeks's description of the Pauline communities also fits the churches in the cities of Asia:

> The extreme top and bottom of the Greco-Roman social scale are missing from the picture. . . . The levels in between, however, are well represented. There are slaves, although we cannot tell how many. The 'typical' Christian, however, the one who most often signals his presence in the letters by one or another small clue, is a free artisan or small trader. Some even in those occupational categories had houses, slaves, the ability to travel, and other signs of wealth. Some of the wealthy provided housing, meeting places, and other services for individual Christians and for whole groups. In effect, they filled the roles of patrons. (Meeks 1983a, 73)

Along with being affluent, many Christians shared fully in urban Roman life. As Paul writes to the Corinthians, if they did not share in that life, they would have "to go out of the world" (1 Cor. 5:10). Ernest Cadman Colwell observed several decades ago that Christians from the beginning had social intercourse with their non-Christian neighbors. Christians were to live a life that could not be criticized or slandered. Non-Christians came to Christian meetings, and Christians ate dinner with non-Christians.[37] At the end of the second century Tertullian states in his *Apology* that Christians have the same manner of life, the same dress, and the same requirements for living; they depend upon the marketplace, the butchers, the baths, shops, factories, taverns, fairs, and other businesses. Christians sail ships, serve in the army, till the ground, and engage in trade alongside non-Christians; and Christians provide skills and services for the benefit of the whole society (*Apol.* 42). Ordinary Christians shared in the common life of cities and empire: "they did not live as separately and aloof as their leaders desired," and for that reason they eventually overcame the "opposition of the heathen masses" (Colwell 1939, 70). There were important differences in moral goals and in the Christians' refusal to participate in idolatrous worship, but as R. M. Grant notes, "the differences must not blind us to the general coincidence between the life-styles and attitudes of non-Christians and Christians alike" (1980, 29).[38]

Non-Christian Attitudes Toward Christians

There is, then, considerable evidence that Christians shared in the urban life of Asian cities. Their style of life required contact with non-Christians in economic, social, familial, and even political arenas. What impression did these Christians make on their neighbors? How did the non-Christians feel about Christians? What kind of opposition was expressed toward Christianity? That is, I shift now from

Christian attitudes and life-styles in relation to urban society to the attitudes of
non-Christians toward Christianity.

Precisely when Christians were recognized as a group separate from Jews cannot
be located. Christians, of course, including the writer of the Book of Revelation,
make a clear distinction between themselves and Jews (Rev. 2:9, 3:9). Certainly by
the time that Tacitus and Suetonius were writing in the early second century, a
distinction was made by Roman writers. Whether the distinction was made during
Nero's reign (54–68) or whether Tacitus and Suetonius referred to it anachronistical-
ly cannot be certainly determined (Tac. *Ann.* 15.44, Suet. *Ner.* 16.2). One can be
certain that by the time of Pliny's encounter with Christians in 112 Christians were
being considered by Roman officials as a distinct group of people, separate from
Jewish communities.[39] Probably by the last decade of the first century, when the
seer was writing, city-dwellers in Asia who were neither Christian nor Jewish knew
that there was a difference between those who went to synagogue and those who
gathered in house churches.

At the same time, however, "the beginnings of Christianity in Anatolia are
bound up closely with the numerous and wealthy Jewish communities" (Johnson
1958, 14, cf. Wilken 1976, 56, Johnson 1972, 182). So we cannot be certain when a
clear distinction began to be made by neighbors, city officials, provincial officials,
and officials of the empire; but probably it emerged at different times among those
groups. Nor can we be certain about what "a clear distinction" would mean for
Christian existence in Asian cities. On the face of it, there would probably be no
legal ramifications for Jewish Christians; for so far as I can see, that group could
still claim whatever Jewish rights they had, so long as Jews did not challenge the
claim of Jewish Christians to "membership" in the synagogue.[40] On the other hand,
house churches and other specifically Christian associations could not claim special
Jewish privileges, for the very visible Jewish synagogues and other associations in
Asian cities belied on the social level the Christian theological claim that Christianity was the true Jewish community (e.g., Wilken 1984, 198).

In Asia opposition to Christianity came primarily from local people, not from
the imperial machinery. That is clear from Pliny's correspondence with Trajan,
written in 112 in Pontus, somewhere between Amisus and Amastris (see Wilken
1984, 15). There is, to be sure, a certain ambiguity in Pliny's query to Trajan: Pliny
pleads ignorance of how to deal with Christians on the grounds that he has never
been present at their trial (*cognitio, Ep.* 10.96.1). That could mean that a body of
legal practice had been established at Rome regarding Christians but that Pliny was
unfamiliar with it — not likely, for if the practice were widespread, Pliny, an experienced lawyer in public service, would know about it. More likely, Pliny inquires
because there is no set legal procedure for dealing with Christians; and he, a Roman
senator sent out to deal with problems in the hinterland, does not want to act
without first making a detailed report to the emperor.[41] Most especially, he wants to
be sure about the "extent of the punishments"; "the grounds for starting an investigation and how far it should be pressed"; distinctions that should be made on
grounds of age; when a pardon should be granted; "and whether it is the mere name
of Christian which is punishable, . . . or rather the crimes associated with the
name" (*Ep.* 10.96.1–2). His procedure is to ask those brought to him whether they

are Christians; if they persist in avowing Christianity through threefold questioning, Pliny orders them to be sent away (for execution) unless they are Roman citizens. For Pliny their greatest offence is their "stubbornness and unshakeable obstinacy" (*pertinaciam certe et inflexibilem obstinationem*) (*Ep.* 10.96.3).[42] Although he has no kind words to speak about Christians, he does say that he "found nothing but a degenerate sort of cult carried to extravagant lengths" (*superstitionem pravam et immodicam*) (*Ep.* 10.96.8). Christians in Bithynia had customarily met to eat together (in addition to early morning worship) until Pliny had banned the meeting of associations (*hetaeriae*). Thus they had sought to comply with Pliny's edicts.[43]

From Pliny's letter it is clear that Christians are brought before him with charges (*deferebantur*). He does not (and Trajan orders him not to) seek out Christians (*Ep.* 10.97.2). Pliny writes that the situation is complicated by the fact that "the charges are becoming more widespread and increasing in variety. An anonymous pamphlet has been circulated which contains the names of a number of accused persons" (*Ep.* 10.96.4–5). As Pliny says, that is the kind of thing that "so often happens," but it nonetheless complicates his task. Pliny complicates our task by not indicating anything about those who bring charges against Christians and why. Perhaps those involved in the religious and economic aspects of the temples and sacrifices brought charges; towards the end of his letter, Pliny may be connecting the influence of "this contagious superstition" (*Ep.* 10.96.9) with the decline of the sale of sacrificial victims. If so, some of the accusers might have been motivated by some of the same forces as the silversmiths in Acts 19 (see Benko 1986, 8). Social class may be a factor in the accusations made to Pliny. Assuming that Christian leaders were involved in trades and crafts, like Aquila of Pontus (Acts 18:2), they would be wealthier than many. They would not, however, be valued as patrons, as was the provincial aristocrat Dio Chrysostom, nor would they be part of a network with imperial or senatorial representatives. They would be more or less unprotected from accusations from either urban dwellers "below" or imperial dignitaries such as Pliny "above."[44]

The Pliny correspondence confirms that the imperial cult was not a central issue in either official or unofficial attitudes toward Christians (see pp. 163–64). For the first time in Roman literature we hear of a formal test for Christianity: Pliny writes that suspects had to repeat "after me a formula of invocation to the gods" and to make "offerings of wine and incense to your statue [*imagini tuae*] . . . along with the images of the gods [*simulacris numinum*]" and to revile "the name of Christ" (*male dicerent Christo*) (*Ep.* 10.96.5, cf. 10.96.6). As Fergus Millar points out, "The Imperial cult thus plays a minor part in this episode" as an element in local religious tradition (1972, 153, cf. Price 1984, 125). Christians had a problem on two counts: they rejected all but the one true God and refused to recognize the divine object of other worship; and they rejected all forms of sacrificing, for Jesus was the one supreme and final sacrifice. They could not sacrifice to any god on behalf of the emperor. That put Christians on a collision course with local religious activity. For Pliny (and Trajan), the real issue centered on renewing local sacred rites; Pliny hopes that such renewal has begun in the towns, villages, and rural districts of Bithynia. Many have turned from Christianity, and he is confident that with a little patience and pressure, that disruptive cult can be checked and the people who were in it reformed (see *Ep*. 10.96.9–10).

Pliny's letter reflects a fundamental suspicion of Christianity. Even though the state did not "hunt Christians down," everyone — Roman official and local citizen alike — seems to agree that Christianity is a troublesome affiliation that deserves punishment (see *Ep*. 10.97). For Roman officials, Christians are placed with Bacchants and Druids as antisocial and a danger to the social fabric of the empire. For local citizens, activity that is specifically Christian is suspect. It was new, foreign, secretive, and exclusive; as such it was superstition intruding into the urban order that supported, and was supported by, traditional, customary, public, and inclusive religious piety (see Wilken 1984, 48–67). Nonetheless, overt conflict between Christians and their non-Christian neighbors requiring official, legal action was rare: "In most areas of the Roman Empire Christians lived quietly and peaceably among their neighbors, conducting their affairs without disturbance" (Wilken 1984, 16). "Maintain good conduct among the Gentiles," writes the author of 1 Peter to Christians in the area of Pontus, "so that in case they speak against you as wrongdoers, they may see your good deeds and glorify God on the day of visitation" (1 Peter 2:12).

Most early Christian literature associated with Asia, Bithynia, and Pontus supports that ideal of living peacefully in the Roman social order. Paul's attitudes expressed in his letter to the Corinthians were no doubt also expressed in Ephesus, where he wrote the letter, which city was for a time a center of his missionary activity. Later the author of 1 Peter exhorts his audience to "be subject for the Lord's sake to every human institution, whether it be to the emperor as supreme, or to governors. . . . Honor the emperor" (1 Pet. 2:13–17). The author of 1 Timothy urges "supplications, prayers, intercessions, and thanksgivings . . . for kings and all who are in high positions, that we may lead a quiet and peaceable life, godly and respectful in every way" (1 Tim. 2:1–2).[45] Those at Thyatira and Pergamum whom the author of the Book of Revelation saw as his opponents probably shared that dominant Christian view. On the face of it, therefore, the Book of Revelation presents a minority report on how Christians relate to the larger Roman society. The seer is apparently advocating attitudes and styles of life not compatible with how most Christians were living in the cities of Asia.

8

Jews in the Province of Asia

The earliest Christians in the cities of Asia were probably converts from Anatolian Judaism. Among all the cities in the province of Asia — somewhere between three hundred and five hundred in number — indicators of a Christian presence are found for the most part where there were also Jews. Those cities are also, for the most part, major cities of the province, for example, Ephesus, Smyrna, Pergamum, and Sardis. Thus, both Jews and Christians in the late first and early second centuries lived in major metropolitan areas. As we have seen, evidence from early Christian literature — Paul's letters, Acts of the Apostles, the Book of Revelation, Ignatius — suggests that the links between Christianity and Judaism were close and that converted Jews made up a significant portion of Christians in the cities of Asia. Jews who converted to Christianity continued in their same occupation; their same social, economic status; and their same relationship to the cities in which they lived and worked. For that reason an inquiry into Judaism will indirectly provide information about social characteristics of some Christians and about the social context of Christianity. This inquiry will focus first on Domitian's attitudes towards Jews and Christians, then upon the social location of Jews in the cities of western Asia Minor, primarily in the cities of the province of Asia. As in the review of Christians in Asia, special attention will be given to the seven cities referred to in the Book of Revelation.

Jews, Christians, and Domitian

Suetonius writes in his *Lives of the Caesars* (after 120, see chap. 6) that Domitian extended the "Jewish head tax" to include not only "those who kept their Jewish origins a secret in order to avoid the tax" but also "those who lived as Jews without professing Judaism" (*Dom.* 12.2). This "head tax" was Vespasian's replacement for the Jewish temple tax after the destruction of the temple at Jerusalem in 70.[1] Suetonius goes on to mention that he was present once when a ninety-year-old man was examined before the procurator and a crowded court to see if he was circumcised. Domitian extended the tax to apostate Jews (perhaps including some Jewish Christians), and to those who were "followers" — though not proselytes — of Judaism, and he did this vigorously (*acerbissime*). According to Suetonius, overspending and economic disarray were the catalysts for this more rigorous administration of the Jewish tax.[2] As we saw in chapter 6, this portrait of the end of Domitian's reign cannot be sustained. But in this case Suetonius suggests nothing particularly malicious on Domitian's part. He simply indicates that Domitian was more rigorous in administering the tax than his predecessors. There is nothing in this statement to suggest that Jews suffered under Domitian; his policy of rigor merely fits with Domitian's general administrative principles of rationality and consistency (see Smallwood 1981, 377).

The notion that Domitian may have used unjust procedures in administering the Jewish tax comes from a legend appearing on the reverse of some Roman coins minted in the early part of Nerva's short reign (96–98) immediately after Domitian's death. The legend reads FISCI IUDAICI CALUMNIA SUBLATA, ("The misadministration of the Jewish imperial treasury is abolished") (Smallwood 1966, no. 28).[3] There is no direct mention of Domitian on these coins; but the legend announces some kind of reform that the Emperor Nerva made in his administration of the Jewish tax. The legend does not indicate the specifics of that reform except that the previous practice is called *calumnia*, some kind of misadministration involving a false charge or wrongful accusation. Applebaum offers the most fitting interpretation of this word, which is also found in Aurelius Victor's *Caesars*. In that work are described *calumniae* by Trajan's procurators that consisted of "extraction of information on income and its sources by threats and bullying" (Applebaum 1974a, 462–463). As Applebaum points out, this aggressive approach to collecting the Jewish tax squares with Suetonius' reference to a procurator's examining a ninety-year-old man to see if he was circumcised.

In Roman sources there are two more passages relevant to Domitian and the Jews. Dio Cassius claims that Domitian killed Flavius Clemens, along with many others, and banished Flavius' wife, Flavia Domitilla, to Pandateria with the charge "of atheism, a charge on which many others who drifted into Jewish ways were condemned" (67.14.2).[4] Dio Cassius also mentions one Glabrio, by name, who was "accused of the same crimes as most of the others, and, in particular, of fighting as a gladiator with wild beasts" (67.14.3). Thus, according to Dio, Domitian condemned fighting with wild beasts, atheism, and Jewish ways. Dio Cassius writes in another place that when Nerva came to power he did not allow anyone "to accuse anybody of *asebeia* [impiety, disloyalty] or of adopting the Jewish mode of life" (68.1.2); one need not connect the two.

One can make all of these references to Jews fit into the distorted portrait of Domitian's last years in the standard sources (see chap. 6). Suetonius refers to the extension of the Jewish tax as a part of the confusion and disorder during Domitian's last years. Dio's account fits the picture of Domitian as a mad megalomaniac eager to condemn those who did not worship him properly. The sharp contrast between Nerva's policies and Domitian's fits with the rhetoric of the post-Domitian era. Nerva proclaims on other coin legends his "new era" of liberty and safety.

Trajan's rescript to Didius Secundus assures in the same vein that he will no longer confiscate property as was done before from avarice. The republican values of humane treatment and recognition of the human "leader" replace the divine claims of Domitian and his past unjust ways. If one takes this propaganda at face value as reporting the social and political situation under Domitian—an assumption difficult to accept (see chap. 6)—*asebeia* ("disloyalty"), the Jewish mode of life, atheism, *calumnia* ("malpractice"), and Domitian's extension of the Jewish tax may be strung together to show Domitian's poor treatment of Jewish converts, if not of the Jews. So Smallwood concludes, "Dio and Suetonius deal with different aspects of the fate of converts to Judaism, Dio with punishment for 'atheism' and Suetonius with delation for tax-evasion, and the two supplement each other in showing how Domitian's development of the imperial cult together with his extension of liability for the Jewish tax put converts into a cleft stick: if they admitted their Judaism, they were liable for punishment as 'atheists'; if they did not, they risked delation to the officials of the *fiscus Judaicus*" (1981, 380).

If, however, the themes of Domitian's excessive demands for imperial worship, the ubiquity of his informers, his indiscriminate charges of disloyalty and treason, and the contrast between Domitian's and later reigns are to be understood along the lines developed in chapter 6, Domitian's conflicting policy of taxing and then killing Jewish converts, which makes little sense in itself, has little evidential support either. The concern to regularize the Jewish *fiscus* fits with Domitian's public policy; the contrasts between Domitian and later emperors, with Domitian at the negative pole, fit the rhetoric of the later period. There is no convincing evidence that Domitian modified the imperial cult by demanding greater divine honors than his predecessors (or successors) or that he arbitrarily brought recriminations against people falsely accused of treason against the state or the person of the emperor. Yet such are required to interpret these Roman texts as evidence for a Domitianic persecution of Jews or Jewish converts. If it can be shown that Domitian oppressed Jews, converts to Judaism, or—less likely—Christians, (who were sometimes confused with converts to Judaism), it must be established by means other than the points of view expressed in Suetonius, Dio, and Nerva's propaganda.

If Roman sources about Domitian tend to be taken at their face value, the views of the Jewish writer Josephus tend to be rejected because of his "special relations" with the Flavians. As Rajak has shown, however, that "special relationship" has been grossly overstated. The Flavians gave Josephus many honors (e.g., citizenship, the Flavian name, Vespasian's old house to work in; cf. Rajak 1984, 194–95), but he had his own agenda in writing.[5] In the *Jewish Antiquities*, which was written in "the thirteenth year of the reign of Caesar Domitian" (Joseph. *AJ* 20.11.3), that is, 93–94, and even more in *Against Apion*, written in 96 a short time before Domitian's death Josephus magnifies "the Jewish race in the eyes of the Graeco-Roman world by a record of their ancient and glorious history" (Thackeray 1967, 51). Or, as Rajak puts it, Josephus is "concerned with apologetics—the presentation of Judaism to outsiders" (1984, 225). This apologetic was concerned to present Judaism in the best light to outsiders and also to make Judaism attractive to the non-Jew. One cannot say that he was seeking converts; but he was certainly aware that Judaism was attractive to some Greeks and Romans, and he traded on that attractiveness

˘ (see Rajak, 1984, 228). Although it is possible that Josephus carried out his apologetic without the notice of Domitian, it is unlikely. More likely, either Domitian was supporting Josephus in an enterprise that could foster converts to Judaism, or Domitian was insufficiently interested in the "Jewish question" to pay much attention to Josephus.

Josephus states that Domitian was aware of him; indeed, that Domitian protected him from false accusers: "And Domitian, who succeeded [Titus], still augmented his respects to me; for he punished those Jews that were my accusers, and gave command that a servant of mine, who was a eunuch, and my accuser, should be punished. He also made that country I had in Judea tax free, which is a mark of the greatest honour to him who hath it; nay, Domitia, the wife of Caesar, continued to do me kindnesses" (Joseph. *Vit.* 76). That statement supports evidence from Quintilian and even Suetonius that Domitian did not abuse the use of informers.[6]

In the twelveth *Sibylline Oracle*, a third-century Jewish text written away from the capital, Domitian's reign is described (in the apocalyptic form of prophecy after the event) as "a great kingdom whom all mortals will love throughout the ends of the earth and then there will be rest from war throughout the whole cosmos . . . from east to west, all will be subject willingly, and . . . upon him heavenly Sabaoth, the imperishable God dwelling in heaven, will bring much glory" (*Sib. Or.* 12.125–132). In this twelveth book of *Sibylline Oracles* from a "provincial tradition" (Geffcken 1902) specific mention is also made of cities' spontaneously becoming subjects of Domitian and of Domitian's favorable treatment of the provinces.[7] *Sibylline Oracles* 12 reflects the strong positive feelings of provincials towards Domitian; Domitian was seen as a "benefactor of all provinces," including provincial Jews (see B. Jones 1979, 61).[8]

Nor are Christian writers thoroughgoing in their condemnation of Domitian. The notion of a Domitianic persecution first appears in the fourth-century work of Eusebius. According to him, John of Revelation was caught in that persecution, "sentenced to confinement on the island of Patmos" (*Hist. Eccl.* 3.18). Eusebius quotes freely from earlier sources, but he is less ambiguous than his sources about the evil character of Domitian. Eusebius tends to follow the official Roman historiography on Domitian, referring to his "appalling cruelty" and his execution "without a fair trial" of "great numbers of men distinguished by birth and attainments" and his banishment of "countless other eminent men." Like the standard Roman pen-portraits, he calls Domitian a "successor of Nero," but Eusebius adds "in enmity and hostility to God" (*Hist. Eccl.* 3.17). He probably draws on Dio Cassius as a source for Domitian's persecutions. Eusebius states "that even historians who accepted none of our beliefs unhesitatingly recorded in their pages both the persecution and the martyrdoms to which it led" (*Hist. Eccl.* 3.18).[9]

In Eusebius' sources and other Christian writings earlier than Eusebius, Domitian is not presented in such totally negative terms. Irenaeus, in connection with a revelatory vision recorded in the Book of Revelation, states simply that "it was seen not a long time back, but almost in my own lifetime, at the end of Domitian's reign" (Haer. 5.30 = Eus. *Hist. Eccl.* 3.18.1, cf. 5.8). He does not mention any persecution at that time. According to second-century Christian writer Hegesippus,

Domitian is one of three emperors who hunt down "the lineage of David."[10] The emperor interviews grandsons of Judas, the brother of Jesus, because they were of the line of David. He sees their work-worn hands and their general poverty, interrogates them about Christ and his kingdom, then releases them. The exact quotation is interesting for our context: "At this [after they described the kingdom] Domitian did not condemn them at all, but despised them as simple folk [or, "viewed them as frugal"], released them, and decreed an end to the persecution against the church" (Eus. *Hist. Eccl.* 3.20.5). Domitian does not, thus, come off as badly as do Vespasian (*Hist. Eccl.* 3.12) and Trajan (3.32).[11] As a Christian from Syria, Hegesippus may be influenced by the provincial view of Domitian reflected also in the *Sibylline Oracles*.

The Christian apologist, Melito, defending Christianity before Marcus Aurelius, argues that the Christian religion and the empire grew up together (flowering in the reign of Augustus!) and that Christianity was a portent for good as Rome expanded (Eus. *Hist. Eccl.* 4.26.7–8). Only Nero and Domitian (the bad emperors in Roman historiography) attacked the Church (Eus. *Hist. Eccl.* 4.26.9). Their ignorance was corrected by Hadrian, Antoninus Pius, and Marcus Aurelius (4.26.10). Tertullian argues similarly in his *Apology*, though he links the emergence of Christianity more accurately to the time of Tiberius (27–37). For him, too, Nero and Domitian are the bad emperors—Domitian "was a good deal of a Nero in cruelty"—who raged against Christianity. Domitian, however, "being in some degree human, soon stopped what he had begun [persecution of Christians], and restored those he had banished" (Tert. *Apol.* 5). Tertullian, thus, like Hegesippus, on the one hand reflects the standard judgment against Domitian but on the other hand qualifies Domitian for the good. Tertullian and Hegesippus may combine two different traditions of Domitian—the standard Roman assessment of Domitian as an evil emperor and the provincial assessment of him as a good emperor.[12]

From the various sources about Domitian's relations to Jews and Christians, two views begin to emerge. There is the view of the standard sources of Domitian's reign—Domitian as the evil emperor—located in the "official" literary traditions and imperial archives at Rome. These sources influence especially Eusebius' account of Domitian's reign, but they also influenced Melito and Tertullian in their apologies to official Roman offices. At the same time, Tertullian equivocates, for he knows another tradition about Domitian, as does Hegesippus. This other, provincial tradition (seen also in *Sibylline Oracles* 12) praises Domitian as a good emperor. In the cities of Asia, the provincials (including Jews and Christians) would more likely have a positive view of Caesar Domitian than would those in Rome.

Jews in the Province of Asia

Although outsiders could distinguish between Christians and Jews by the last decade of the first century, in Christian life they remained entangled historically, socially, and theologically. Many Christians in urban Asia converted from Judaism, which linked the two traditions. Furthermore, the presence of the synagogue was living testimony to the continued life of Israel apart from Christianity, and that social presence created theological problems for Christian claims that the church had supplanted the synagogue (see Wilken 1984, 112–117). The relationship between Christianity and Judaism is a prominent theme in Christian literature from Asia (cf. Acts, Paul's letters, Ignatius, Melito), and the Book of Revelation is no exception. The seer's reference to Jews at both Smyrna and Philadelphia as members of "the synagogue of Satan" has both theological and social implications. He is rejecting the Jews as being the "people of God"; but he is also condemning Jewish accommodation to Greek city life, just as he condemns Christians who do the same (Rev. 2:9, 3:9, cf. 2:14, 21). Because of these links between the church and synagogue, the status of the Jew can give indirect evidence for the placement of Christians in the social order.

Jews had been settled in this geographical area as early as the end of the third century BCE, when Antiochus III, a Seleucid king, ordered Zeuxis, one of his subordinates, to relocate some two thousand Jewish families from Babylonia to Phrygia as a military colony (*katoikia*) to help control political revolts that were occurring in Phrygia and Lydia (Joseph. *AJ* 12.147–49). Antiochus gave orders to grant each family "a place to build a house and land to cultivate and plant with vines" (*AJ* 12.151). According to Josephus they are also allowed "to use their own laws" (*AJ* 12.150) and to provide grain for "their servants" and "those engaged in public service" (*AJ* 12.152).[13] Antiochus also orders that thought should be taken so that the Jews "should not be molested by anyone" (*AJ* 12.153).

The protection given to this Jewish *ethnos* (Joseph. *AJ* 12.153) seems to be maintained down through the hellenistic and Roman periods (see Applebaum 1974a, 420). Josephus once again provides the evidence. In various documents from the time of Julius Caesar, Josephus mentions such privileges as the right to assemble for worship, to obey the Jewish Law, to hold funds, and to build synagogues (see *AJ* 14.190–216; Smallwood 1981, 134–35). Dolabella, after 44 BCE, on orders from Rome, granted to Asian Jews exemption from military service and permission to maintain their native customs (*AJ* 14.223). Josephus includes decrees to Ephesus (*AJ* 14.225–27, 228–29, 234, 236–40, 262–64), Sardis (*AJ* 14.235, 259–261), Laodicea (*AJ* 14.241–43, cf. a reference to Tralles at 14.242), Miletus (*AJ* 14.244–46), and Pergamum (*AJ* 14.247–55).

Augustus later (c. 12 BCE) renewed those privileges to the Jews of Asia, including their right to send "sacred monies" to Jerusalem (the temple tax) (see *AJ* 16.166), and the edict was publicly displayed in Caesar's temple (*AJ* 16.162–65). Marcus Vipsanius Agrippa, a longtime close associate of Augustus, during a mission to the East also sent an edict to Ephesus that the sacred Jewish temple tax should be protected (*AJ* 16.167–68, cf. 12.125–28, 16.27, 60). Gaius Norbanus Flaccus, proconsul of Asia, transmitted to the city leaders of Sardis Caesar's edict that they should not impede the collecting or the sending to Jerusalem of the temple tax (*AJ* 16.171). Finally, Julius Antonius writes to the people of Ephesus to permit Jews "to follow their own laws and customs, and to bring the offerings . . . travelling together under escort [to Jerusalem] without being impeded in any way" (*AJ* 16.172). From those and other documents about Asian cities, one may conclude that Jewish rights were challenged in the reign of Augustus but that they had been firmly reestablished by the end of his reign (see Applebaum 1974a, 443).

These rights were reiterated by Claudius (*AJ* 19.286–91) and again, after the Jewish War, by Vespasian and Titus (*AJ* 12.119–24). As Applebaum points out, Jewish privileges in the empire came about through a "series of city-laws of local application, which, although they possessed common principles, differed in detail from one city to the next" (Applebaum 1974a, 457). Jewish status outside Judaea was "ethnic rather than national" and "the question of whether Judaism was a 'permitted religion' does not seem to arise at all."[14]

Phrygia

Jewish communities were widely distributed throughout western Asia Minor (cf. Philo, *Leg.* ˙245). In Phrygia (the central, eastern portion of the province of Asia and part of the province of Galatia east of Asia), where Antiochus III had sent Jews at the end of the third century BCE, there is evidence for Jews in several cities. Apamea must have been the center of a large Jewish settlement; for Flaccus, a governor of Asia around the middle of the first century BCE, had confiscated from Apamea a temple tax of almost a hundred pounds of gold. Apamea was probably a collection center for the money to be sent to Jerusalem, but it was at least as important as a collection center (see Stern 1974, 143). Other evidence for a Jewish presence is found in three third-century-CE inscriptions from Apamea involving curses against violators of Jewish grave sites (MAMA 6.231; CIJ 773, 774). More interesting, Apamea seems to have adopted the Noah story (probably through Jewish rather than Christian influence). Coins minted there from the first half of the third century CE show Noah and his wife stepping from the ark along with the caption *Noah*. Mount Ararat, where Noah's ark came to rest, was identified in Phrygia somewhere near Apamea. Here is probably a case of Jewish cultural traditions mixing freely with Greek and Phrygian culture, which indicates acculturation of the Jews in this area (cf. Kraabel 1968, 121–22; Schürer 1986, 28–30).

Laodicea, one of the seven cities in the Book of Revelation, was also a collection center for the Jewish temple tax; there, however, Flaccus confiscated only about twenty pounds of gold (see Cic. *Flac.* 28.68). Laodicea was both an important commercial center (see Rev. 3:17) and a center for organized Judaism. Josephus records a letter from magistrates of Laodicea to Gaius Rabirius, proconsul of Asia, confirming that they would honor his instructions to allow Jews to observe the Sabbath and other rites and to recognize the Jews as "friends and allies" not to be harmed (*AJ* 14.242, cf. Applebaum 1974b, 477).[15] The seer of Revelation does not mention this Jewish community in his condemnation of the wealthy Christians there.

Several inscriptions, mainly epitaphs from the second and third centuries CE related to Jews, have been found at Hierapolis, a city located a few miles north of Laodicea in the Lycus Valley (see *CIJ* 775–80). Jews took over the local Phrygian custom of decorating graves with flowers but adapted the custom so that graves were decorated on the Jewish holidays of the Festival of Unleavened Bread or the

Pentecost (*CIJ* 777, see Kraabel 1968, 128). Jewish names—sometimes double names—include Hebrew, Greek, and Latin. In *CIJ* 777 local trade associations— which included both Jews and Gentiles—are given the responsibility of decorating with flowers the Jewish graves of Publins Aelios Glykon Damianos and his family—the chairman of the purple-dyers for the Festival of Unleavened Bread and the tapestrymakers for Pentecost. Finally a prosperous man who is probably a Jew notes on a grave prepared for him and his family that he made seventy-two business trips to Rome from Hierapolis (see *IG Rom*. 4.841; Kraabel 1968, 135).[16]

Acmonia, about a hundred miles north of Hierapolis but reached by road through Philadelphia, had a large Jewish community, which has left us an abundance of epigraphy. Sometime in the second half of the first century CE a building for a synagogue was donated to the Jews by Julia Severa, a magistrate there in the time of Nero. She was high priestess of the *theoi sebastoi* and a Gentile sympathizer to Judaism (see Applebaum 1974a, 443). Her activity in other religions was not a hinderance to receiving her gift (*CIJ* 766). A bilingual Greek-Hebrew inscription from Acmonia may be a building fragment from that synagogue (see Kraabel 1968, 92). In the third century a wealthy Jew by the name of Aurelios Phrougianos (248– 49 CE)—who prepared a tomb for himself with warning curses from Deuteronomy—held several municipal offices, including superintendent of the markets, provider of grain, and sheriff (*CIJ* 760); and from the same period another Jew, Titus Flavios Alexander, lists the offices of "magistrate, officer of the peace, provider of grain [two terms of office], senate president, superintendent of the markets, and chief magistrate."[17] From these and other third-century inscriptions from Acmonia it is clear that Jews took the highest responsibilities and offices in civic life and that they had the economic means and appropriate social status to fulfil these obligations.[18] At Acmonia, Jews are well integrated into the municipal life. Kraabel concludes, regarding Jews and pagans, "No group appears to be set apart, but all mix in the social and economic life of the city" (1968, 115).

From elsewhere in Phrygia, but outside the circle of the churches of Revelation, come a few epigraphic remains of a Jewish presence: at Synnada, about twenty miles east and south of Acmonia, a fragment of a first-or second-century inscription referring to a son of Julios, an *archisynagogos* (*CIJ* 759); fifteen miles north and east, at Docimium, a stone with a carved menorah, perhaps a fragment from a synagogue (Kraabel 1968, 80); from Eumeneia, about ten miles northwest of Apamea, may come at least one Jewish inscription, a curse against anyone violating a grave; he will receive a curse against his children's children (*CIJ* 761).[19]

Lydia

In Lydia, a region in the river valleys of Phrygia, the most important Jewish community was at Sardis.[20] Recall that the author of Revelation condemns wealthy Christians at Sardis and does not mention Jews there. In a document from the first century BCE Josephus refers to rights given to the Jews at Sardis by Lucius Antonius, a governor of Asia (see Magie 1950, 1256). In a letter addressed to the city council at Sardis, Lucius Antonius directs the council to allow Jews to have "an association of their own in accordance with their native laws and a place of their own, in which they decide their affairs and controversies with one another" (Joseph. *AJ* 14.235, Applebaum 1974b, 442). A little later Josephus includes a decree passed by the council (c. 50 BCE, cf. Applebaum 1974b, 477–78, Kraabel 1978, 16–18) to allow Jews those liberties, and the market officials are charged to bring in suitable food for the Jews (Joseph. *AJ* 14.259–61). Josephus also records a decree from Gaius Norbanus Flaccus (the confiscator of Jewish funds), proconsul, to the magistrates and council of Sardis (c. 30 BCE) ordering that the Jews should be allowed to send money to Jerusalem (*AJ* 16.171; Millar 1966, 161). Thus, Jews at Sardis are said to have been able to practice their ancestral laws, meet for cultic and communal activities, erect communal buildings, conduct litigation among themselves in their own courts, collect funds for Jerusalem, and have their own foods in the market (Applebaum 1974b, 478–79).

A spectacular archaeological find of a Jewish synagogue at Sardis may also provide evidence for the socioeconomic status of Jews and their integration into civic life there. Archaeologists have unearthed a large social center that included a gymnasium, shops, *palaestra* complex, and a Jewish synagogue (see Hanfmann 1972, 1975). Although the synagogue was walled off from the rest of the facilities, it opened out onto "Main Street" (see Kraabel 1978, 233). Some of the furnishings of the synagogue are striking examples of Jews' combining their own with Roman and indigenous Lydian traditions. A large table was probably the centerpiece of worship; decorations on the tabletop show it to have been originally from the Augustan period, when it was used for some other purpose. The tabletop is supported on either end by two slabs, each of which is "decorated with a powerful Roman eagle clutching thunderbolts" (Kraabel 1978, 22). The table is flanked by self-standing lions, "originally Lydian, sixth-fifth century BC" (Kraabel 1978, 22). One of the Jewish donors to the synagogue is identified as from "the tribe of Leontioi," a group within the Jewish community at Sardis (see Robert 1964, 46). The lion, however, also plays a significant role in the history of Sardis; and lions are found in connection with the altar to Cybele uncovered in excavations of Sardis (see Kraabel 1978, 22–23). Kraabel concludes, "It seems likely that the Sardis Jews are not simply 'reusing' the lion-statues in their synagogue, but actually associating themselves in some way with this traditional Sardis image" (Kraabel 1978, 23). At the same time, the synagogue has the niche for the Torah scrolls and other appropriate Jewish identifications. Thus the mixture of materials, symbols, and images of Roman, Lydian, and Jewish traditions in the synagogue indicates a vital Jewish community integrated into Lydian and Roman culture and society. Kraabel sees here "the expressions of self-confident Jewish communities, existing in Asia Minor . . . for over a half a millenium. . . . These are bold acts, not timid ones, and should not be misrepresented as 'syncretism' or 'apostasy' simply because they appear strange in the twentieth century" (Kraabel 1978, 24).

From other inscriptions found at Sardis there is further evidence for a Jewish community integrated into Sardian social life. On at least nine inscriptions Jews identify themselves as members of the city council (*bouleutes*), citizens of the city, and citizens of the empire. Three others belonged to the Asian provincial administration, one of whom was a procurator. Although these are third-century-CE inscriptions, Kraabel argues that they confirm that "Jews had been in Sardis a long time" and had accumulated significant wealth and status (1968, 219–20). Sardis reveals a Jewish community quite integrated into the social, economic and political life of a major Anatolian city, while retaining an internal cohesion and a clear understanding of itself as a part of Diaspora Judaism" (Kraabel 1968, 9–10).

To what extent this synagogue contributes to our understanding of Jewish existence in the first century CE depends on the date of the synagogue and how much continuity existed in Jewish life at Sardis through the first three centuries of the Common Era. The synagogue did not become a part of the gymnasium complex before the third century. From an analysis of coins found under the mosaic floor of the synagogue, it is likely that remodeling for the synagogue did not occur before 270, nearly two centuries after the writing of the Book of Revelation. By the end of the third century, Jews at Sardis were integrated socially, politically, and economically into the city, the province, and the empire. Can one assume, as Kraabel does, that such integration was present over the previous half millenium?[21]

Marianne Bonz argues (against Kraabel) that the economic prosperity and social acceptance of Jews at Sardis were tied to the economic policies of Septimius Severus (193–211) and Caracalla. Because of extravagance and mismanagement, Caracalla needed to broaden the tax base that included services (liturgies) for the cities, and Jews were brought in as a part of the necessary adjustments.[22] Because of the economic crisis in the empire during the third century, the costs of municipal offices went up and interest in them went down (see A. Jones 1940, 182–91). There is insufficient evidence, however, to link completely the social status of Jews at Sardis to those economic and social upheavals. More Jews may have been harnessed into giving liturgies to the city in the third century, and perhaps the economic crisis contributed to the conditions that allowed a synagogue in the gymnasium at Sardis. But there is sufficient circumstantial evidence from Josephus to conclude that prior to this time Jews lived an integrated social life at Sardis.

Philadelphia is situated about thirty miles southeast of Sardis. John of the Apocalypse praises Christians there but condemns the Jews as belonging to the "synagogue of Satan" (Rev. 3.9). A third-century inscription found at nearby Deliler provides evidence for a synagogue; in the inscription Eustathios and his fiancée dedicate to the synagogue a basin for water ablutions (*CIJ* 754). Ignatius, bishop of Antioch, may provide evidence for Jews at Philadelphia when he writes, "But if any one propound Judaism unto you, hear him not: for it is better to hear Christianity from a man who is circumcised than Judaism from one uncircumcised" (*Philad.* 6).[23]

A few other cities in the region of Lydia certainly included Jewish communities. At Thyatira — where Jezebel the prophetess lived (Rev. 2:20–21) and Lydia, "seller of purple cloth," was born (Acts 16.14) — a second- or third-century inscription probably provides evidence for a synagogue.[24] At Hypaepa, between Sardis and Ephesus, an inscription from about 200 was found, the remains of which reads simply, "of the younger Jews" (*CIJ* 755, Ἰουδαίων νεωτέρων). It refers to either a group in the local synagogue or more likely to "Jewish ephebes in the local palaestra" (Kraabel 1968, 181, cf. the inscription from Iasos in Caria in Kraabel 1968, 16). Finally, at Magnesia in the shadow of Mount Sipylos, Strato, a Jew, son of Tyrannos (*CIJ* 753) placed a funerary inscription.

The Coastal Region

Finally, we may note some of the cities in the coastal region along the Aegean Sea where there is evidence of a Jewish presence. A Jewish community is well attested at the economic and political center of Ephesus (one of the seven cities of the Apocalypse). Josephus refers to decrees from the latter half of the first century BCE reconfirming Jewish privileges.[25] Two funerary inscriptions from the second or third centuries CE were found at Ephesus: one set up by an "official physician" (ἀρχείατρος) by the name of Julius or Iolios and one was set up by Markus Mussios. Both charge the Jews of Ephesus to care for their graves.[26] As we saw in chapter 7, Jewish magicians are referred to in Acts 19:13–19. A few protective amulets have been found in and around Ephesus, indicating that there were "Jewish elements in the magic of the Ephesus area" in the first century (Kraabel 1968, 59). That offers further evidence for Jewish acculturation in Asia. Sometime during the middle of the second century, the Christian apologist Justin Martyr debated there with Trypho, "the most distinguished Jew of the day" (Eus. *Hist. Eccl.* 4.18.6). A few distinctively Jewish household and personal items have been found at Ephesus, but for the most part the Jews there shared in the common culture. Just as they bore the same name as the indigenous citizens (cf. Joseph. *Ap.* 2.39), so they "displayed the dress and the vocabulary of their Ephesian neighbors" (Kraabel 1968, 60).

Epigraphic and literary evidence from Smyrna indicates an active Jewish community. As we have seen, the seer of the Apocalypse refers to the "synagogue of Satan" there as well as at Philadelphia. A few years later, according to Ignatius, the Christian community at Smyrna was made up of both Jews and Gentiles (I *Smyr.* 1).[27] In an inscription from Hadrian's reign (c. 125), specifying contributions made to public buildings, a small contribution is made by "those formerly Judaeans" (οἱ ποτὲ Ἰουδαῖοι, *IG Rom.* 4.1431). The phrase may refer to apostate Jews, but Kraabel has made a good case for interpreting it to mean "formerly from Judaea" (1968, 28–32). Thus, Jews supported civic life in Smyrna just as they did elsewhere in Asia. Later inscriptions indicate the presence of a synagogue at Smyrna (*CIJ* 739–40). There are also "funerary inscriptions of two officers of the Jewish community" there (*CIJ* 741; and Dittenberg. *SIG* 1247; Kraabel 1968, 41). In another inscription a woman benefactor by the name of Rouphina is honored as "leader of the synagogue" (*archisynagogos*) (*CIJ* 741).[28]

Jews are also evident at Pergamum, where, according to John, is "Satan's throne" (Rev. 2:13). Flaccus confiscated a Jewish temple tax there around the middle of the first century BCE.[29] Somewhat earlier, Josephus mentions Pergamum in connection with the confirmation of Jewish rights (see *AJ* 14.247–55).[30] Two objects have been unearthed by archaeologists at Pergamum: one from a synagogue showing "a well-carved menorah" and other distinctively Jewish symbols; the other, a third-century altar with the inscription "God, Lord, the one who is forever" (see Kraabel 1968, 180).

Evidence from Miletus (the town closest to the island of Patmos, where John had his visions) probably shows another Jewish community well-integrated into urban social life. An inscription carved on a row of theater seats reads "the place of the Jews who are also called *theoseboi*" (*CIJ* 748, cf. *CIJ* 749). Jews entered fully into life at Miletus without losing their identity. As Kraabel aptly says, this inscription says "much about their social life, little about their piety" (1968, 16, but cf. Schürer 1986, 167-69). Earlier evidence of Jews in Miletus comes from Josephus; he cites a document (c. 150-50 BCE) that refers to a proconsul ordering the magistrates, council, and people of Miletus to stop attacking the Jews and forbidding them from observing their customs (*AJ* 14.244-46).[31]

Jews in Asia: A Composite Portrait

Epigraphic, numismatic, and literary evidence for a Jewish presence in the province of Asia is fragmentary and happenstance. Nonetheless, we can develop a portrait of Diaspora Judaism in this area of the Roman Empire, drawing upon evidence from roughly the second century BCE to the fourth century CE. This portrait is important for our study of the Book of Revelation, for it shows how Jews in Asia lived, worked, and related to their urban setting. We assume that Jewish converts to Christianity continued to relate to their urban setting along the same lines; that is, their conversion to Christianity did not cause or require of them a drastic change in their occupation; in their social, economic status; or in their engagement in urban life.

With the transition to the empire in the latter half of the last century BCE came certain challenges in the cities to Jewish rights and privileges. Josephus is a primary source for these challenges. By the turn of the century, however, Jewish communal rights were reestablished and attacks against them came to an end. Throughout the first century of the Common Era, even during and immediately after the Jewish War in Palestine (66-70), Jews in Asia shared in the prosperity of the common, urban life and were not persecuted or punished for their Palestinian connections. Nor were there successful attempts at putting in jeopardy the rights of Jews to observe their legal tradition, to gather at the synagogue for worship and study, or to be exempt (albeit tacitly) from the state cult and military service.[32]

Guarantee of those rights did not, however, create in the cities of Asia a Jewish ghetto, a sheltered community that carefully avoided or circumscribed social intercourse with non-Jewish neighbors. Jews identified with the cities in which they had lived for centuries; in Sardis, they called themselves Sardians, in Ephesus they were Ephesians, and in Antioch, Antiochenes (see Joseph. *AP.* 2.39; Kraabel 1978, 24).[33] From inscriptions at Acmonia, Sardis, and Ephesus, we know of Jews who served municipal, provincial, and even imperial offices — and all this while active in the synagogue; honored members of the synagogue were not troubled by the "religious observances connected with citizenship and office-holding" (Kraabel 1968, 221). At Smyrna, Jews contributed, however modestly, to the development of the city. Jews at Hypaepa, Iasos in Caria, and elsewhere participated in the gymnasium (Applebaum 1974a, 447). At Miletus they had "reserved seats" for the theater. At Acmonia they received funds for the synagogue from Julia Severa who was also affiliated with "pagan" cults. At Sardis, Jewish shops were set up along the south wall of the synagogue beside a public restaurant and shops by non-Jews.

Jews were active members of guilds and trade unions; at Hierapolis two guilds were responsible for decorating the graves of Jews on the Feast of Unleavened Bread and the Pentecost.[34] Probably Jews had "penetrated all the peninsula's normal occupations" (Applebaum 1976, 715). Applebaum cites evidence for Jews in farming, vine growing, and ownership of land. Jews were leather workers, manufacturing tents and hobnail boots; metal workers; and perfumers. Most common of all, Jews were involved in the manufacture of textiles — dyers, carpetmakers, and makers of woolen wares (see Applebaum 1976, 715-19).

Jews in the Asian cities not only identified socially with their urban environment; they also shared in the cultural traditions. At Ephesus only a few distinctively Jewish artifacts have been discovered. At Laodicea the local customs of decorating graves with flowers was taken over by Jews. At Apamea the flood story of Noah was conflated with Greek and Phrygian flood stories, and at Sardis the symbolism of lion and eagle in Judaism mixed with that in Rome and Lydia.

In sum, the Jews in the cities of Asia neither withdrew into ghettos nor assimilated and lost their identity as Jews. They engaged in what Applebaum calls a "mutual rapproachement" (1974a, 443). Throughout the first two centuries of the Common Era, Jews lived stable lives socially and economically in those cities (see Applebaum 1976, 702). As Kraabel summarizes, in Asia Minor one kind of Diaspora Judaism emerges, "a picture of a number of Ionians, Phrygians and Lydians, each of whom participates in the life of his own city, speaking its language, fitting into its commercial and social life and its government, honoring its traditions, and all the while remaining within the race and the faith of the Jews" (1968, 13). In each of the seven cities mentioned in the Book of Revelation, Jews lived according to their traditions and as active, recognized participants in the municipal life. Those who converted to Christianity continued to live the same kind of life in the seven churches of the Apocalypse.

9

Urban Life in the Province of Asia

The last two chapters have shown that Christians and Jews in Asia were not isolated from economic and political social structures in the cities where they worshipped and worked. In this chapter I shall consider more systematically some of the features of urban life only alluded to in the previous two chapters. The focus here will be on the organization of cities and the economic and political connections between the cities in Asia and the larger empire. Implicit throughout the chapter is the thesis that political, economic, and social relations within the cities and between the cities and the empire were stable. Those relations provide no basis for assuming crises between empire and province or among classes within the cities of the province. Life in the cities was fairly prosperous and the social structures provided benefits for all; though, as always, the wealthy received more benefits than the poor. In other words, life in the provinces under Domitian was carried on much as it had been and would be for at least another century. Where possible, I provide examples of this life from Ephesus, Smyrna, Pergamum, Thyatira, Sardis, Philadelphia, and Laodicea, the circle of seven cities in the Book of Revelation.

Land Control

Cicero declared that "in the richness of its soil, in the variety of its products, in the extent of its pastures and in the number of its exports it [Asia Minor] surpasses all other lands" (*Leg. Man.* 14). Among the provinces in Asia Minor, Asia was the richest. That province contained four great river valleys flowing from the eastern highlands westward into the Mediterranean: from north to south they are the Caicus, Hermus, Cayster, and Maeander, each separated from the next by a range of mountains. These hills and valleys provided the natural conditions for a rich variety of agricultural resources: grains, wine, olives, fruits and nuts, aromatic flowers for perfumes and ointments, textile plants, forest products, and animal pasture for sheep, goats, and swine. The land also offered lead, copper, iron, salt, and marble, which could be mined and quarried (see Broughton 1938, 607–26).

By the time of Augustus (then called Octavian, c. 30 BCE), a mixture of peoples — some indigenous and some invaders — had left their mark on Asia Minor, and ethnic boundaries were blurred, in part by the way the land was organized and controlled. In the Seleucid period (c. 312–64 BCE) the land was incorporated into large estates, with villagers paying dues to the landlord and the landlord, in turn, paying taxes to the government.[1] Retired soldiers received royal grants of land and settled on them, bringing another new element into Asia; this land was either owned by the king or developed into military establishments (*katoikiai*) with corporate power (cf. Broughton 1938, 632–37). In the period of the Roman Empire, villages could have the same internal organization as cities and were also represented at provincial gatherings, like cities (see Price 1984, 82). More often than not, however, villages with their surrounding land were incorporated into city territories where the land was either owned publicly by the city itself or privately by citizens who lived in the city (Broughton 1938, 637–40).

Temples also owned village lands from before the time of the Seleucids through the Roman period (Broughton 1951, 243). Some temples gained great economic power. For example, the temple of Artemis at Ephesus possessed quarries, pastures, salt pans, and fisheries and extensive estates in the Cayster Valley. Wealthy temples also served as banks, acquiring land through mortgages and business investments.[2]

From the days of the republic, Roman and Italian businessmen were involved in land investments in Asia (Broughton 1938, 543–54). By the time of the Flavians and Antonines, however, these businessmen had become fully provincialized, and there was little or no distinction between them and wealthy provincials. Finally, the imperial family held private land in Asia. For example, Livia, Augustus' wife, held land at Thyatira (*IG Rom.* 4.1213, 1204; Broughton 1938, 648), and Domitian had an imperial estate around Pisidian Antioch (see Pleket 1961, 308). Vespasian probably reorganized all the Julio-Claudian estates as these imperial possessions grew larger and their administration became more complex (see Broughton 1938, 652). Large estates required supervisors and managers, who attained significant status, especially if they worked for an important senator or on an imperial estate. Thus, by the imperial period, land in the province of Asia was incorporated into military settlements; temple estates; public city properties; and private estates of provincials, Romans, and the imperial family. Possession of land assured income from the sale of its natural resources, and it was also seen as a safe long-term investment.[3]

Urban Development

The other significant factor in the prosperity of Asia Minor was the growth and maintenance of the cities. Pompey had drawn up charters for the cities of Asia prior to the empire; these charters continued to guide the cities throughout the second century CE (see Broughton 1938, 533). Cities were the places of regional and local trade (aside from village fairs), and they were the places where most industry developed, intensive labor was needed, and the professions recognized. As we have seen, cities also provided a favorable habitat for nascent Christianity.

Distribution and Interconnection of Cities

At the beginning of the empire the coast of the province of Asia was completely occupied by city territories; in the interior of the province, cities were to be found primarily on the major roadways, with fewer cities toward the central plateau.[4] At the time of Octavian's visit to Samos after the battle of Actium (31 BCE), all seven cities mentioned in Revelation were in existence (see Broughton's list of cities, 1938, 700–701). Indeed, they were among the largest cities of the province. Evidence of urban population is scant, and estimates are based upon extrapolations from various sources, but Broughton estimates that in the middle of the second century Pergamum had a population of about 200 thousand, while Smyrna and Ephesus were probably larger (1938, 812–13). Next to Smyrna and Ephesus, Sardis and Thyatira were among the larger cities in the Lydian region. And in the Phrygian region, Laodicea and Apameia were "fairly large" (Broughton 1938, 815). Of the seven cities of Revelation, only Philadelphia is not to be counted among the major cities of Asia.

The cities of Asia were connected by a network of roadways, which emperors improved, remade, and expanded.[5] A web of such roadways connected and surrounded the seven cities of the Apocalypse. Out of Ephesus a northwest roadway went to Smyrna and then on north through Cyme, Elaia, and Pergamum (continuing to Cyzicus on the Propontis). Out of Pergamum another road went east to Germe and then south to Thyatira where it junctioned with another artery coming down from the north. That road connected Thyatira with Sardis. Another road angled southwest out of Thyatira to Magnesia ad Sipylum. Sardis was a major intersection of several roads: one came from the west with a branch to Magnesia ad Sipylum and another branch to Smyrna; one wound southwestward to Ephesus, one east to Acmonia, and southeast to Philadelphia. At Philadelphia one road went east and north, following the Hippourios River joining the Sardis-Acmonia roadway, and another went southeast to Laodicea and the Lycus Valley. At Laodicea that road junctioned with a major east-west road that went eastward through Asia Minor and westward through Tralles and Magnesia on the Maeander up to Ephesus, which completed the loop of the seven cities mentioned in the Apocalypse. Thus, all seven cities were strategically located on roadways connecting the hundreds of cities of Asia Minor. Under the Flavians, road building extended into the interior of Asia Minor, with greatest attention given to the central and eastern sections so that the eastern frontier could be supported militarily.[6]

Roads were used by travelers and by small caravans transporting textiles, precious metals, ointments, and other valuables small in bulk. Movement of heavy materials was very expensive and rarely done. There is little evidence for inns along the roads. Individuals arranged to stay in rooms provided by people in their religious or professional networks. Paul, for example, stayed with Christians. Security for the roads was left to the local authorities and, according to Broughton, was successful for the first two centuries of the empire (1938, 866–68). Roads were also important for the imperial post (*cursus publicus*) organized by Augustus (cf. Domitian's order to his Syrian procurator, discussed below).

Among the many cities of Asia—perhaps as many as five hundred—there was lively competition for honors and status.[7] Ephesus was capital, seat of the governor (proconsul) and center for both the senatorial and imperial treasuries (see Broughton 1938, 708). Augustus granted also to Ephesus (and Nicaea) the right to dedicate sacred precincts to Rome and Caesar, for, as Dio Cassius says, "these cities had at that time attained the chief place in Asia and Bithynia respectively" (51.20.6, see Broughton 1938, 709).[8] Another kind of urban honor involved the place where Roman officials would hold court in their judicial itinerary.[9] The province was divided into assize districts, areas around a designated city. These assize districts changed over time and evidence for them is not extensive, but probably in the time of Augustus the assize cities included Ephesus, Smyrna, Pergamum, and Sardis as well as Alabanda, Adramytteum, Apamea, Cibyra, Synnada, and Philomelium (see Habicht 1975, 70). Sometime no later than the second century, Philadelphia became an assize center separate from Sardis, and in the third century Caracalla granted Thyatira assize status separate from Pergamum (see Burton 1975, 94).[10] A final status symbol for a city was to have a shrine or temple with the right of asylum for evildoers. Among shrines in the cities of Asia with that status Tacitus lists Artemis of Ephesus, Asclepius of Pergamum, Aphrodite Stratonicis of Smyrna, Artemis of Sardis, and nine others.[11] Issuance of coinage also provides evidence for the importance of cities. Most of the cities along the coast began to issue coins at the beginning of Augustus' reign. Coins from cities in the Maeander, Cayster, and Hermus valleys suggest that at the beginning of the empire trade routes into Phrygia were again in operation (Broughton 1938, 713). Finally, building programs, gifts, and establishment of foundations by private families indicate the wealth and status of cities in Asia.[12] From this brief review, it is clear that the cities mentioned in Revelation are major cities in the province of Asia, linked with each other and with the larger empire.

The Organization of Cities

City organization in Asia followed a pattern found elsewhere throughout the eastern part of the Empire. Roman powers did not try to change the basic democratic structure of the old Greek city with its council, administrative magistrates, and assembly of citizens (see A. Jones 1940, 170–91). Nonetheless, several changes were made within that structure. Furthermore, although the empire remained under Vespasian more or less "a congeries of self-governing communities" (Magie 1950, 639), the self-administration of those communities was overseen by Roman authorities.

Through changes in requirements for holding office, the cities of Asia under Rome came to be governed more like the senate in Rome than like a Greek democracy (A. Jones 1940, 171). Members of the city council had to own property (A. Jones 1940, 338). More important, the council became less and less responsible to the assembly of citizens. Membership in the council became permanent rather than subject to election, and gradually the council members came to control its membership. For the most part, the council consisted of past and present magistrates, and the council determined the list of names to be submitted to the people (where this was done) for magistracies.[13] With such power the council "inevitably became the governing body of the city" (A. Jones 1940, 171). The assembly simply confirmed what the council recommended.

The administration of the city was often headed by one official whose name was used to date public documents. *Eponymous* ("named after") was normally placed before his title.[14] An executive board of magistrates—three to five in number—watched over city government, managing the income and expenses of the city, enforcing enactments of the council and the assembly, and acting as judiciary to try those who violated such enactments (see Magie 1950, 644).

At least seven individual officials were involved in running city government:[15]

1. A clerk (*grammateus*) presented resolutions to the assembly, oversaw erection of statues decreed by council or assembly, and tended to many other details of civic administration, including endowments and distribution of money (Magie 1950, 645). In Acts of the Apostles 19:35 the clerk of Ephesus speaks to the people when Paul comes into conflict with Demetrius the silversmith.

2. A superintendent of the markets (*agoranomos*) oversaw the maintenance, and even the construction, of buildings in the marketplace, established prices, determined accuracy of weights, and saw to the supply of food and oil. In times of recession or shortages, this office could be very expensive, for the *agoranomos* would personally have to provide for the city's needs (Magie 1950, 645). For the overseeing of grain many cities had, in addition to an *agoranomos*, a provider of grain (*sitones*), who had to make grain available to citizens at a reasonable price, and sometimes a special administrator for the supply of oil to the city, so important for cooking, heating, and the baths (Magie 1950, 646).

3. A superintendent of streets and sanitation (*astynomos*) was responsible for such things as checking the stability and placement of house walls lest they endanger or encroach upon public thoroughfares, keeping drains and fountains in good condition, and providing the city with a water supply (though a special commissioner might be set up for that, see Magie 1950, 646).

4. A treasurer (*tamias*) paid out money as directed by the council, and at Pergamum had responsibility for public slaves.

5. An officer of the peace (*eirenarches*) kept law and order, arrested bandits, and testified at trials in connection with arrests. That officers of the peace, not sheriffs, were involved in arresting Christians is probably evidence that Christians—at least prominent Christians—lived in the cities and not in rural areas.[16]

6. A sheriff (*parafylax*) had responsibilities for the larger territory around the city, protecting its land and its boundaries (Magie 1950, 647–48).[17]

7. A city attorney (*ekdikos*) acted on behalf of the city in both internal and external affairs (e.g., in a legal issue with the Roman government, see Magie 1950, 648–49).

The administering of city government under the empire developed into a patronage system. Only the wealthy could afford to hold office, for not only were fees paid in order to be "elected," but also the officer had to spend his or her own money to execute the responsibilities of the office.[18] As we have seen, wealthy Jews in the cities held these offices. In return, recipients granted honor and prestige to their patrons. During the time of the Flavians, when opportunities in the imperial system were available for wealthy provincials, they happily took these civic magistracies as stepping stones to imperial service, equestrian status, and, eventually, the prestige of being a senator at Rome. In the later second and third centuries, as costs of office went up and interest in local politics weakened, the wealthy were not so eager to support their cities (see A. Jones 1940, 182–91).[19]

The wealthy were required not only to support the various executive and legislative magistracies but also to donate liturgies (services) like helping in transporting grain, paying the imperial tax, erecting a building, or paving a street (see Broughton 1938, 802–3). The distinction between a magistracy and a liturgy was not absolute or universal. In general, however, a magistracy carried greater weight than a liturgy: it entitled one to membership in the council, whereas a liturgy did not; those who became Roman senators were exempt from paying liturgies but had to serve magistracies; a man found guilty of an offense might serve a liturgy but not a magistracy; only citizens could serve magistracies, while any resident could be required to pay a liturgy. Liturgies were sometimes also donated by a god or goddess, the funds being then taken from the treasury of that cult.

Two of the most important liturgies were the running of the gymnasium (*gymnasiarchia*) and the overseeing of contests at festivals (*agonothetes*) (see Magie 1950, 652). The former had responsibility for one or more gymnasium in the city — for the building itself and for expenses in running the gymnasium. Since the gymnasium was "the chief centre of the social life of the community," its upkeep could be very costly, for it included not only baths and exercise rooms but also lecture halls and sometimes a library (Broughton 1938, 806–7, cf. Magie 1950, 62, A. Jones 1940, 220–26). Vespasian gave special privileges to gymnastic trainers and athletic associations, so that they did not have to pay taxes nor provide lodging for officials and troups (see Magie 1950, 572). Large cities had more than one gymnasium, with their clientele divided by age groups (boys, ephebes, young men, elders) or, less often, by sex. In most cities of Asia, young men's associations met at their gymnasium for social life, engaging in contests and studies, and carrying on business related to the club; so, too, elders' associations met at their own gymnasiums for much the same reasons. Both of these associations had recognized corporate status, as they could receive endowments, had their own officials (patterned after the city's), and could prosecute as a corporate entity.[20] The overseer of contests (*agonothete*) was responsible for the contests at festivals — enrolling contestants in musical, dramatic, and athletic events; organizing the events; and awarding prizes (see Magie 1950, 653; A. Jones 1940, 233–34). He was responsible for making sure that the contests were held in appropriate splendor, even if he had to pay for them himself. That officer worked closely with the one who headed up the festival, normally a *panegyriarch*. Providing contests was very costly, so sometimes the local priest to the emperor (always one of the wealthiest of the city) took the office (see Magie 1950, 653).

City Finances

Cities in the province of Asia were crucial to the economic well-being of the province as a whole. How they gained and spent money thus contributes to an understanding of life in the province. As we have seen, cities owned land — sometimes a considerable amount — around the municipality itself. Revenue was derived from taxes on that land if owned privately by a citizen, from rents and income if it was public land. Some income was also received from sale or rental of concessions and space for marketing goods, market taxes, charges for public utilities, and fines and from harbor dues if the city was located on the coast. Direct taxes on citizens were discouraged as burdensome, indirect taxes being preferred. Public baths and public latrines were free in the towns of Asia. The most important source of income was no doubt the payment made to the city by magistrates and priests when they assumed office. The wealthy often gave beyond the required amount, establishing gifts and foundations, providing gymnasiums, religious festivals, public buildings, city banquets, and support for the food supply. These gifts were motivated by civic pride; by a desire to be noticed by the imperial system; and by a need for "social insurance" lest the poor riot in protest against the irresponsibility of the wealthy.[21]

On the debit side of the ledger there were few personnel expenditures, for no salaries were paid to the high magistrates — as we have seen, they themselves had to pay — and often public slaves did menial tasks. Only a few lesser officials received any salary or fees (see Broughton 1938, 804). Public works were by far the largest expense to a city — the building of aqueducts, gymnasiums, markets, streets, temples, and municipal buildings. Even though these were frequently supported by imperial aid and private benefactors, they remained a significant expenditure for the cities. Cities also provided festivals, popular amusements, banquets, food distributions, grain supply, and the games and celebrations connected with the imperial cult, which demanded large sums. The gymnasiums were especially expensive. Another expense involved sending embassies to Rome to make appeals to the emperor about issues involving the city. The solvency of the cities depended on the generosity of wealthy families and on general prosperity. Broughton states that if such generosity ceased or foundations did not yield income, "the margin of comfort and well-being would be wiped out" (1938, 809).

Industry and Labor

Production of textiles was the most important industry in Asia Minor, with Lydian embroideries, red dyes from Sardis, wool traded at Thyatira and worked at Laodicea and Colossae, and hemp at Ephesus. Pergamum produced parchment; there were tanners, leatherworkers, and coppersmiths at Thyatira; silversmiths and goldsmiths at Smyrna (Broughton 1938, 817–30). As the interior of Asia Minor was opened up, industrial activity shifted from the islands in the Aegean and the coastal cities to inland cities such as Thyatira, Laodicea, and Aphrodisias (see Broughton 1938, 839). Except for textiles (and shipbuilding in the north), industrial production served local and regional needs. Because of difficulties in transporting bulky items, workmen with skills were transported, rather than their products (see Broughton 1938, 868). Skilled workers such as the Christian apostle Paul were thus natural itinerants. Industrial development took the shape of individual craftsmen working in small shops for local consumers (Broughton 1938, 839). Every village had a marketplace where fairs took place as well as regular marketing. Fairs brought not only the buying and selling of goods but also entertainment, temple prostitutes, soldiers, and religious celebrations (Broughton 1938, 870–71; MacMullen 1970). Movement of commerce is hard to trace, but the presence of coins from other cities and epitaphs of visitors from another city can sometimes provide clues about the movement of goods. For the most part, traders moved around within Asia Minor, which was a self-sufficient area (Broughton 1938, 876).

The various industries in Asia provided ample opportunities for labor; slavery was insignificant in both agriculture and industries. Farming was done by owners of small plots of land or by tenants who rented the land. In busy seasons of planting and harvesting, laborers could pay off their debts through agricultural work (Broughton 1938, 691–92). Individual artisans, members of the craft guilds, and their known helpers were almost all free men, not slaves or freedmen. This is true, so far as we know, of Sardis, Philadelphia, Thyatira, Ephesus, and Smyrna. Perhaps slaves or freedmen would tend not to appear in inscriptions, and it may be that they had a place as assistants to guild members, but there is no indication that slaves played an important role in the industries of Asia. The occurrence of strikes is a further indication that the labor market was free. Of course, there were slaves. An inscription from Thyatira honors a slave dealer there (*IG Rom.* 4.1257), and according to Galen there were thousands in Pergamum. Another inscription from Thyatira requests relief from the 5-percent tax on manumission of slaves (*IG Rom.* 4.1236).[22] Slaves must have been used primarily as domestics, personal agents, clerks and secretaries, civil servants, and menial laborers. Slaves also worked the mines and quarries. So Broughton concludes, "Industrial production in Asia Minor depended mainly upon the labor of free men, a notable contrast with conditions in Italy of the same period" (1938, 841).[23]

Those involved in industry and trading formed guilds according to their craft.[24] At Pergamum there were dyers, at Smyrna silversmiths and goldsmiths, at Ephesus bakers, at Thyatira wool dealers, at Sardis builders, at Philadelphia woolworkers, at Laodicea fullers and dyers. Apparently these associations emerged as significant organizations only in the imperial period, the great majority in the second and third centuries CE. Little is known about the organization of these guilds. There was a presiding officer, other officers such as a treasurer, sometimes a steering committee, and supporting patrons. The guilds could receive gifts and trusts, and they could be fined. There is no evidence that the guilds trained workmen, as in an apprentice system. Sometimes the workers and/or their guilds would strike, hold slowdowns, or leave one job for a more lucrative one. What Broughton observes about a bakers' strike was true of strikes in general: "Striking was a recognized means of exerting pressure and not punishable unless accompanied by seditious action" (1938, 848). At Miletus a group of workmen consulted the oracle about whether they should remain on one job or go to another. At Pergamum builders were delaying completion of work contracted; they were causing such difficulties that the proconsul interfered. From Sardis (459 CE) there is an agreement between the general contractors of a building and the union of artisans doing piecework on the building. The union makes a contract binding its members to work on certain terms. All of these documents provide evidence that workers were free agents who could band together in unions to negotiate with their employers (see Broughton 1938, 846–49).

The professions included architects, physicians, teachers, lawyers, actors, and performers. Teachers were freed from certain municipal burdens and, in a few cases, received a salary from the city; but public teaching had little status, and true wealth and fame derived from private teaching of wealthy pupils. At the shrine of Asclepius at Pergamum a kind of university charter has been discovered supporting medicine and rhetoric. Sophists of note taught and lectured at Smyrna, Ephesus, and Pergamum. At Smyrna there was a school of medicine founded sometime in the first century BCE; the Asclepion of Pergamum continued in importance, and at Ephesus there was some kind of association of physicians. Medical science progressed there by means of contests in writing medical treatises and solving medical problems. Vespasian supported the work of physicians, and educators more generally, when he exempted them from any obligation to provide lodging for traveling officials or troops and from payment of taxes (see Magie 1950, 572). Law was among the professions in the cities of Asia, but it required Roman training and developed late. Entertainers and performers were held in high honor. Successful performers were given citizenship and honorary council memberships.[25] Guilds of these entertainers and performers also had wealth through trust funds bequeathed to them.[26]

Wealth and Class Conflict

The great wealth of Asia was, of course, not divided equally among all its inhabitants. The Roman social historian Rostovtzeff classifies the urban population according to wealth. From the wealthy bourgeoisie, upon whose grants, foundations, and liturgies the cities depended for their well-being, came the governing officers of the city; and they were the class that negotiated with Rome. Below them came what Rostovtzeff calls the petty bourgeoisie: "the shopowners, the retail-traders, the money-changers, the artisans, the representatives of liberal professions, such as teachers, doctors, and the like. . . . the salaried clerks of the government and the minor municipal officers." Below the petty bourgeoisie were "the city proletariate, the free wage-earners and the slaves employed . . . in the households." These classes were not static, and there was in this time of prosperity an upward movement: "We have no means of drawing a line between the higher and the lower *bourgeoisie*, as the former was certainly recruited from the latter" (Rostovtzeff 1957, 190).

Since Rostovtzeff is usually appealed to by those who assume that economic crisis was one of the crises occasioning the Book of Revelation, his approach to social and economic history deserves attention. In his analysis of Asian society Rostovtzeff tends to assume a model of social conflict: class conflict between the poor and the wealthy in the cities of Asia and political conflict between Asian provincials (rich and poor) and Rome (1957, 117).[27] Rostovtzeff even combines the two conflicts: "The social movement in the cities, especially among the proletarians, necessarily assumed an anti-Roman aspect, since the Romans as a rule supported the governing classes, the manifest oppressors of the proletariate" (1957, 117). Elsewhere, regarding Bithynian cities, Rostovtzeff writes, "Attempts at a social revolution and a bitter struggle against the governors were the main features in their life" (1957, 587). In fact, evidence for these conflicts comes only from a few literary sources such as Dio Chrysostom and Plutarch (see Rostovtzeff 1957, 586–87). In contrast to Rostovtzeff, MacMullen notes that "only one doubtful instance is known of poverty and anti-Romanism conjoined, to be set against a mountain of indirect proof of the popularity of the empire among the lower classes" (1966, 189).

Two specific cautions need to be made in using Rostovtzeff's analyses for interpreting the Book of Revelation: (1) serious social distinctions and class conflicts emerged after the end of the first century CE;[28] (2) slaves were an insignificant factor in the economics of agriculture and industry (as Broughton shows), so their presence cannot be used to assume — as Rostovtzeff assumes — that wages were "hardly above the minimum required for bare subsistence" (1957, 190). Broughton offers a more balanced analysis of the economic situation in Asia: "We cannot exclude the possibility that the lower classes in the towns had benefited little [from the prosperity since Augustus], that the prosperous facade concealed much unrest, and that a great deal of the generosity of the wealthy in the towns is to be classed as social insurance. Against this view, however, we have to put the insufficiency of our evidence, which may largely deal with individual and sporadic occurrences of unrest, the genuine increase in industrial and commercial activity, and the wide extension and adoption of city organization itself as Asia Minor within the same period. These processes could hardly have progressed so far without allowing some of the benefits of the imperial peace to filter down to the lower classes" (1938, 812). Greater regularization and imperial control of provincial politics had advantages for the lives of many provincials, especially those who were less prosperous and not active in the political arena. As Oliver points out, the imperial government was concerned to protect the cities from local aristocracy, especially if it abused its connections with Rome (1953, 953–58). Augustus himself issued an edict that Roman citizens — with few exceptions — had to take responsibility for services (liturgies) in their cities. The Roman court was open to the appeal of cities experiencing aristocratic oppression or irresponsibility on the part of the wealthy.[29] Oliver thus concludes by agreeing with Aelius Aristides "that Roman rule satisfies both rich and poor" (Oliver 1953, 958).

A key issue centers on how to interpret urban unrest. Most of the evidence suggests that protests, strikes, and riots brought the results desired by the protesters. Chrysostom recommends that in response to the riots at Tarsus the lower classes be admitted to full citizenship without requiring the normal fee.[30] Dio Chrysostom and his family were threatened in their home town of Prusa with stones and firebrands, apparently because they, with their great wealth, were not adequately subsidizing the grain supply, and as a result bread was too expensive. The "poor" demanded that instead of investments in porticoes and workshops to be rented out to tenants, Dio Chrysostom should give a liturgy in support of the grain market (*Or*. 46; Magie 1950, 581). In response, Dio did agree to assume the burden of the grain supply.[31] As we have seen, a city's prosperity depended upon the liturgies, or services, of wealthy citizens. Without that service the cities could not survive. Urban unrest does not necessarily point to "a reservoir of ill-will among the lower classes ready to be converted into open violence" (Macro 1980, 691). It may simply indicate that when the wealthy were not fulfilling their obligations to the city, the poor had an effective social tool to right the imbalance. Correctives did not have to await action from above; they could be made by acts from below.

Networks between Asia and Rome

Strictly speaking, Asia was a senatorial province governed by a proconsul appointed by the senate. That arrangement should not be construed, however, to mean that the network of relations was primarily between Asia and the Roman senate; for the emperor also related to the province in several different ways. Indeed, Fergus Millar (1966) has shown that in practice there is no clear distinction between senatorial and imperial provinces. Both emperor and senate could make regulations affecting all the provinces, and communities in senatorial provinces might send representatives to either the senate or the emperor. Letters, queries, and directions also flowed between emperor and Asian proconsul in the Flavian period.[32] In connection with his discussion of provincial control, Millar makes a general point that is relevant for our understanding of the network of relationships between Rome and Asia: "What we see is not an arrangement of compartments, of administrative hierarchies, but an array of institutions, communities and persons, the relations between which depended on political and diplomatic choices which could be made by any of the parties" (Millar 1966, 166). For an established province such as Asia, queries and directives took different turns as they passed through city fathers, provincial assembly, procuratorial and proconsular offices, senators, and emperor. Within this network, communication could be initiated at any point.[33]

Prosperity and Integration under the Flavians

Octavian (Augustus), the first princeps of the empire, brought peace and stability to it, including the province of Asia. He assured economic recovery in Asia by canceling all public debt and by minting a considerable number of coins at Ephesus and Pergamum. Asian prosperity continued through the era of the Flavians (Vespasian, Titus, and Domitian) (see Magie 1950, 566).[34] Under Vespasian the principate stabilized once again after the unrest following the death of Nero. The year of conflict after Nero's death—in which Galba, Otho, and Vitellius one after another laid claim to the throne—left the Roman treasury bankrupt by the time Vespasian came to power. In contrast to the period before Augustus came to power, however, the conflict of 69 was fought in the West and did not directly affect the provinces of Asia Minor. Vespasian, acclaimed emperor while in Alexandria (Egypt), sailed toward Rome along the coast of Asia Minor and received oaths of allegiance along the way (see Magie 1950, 566). The recovery under Augustus and his dynasty continued under the Flavians, Trajan, and Hadrian, as the cities of Asia flourished both economically and in the status and privileges they held.[35]

In the Flavian period, provinces—even senatorial ones—came more and more under the control of the emperor. As regards Asia, Vespasian probably appointed Titus Clodius Eprius Marcellus proconsul there for three successive years. As Magie indicates, his appointment in 70–73 would not have been popular with the senate, but he was an able and wealthy man on whom Vespasian depended (1950, 569).[36] Domitian showed the same forthrightness when he appointed Gaius Minicius Italus as his imperial procurator in Asia, who acted as governor, after Gaius Vettulenus Civica Cerialis, then proconsul of Asia, was put to death for conspiracy (c. 87–88) (see B. Jones 1979, 25–26; Dessau, *ILS* 1374). Apparently that situation was one that "necessitated vigorous action"; and even though it may have been viewed as "high-handed" by the senators, it "had no real significance" (Magie 1950, 578). Brian Jones says that over a two-year period "Asia was governed by the patrician C. Vettulenus Civica Cerialis, the imperial procurator C. Minicius Italus and the Italian senator L. Mestrius Florus—a procedure that was no doubt regarded with marked distaste by the traditionalists" (1979, 28).[37] This assessment is most likely correct, since the proconsulship of Asia or Africa was traditionally viewed as the crowning achievement in a senatorial career. Whatever may have been the attitude of certain senators in Rome, however, the Flavian emperors saw to it that excellent governors were appointed to the eastern provinces. Domitian promoted to higher office men whom Vespasian had earlier marked for promotion (see Magie 1950, 578).[38]

Under the Flavians another trend in appointments may also have been unpopular with the senate, namely, the increased reliance on eastern provincials. Brian Jones points out that "Domitian was the first emperor to appoint easterners to imperial praetorian provinces" (1979, 2), but that probably followed Vespasian's pattern of promoting easterners.[39] Broughton attributes the increase in the number of private estates during the Flavians and Antonines in part to the "rise of native families to imperial prominence" (Broughton 1938, 666, 745). As we have seen, opportunities in the imperial service stimulated wealthy provincials to serve their cities in order to gain recognition by the emperor and to be promoted into the imperial ranks.

Vespasian and his sons also systematized and organized the provinces, especially with respect to finances. When Vespasian came to power, the treasury was bankrupt and he needed a maximum amount of revenues.[40] He revoked several grants of freedom and immunity from tribute made by Nero, and he reduced the number of

exemptions allowed (Broughton 1938, 740; A. Jones 1940, 129).[41] In continuity
with policies as early as Julius Caesar, Vespasian shifted tax gathering from publi-
cans or tax farmers to a government agency headed up by a head procurator at
Rome and a procurator with a staff of the emperor's freedmen in each of the
provinces responsible to Rome (see Magie 1950, 567–68). Probably the procurator
of Asia was responsible for collecting the *fiscus Asiaticus*, a treasury of the emper-
or, first attested in the time of Domitian.[42] Apparently the procurator who oversaw
the marble quarries around Synnada also was responsible to the head procurator at
Rome (Magie 1950, 568). This development of the office of procurator also probably reflects Vespasian's reorganization and expansion of imperial estates in Asia
(see Broughton 1938, 652–54; Magie 1950, 1425–26). The expansion of this imperial
office in a senatorial province further illustrates the complexity of the network of
relations between Rome and Asia.[43]

The Flavians were also concerned about unnecessary expenditures by cities (see
A. Jones 1940, 135). As early as Vespasian's reign, a city had to obtain permission
from the governor to increase taxes. Sending delegations to Rome was also an
enormous expense born by cities. In order to control those costs, Vespasian limited
delegations to the emperor to three members (A. Jones 1940, 135). Because of city
rivalry, cities would sometimes spend beyond their means, especially on buildings,
and get into financial difficulty. To prevent such foolishness, Domitian created the
office of *curator civitatis* to help cities keep solvent and not spend beyond their
means.[44] These auditors watched over investments, communal funds, public lands,
and public buildings. They had veto control over virtually all aspects of a city's
finances (see Magie 1950, 598). Under Trajan these offices developed extensively
(Broughton 1938, 744).

Such regularization and organization of the provinces and their cities took away
from civic and provincial autonomy, but those measures helped make the cities of
Asia a vital part of the empire. At this time, "making it" in the world required
engagement with the empire and entering into its bureaucracy. In the long run, local
politics became less interesting to the wealthy because in that sphere there was little
opportunity for initiative and significant contribution. The major decisions were
made by the Roman administration.

The Imperial Cult

The imperial cult constituted a part of the network between Rome and Asia. It was
an equally important means for representing the emperor to those in the province
and those in the province to the emperor. Emissaries to the emperor were often
imperial priests in Asia, and requests for privileges were associated with offers of
cult (see Price 1984, 243; Millar 1977, 365). In the provinces rituals, images, temples, and shrines associated with the imperial cult contributed significantly, if not
decisively, to the people's understanding of their relationship to the emperor.
Cult to the emperor was established through a reciprocal process between Rome
and provincials. Honor involved both sides: it was an honor for the city to have

a temple or some designation of imperial cultic status, and it was an honor for the emperor. Both emperor and senate accepted the existence of the imperial cult, and Rome could exert pressure to establish and maintain these cults, as did the provincials.[45]

S. R. F. Price considers the network of relations between Rome and Asia as a network of power that found expression in religion as well as in political diplomacy and decrees. Religious expressions of power are no more a gloss on politics than is politics a gloss on religion (see Price 1984, 235). Social structures such as rituals of sacrifice, political negotiations, taxes, or civic claims are equally channels of power. All these relations together make up the network. No relation is merely symbolic or simply expressive of another. Religious expressions are really about religion, however much they may also express political obligations.[46]

Distribution of the Imperial Cult

In many aspects Augustus' reign was the high point of activity for the imperial cult in Asia. Language praising him is lofty and is similar to that offered to the gods, and a number of temples and sanctuaries were built during and immediately following his reign.[47] Priests of Augustus were found in more than thirty cities of Asia Minor, and more of his relatives received cult than was the case with any other emperor.[48] Although Tiberius has the reputation of having rejected cult in his honor, he had more than ten priests in Asia Minor (Price 1984, 58; Magie 1950, 1360). Under Claudius there arose, alongside priests to the emperor himself, priests of the Augusti or Sebastoi in general, who had responsibility for the cult of former and present emperors (see Magie 1950, 544; Price 1984, 58). Imperial cult continued under the Flavians and Antonines.[49] Domitian was no more and no less insistent on divine prerogatives than other emperors and, so far as one can tell, he did not change in any fundamental way the social expressions of imperial religion (see chap. 6).[50]

Christians living in the seven cities mentioned in the Book of Revelation probably found the imperial cult an objectionable social, religious institution, but it was just as objectionable under Claudius as under Domitian. Change in emperors throughout the first century did not affect the presence of the cult in Asia. The cult was most heavily represented in Lydia, southern Phrygia, Caria, Lycia, and then along the southern coast of Asia Minor into Cilicia (Price 1984, xxii–xxv). In Price's evidence for the geographical distribution of the imperial cult, all seven cities of Revelation are represented. Five of the seven cities had imperial altars (all but Philadelphia and Laodicea), six had imperial temples (all but Thyatira), and five had imperial priests (all but Philadelphia and Laodicea).[51]

Price, who has written the definitive work on the imperial cult in Asia Minor, also traces out the distribution of the cult in social and cultural terms. He concludes that it is found only in organized urban settings (cities or villages), not in rural areas, which lacked communal organization (1984, 79–86); and it expresses a "Greek idiom" rather than indigenous, non-Greek customs or Roman expressions of piety (pp. 87–91).[52] In sum, "only when cults acquired a communal organization and borrowed sufficient traits from the cults of the dominant Greek culture did they give a place to the emperor" (p. 98).[53] In contrast, rural areas in which neither the Greek language nor the Greek pantheon was present show no signs of the imperial cult (pp. 92–98).[54] The churches of the Book of Revelation were located geographically, organizationally, and culturally where the imperial cult was most heavily distributed.

The Imperial Cult and the Provincial Assembly

A provincial assembly—called a *koinon* (commonalty)—had existed in Asia before the empire. For example, in the republican period it had protested to the senate the extreme demands of tax farmers; and in 42–41 BCE Marc Antony confirmed through that body certain privileges to athletes (see Millar 1977, 385–86, 456). During the imperial period that body became part of the network between the emperor and the cities of Asia. In 12 BCE the provincial assembly inscribed in the temple at Pergamum Augustus' edict making clear the rights of Jews in Asia (see Millar 1977, 391; Joseph. *AJ* 16.6.2); and the provincial assembly sent the orator Scopelianus of Smyrna to make an appeal to Domitian to rescind the edict regarding the planting of vines (see Rev. 6:6). Sometimes the assembly would even bring charges against a proconsul of the province who abused his office (see Magie 1950, 451).

More and more, however, the interaction between emperor and the provincial assembly in Asia took forms involving the imperial cult. In 29 BCE Augustus granted the request of the provincial assembly of Asia to build a temple to him and Roma at Pergamum—not a city temple, but a temple built and administered under the auspices of the provincial assembly.[55] About the same time, the assembly offered "a crown 'for the person who devised the greatest honours for the god' (sc. Augustus)" (Price 1984, 54). Eventually, with the cooperation of the Roman governor, the assembly recognized the greatest honor as being to begin the year on Augustus' birthday. The decree stated, "We could justly hold it [Augustus' birthday] to be equivalent to the beginning of all things. . . . And since no one could receive more auspicious beginnings for the common and individual good from any other day than this day which has been fortunate for all, . . . therefore it seems proper to me [the Roman governor] that the birthday of the most divine Caesar shall serve as the same New Year's Day for all citizens."[56]

The assembly of over a hundred delegates met annually to conduct business, find ways to represent their interests in Rome, and carry out activities of the imperial cult (see Magie 1950, 448). At first they met in Pergamum, then at Smyrna (c. 29 CE), Ephesus (where a third temple was built), and other cities such as Sardis, Philadelphia, and Laodicea (see Magie 1950, 448, 1295). The prestige of holding the provincial assembly and maintaining an imperial temple flamed the rivalry and competitive spirit among the cities of Asia (see Price 1984, 62–65, 126–32). By the end of the first century, cities advertised with pride their status as *neokoros* (temple warden), that is, their claim to an imperial temple with provincial status (see Price 1984, 67). Sometimes a city would be twice or even thrice *neokoros*, that is, it would have two or three imperial temples with provincial status.[57] The senate and emperor made the final decision about which cities would be granted the title of *neokoros*, so it was a title that made the city visible to the powers of Rome. Furthermore, this is another indication that the imperial cult "fell under the regular processes of Roman administration" (Price 1984, 70).

The annual president of the assembly was called the chief priest (*archiereus*, or perhaps sometimes *Asiarches*; see Magie 1950, 449–50, 459, 531). He was responsible for carrying out the business transacted at the meetings, especially for executing formal communications to city or emperor, as well as for cultic duties. If *Asiarches* supported Paul in the riot at Ephesus, he had some well-placed friends (see Acts 19:31). Other officers included "advocates" (usually to the emperor); a secretary (*grammateus*); and a treasurer who managed the assembly's money and investments and, perhaps, the minting of coins by the provincial assembly. These provincial offices, especially that of the chief priest, were stepping stones into senatorial status at Rome.[58] The son of a provincial priest would typically gain equestrian status, and the priest's grandson, senatorial, and perhaps consular, status (see Bowersock 1973, 182). These offices were thus filled by the provincial elite, the wealthy, and those who were seeking status beyond their provincial realm.

Imperial Festivals

One costly responsibility of the provincial priest was to provide a festival in connection with the imperial cult. Individual cities also held festivals celebrating certain critical times in the life of the emperor (see Magie 1950, 470). In other words, imperial festivals were an important expression of the cult in urban provincial life. Sometimes the emperor shared cult in a festival for a local deity. For example, during a festival for Artemis at Ephesus, images of the emperor and some of his family processed, along with images of Artemis, from the temple of Artemis to a theater (see Price 1984, 103–4). The major imperial festivals, however, were held only for the emperor. City festivals involving widespread competition among athletes and musicians were held usually every four years. Annual festivals were also held in many cities, most often on the emperor's birthday (Price 1984, 105; Magie 1950, 448). His birthday was sometimes even celebrated on the first day of each month (see Pleket 1965, 341; *IG Rom.* 4.353). Provincial festivals were held annually at first but later more often, when several cities had the status of *neokoros* (see Price 1984, 104). Imperial festivals might also be held to celebrate the accession of an emperor, success in war, or in honor of other members of his family. Since these festivals lasted more than one day, one can see that several days in a year could be devoted to the "imperial days" when festivities, cultic observances, feasts, and distributions were held in the life of a major city.

These festivals attracted large numbers of visitors from all ranks and professions. Dio Chrysostom mentions governors, orators, prostitutes, craftsmen, and tinkers, among others.[59] Special tax breaks were given to peddlers and craftsmen selling wares (see Broughton 1938, 870). Public buildings such as the gymnasiums would be overcrowded, and sometimes the people became unruly (see Price 1984, 107). No doubt shopkeepers and craftsmen increased business during these festivals, and undoubtedly they brought prestige to the city involved. Nonetheless, supporting the festivals was costly. Provincial festivals were supported by contributions from the whole province. Sometimes the emperor himself donated huge sums of money. Provincial dignitaries and wealthy officeholders, such as the high priest of the provincial assembly, would underwrite the cost of a festival or at least some part of it, such as the expenses of the gymnasiums or distributions to the festival-goers.[60] Trust funds were also set up to help assure the continuation of a festival at a particular city (see Price 1984, 102-3, 109, 112-13).

For these festivals, imperial temples and sanctuaries were wreathed with flowers. Animals were sacrificed at various altars throughout the main locations of the city, for example, the council house, temples of other deities, theaters, the main square, stadiums, and gymnasiums. These political, religious, and public buildings were linked together by processions of dignitaries, garlanded animals being led to slaughter, and bearers of icons and symbols of the emperor. As the procession passed by, householders would sacrifice on small altars outside their homes.[61] The whole city thus had opportunity to join in the celebration.[62] In provincial festivals the procession would be carefully organized. The provincial high priest would march with crown and purple garb, surrounded by young, male incense-bearers (see Price 1984, 129; Dio Chrys. *Or.* 35.10). Representatives from cities and villages of the province would march in the procession (Price 1984, 128). In Asia, delegates from Pergamum, Ephesus, and Smyrna shared the head of the procession as the three major cities in the province (Price 1984, 129). Unfortunately, we do not know whether or how Jews or Christians may have participated in such processions.[63]

Temples, Shrines, Altars, and Statues

Different kinds of buildings, shrines, and sanctuaries provided the physical setting of imperial cult. There were freestanding buildings similar in structure to temples to the gods (not heroes, see Price 1984, 156, 163, 165). In and around them might be statues of emperors, imperial iconography, and cult tables for sacrificial offerings (p. 156). Sometimes in a temple to a god or goddess the emperor was recognized through inscriptions, statues, or a separate shrine (p. 147). When the emperor shared cult in a temple to a deity, the emperor was rarely made equal to the god (p. 149). In the temple, statues of the emperor were subordinated to statues of the deity (p. 147); and through various techniques the emperor was made to appear a participant in worshiping the deity rather than an equal (p. 149).[64] Imperial shrines have also been found in a building housing the city council or headquarters for a group of merchants (p. 134). Such shrines link the city or organization of merchants with the emperor. In a large city such as Ephesus there were everywhere physical reminders of the emperor. That city had imperial temples to Roma and Julius Caesar, to Augustus, to the Augusti (see Magie 1950, 572), to Domitian (see Price 1984, 157), and to Hadrian; shrines affiliated with Augustus and Hadrian in connection with an Artemision; a royal portico with statues of Augustus and Livia; and an Antonine altar (Price 1984, 254-57). In addition there were imperial statues in public buildings, on the streets, on fountains, and on city gates (see Price 1984, 135-36).

Although these physical presences of the emperor were found throughout a city, they were usually placed in prominent positions, either elevated geographically or at the center of civic life (e.g., Price 1984, 137).

Although not all statues of emperors found throughout a city were objects of cult, those in a sacred place or those used in the special time of festival had a ritual significance.[65] In those cultic settings, icons of the emperor could embody and evoke divine attributes as the emperor in person could not have. For example, one type of imperial statue was constructed overtly after statues of the gods.[66] The emperor in person would never have so evoked the thought of the gods.[67]

Ritual sacrifices of incense, wine, and sometimes a bull were made in connection with imperial images (Price 1984, 188). These sacrifices, like most sacrifices related to the emperor, were for the most part made on *behalf* of the image of the emperor, not to it (see Price 1984, 188). So an imperial high priest at Aphrodisias "sacrificed to the ancestral gods, offering prayers himself on behalf of the health, safety and eternal duration of their [the emperors'] rule."[68] And a woman of the same city and office "sacrificed throughout all the years on behalf of the health of the emperors."[69] Small statues, metal busts, or painted portraits were also carried in processions at the time of imperial festivals by special officials (*sebastophoroi*) (see Price 1984, 189–90). Finally, statues also apparently played a part in the imperial mysteries, in which an official (called a *sebastophantes*) revealed an imperial image at certain critical moments during the ceremonies.[70] In a famous inscription from Pergamum regarding the imperial provincial choir (*IG Rom.* 4.353), it is said that the choir sang at several imperial festivals including the Μυστήρια — a temple ritual involving sacrificial incense, candles, lamps, and sermons, as well as hymns.[71] Pleket suggests that the lamps were used to illuminate the image of the emperor in these rites.[72]

Statues of the emperor were also significant apart from their connection with religious cult. Imperial statues served as places of refuge throughout the empire (Price 1984, 192). Slaves, for example, could flee from their masters by taking refuge at a statue. A prospective buyer of a slave might thus be reassured that the slave was "neither a gambler, nor a thief, nor had he ever fled to [Caesar's] statue."[73] Statues also served as a lawful place for paying fines and depositing petitions (Price 1984, 193). Sacrifices might be made to imperial statues by those entering marriage; small statues were set up in private homes; graves were protected by appeal to the imperial presence, or statues were placed in grave sites; and slaves were set free in front of a statue of the emperor (see Price 1984, 119). Statues of the emperor thus had a variety of religious and political-legal significances in the provinces. They were permanent, fixed reminders — or, better, representations — of the emperor and his power.

Christians and the Imperial Cult

As evocations of power imperial statues could also reveal divine portents. When Caesar defeated Pompey at Pharsalus, the goddess Victory at Tralles turned towards the statue of Caesar (see Price 1984, 195). The power and cult of the statue may be at play in the Book of Revelation. For example, in Revelation 13:11–18 the beast from the land mirrors the earlier beast from the sea, somewhat as the emperor's statue mirrors the emperor. The beast from the land also requires that worship and cult be given to the first beast, as the cultic officials in Asia required observances to the emperor. He caused divine portents or signs (Rev. 13:13, cf. 16:14) and required that all people worship the image of the beast. If the worship of the beast and his image in the Book of Revelation is linked to emperor worship, then a fundamental conflict in the book centers upon the true worship of God and his Christ versus the false worship of the emperor and his cult. Price suggests that the "colossal cult statue" of Domitian at Ephesus may specifically have been in John's mind (Price 1984, 197).

Whatever the exact connection between the "image" of the beast in Revelation and imperial images,[74] the importance of the imperial cult for early Christianity should not be inflated (see Price 1984, 15). The greater issue revolves around Christians' relation to adherents of traditional religious cults rather than their relation to the cult of the emperor (Price 1984, 125).[75] For example, as we have seen, sacrifices were made in traditional cults on behalf of the emperor. That was expected and required. But Christians rejected all forms of sacrifice: they did not sacrifice to their God either on behalf of the emperor or for any other reason. For Christians, then, sacrificing itself was at stake, not obeisance to the emperor. Price cites only four instances where Christians were asked to sacrifice to the emperor—two of them occurred when the Christian refused to sacrifice to the gods, the others were either on the emperor's behalf or to his image placed among the statues of the gods (see Price 1984, 221). For the most part, the emperor in the imperial cult was subordinated to the gods, so that the imperial cult could be assimilated to the cult of the gods. For Christians, however, who did not accept the traditional Greek gods and saw them as antithetical to their own religious claims, the imperial cult was rejected as a correlate to the rejection of traditional cults. The forms of traditional Greek religion were central, the imperial cult was secondary to that.

Domitian and the Provinces

Under Domitian the provinces flourished. Domitian, like his brother and father before him, built and maintained roads in Asia Minor, established cities in the interior plateau, and created new offices to oversee municipal administration. More specifically, Domitian's reign brought new privileges, a heightened status, and economic prosperity to the cities of the province of Asia. In Asia the cities' prosperity was reflected in the amount of building going on, in the minting of coinage, and in the private gifts and foundations available to support civic endeavors.[76] Domitian's fifteen-year reign from 81 to 96 also continued a trend of imperial control of provincial affairs, as Domitian continued Vespasian's systematization and organization of the provinces, especially with regard to finances. On the whole, this imperial attention under Domitian resulted in better administration of the provinces.

There is no indication that Domitian displayed "exaggerated pretensions" toward the provinces in either the imperial cult or in imperial edicts. Magie even proposes that Domitian had a reputation for supporting and giving favors to the provinces (1950, 577), which would explain the positive view of Domitian in provincial Jewish and Christian writings. The greater integration of the provinces into the empire during Domitian's reign and the imperial benefits given to them resulted in a greater justice and equity among provincials. Domitian's concern to deal justly and fairly with all provincials no doubt stemmed in part from a concern for the empire as a whole: healthy provincial life was crucial to the empire. But Domitian also seemed to be concerned about the poor and the weak, and demanded honesty and justice from his procurators and provincial governors in dealing with those people.

Two inscriptions nicely illustrate Domitian's attitude toward the provinces. One comes from Pisidian Antioch, circa 92–93, during a time of grain scarcity.[77] At the request of civic leaders in Antioch, a legate of Domitian intervened to institute emergency measures to deal with the high price of grain. According to these measures, all who had grain were required to make it available on the market (except what was needed for seed and family needs).[78] Further, a sale date for that grain was fixed, and a maximum price was allowed. A general principle was also appealed to in the procedures being followed: "It is most unjust that hunger of one's own fellow-citizens should be the basis for profit to anyone" (D. Robinson 1924; Ramsay 1924–25). At the same time those who had grain could charge up to double the normal amount. Thus, wealthy landowners who had grain were allowed to make more money than in normal times, but those who needed to buy grain were also protected.

Domitian's attempt to regulate the planting of vines—enacted about the same time as this scarcity at Antioch—may also be related to the problem of grain shortage that plagued Italy as well as Asia Minor. In order to make more room for the planting of grain, Domitian ordered that at least half of all existing vineyards be plowed up. In response to this regulation, a delegation from the Asian province (probably from the provincial assembly) traveled to Rome to exempt Asia from this edict. The delegation was successful, and Asia was made exempt.[79] Some interpreters of Revelation see these policies of Domitian reflected in Revelation 6:6, "A quart of wheat for a denarius, and three quarts of barley for a denarius; but do not harm oil and wine," which accepts a high price for grain but rejects any harm to oil and wine (see Court 1979, 59–60).

A second inscription comes from Hama, Syria. It is a copy of a letter from Domitian to his procurator, Claudius Athenodorus, apparently with regard to the imperial post:[80]

> From the orders of imperator Domitianus Caesar Augustus, son of Augustus, to Claudius Athenodorus, procurator: Among those special issues needing considerable attention, I know that the privileges of cities received care from divine father Vespasian. Having looked them over carefully, he ordered that the provinces not be burdened with either the contracting of beasts of burden [for the royal post] or the annoyance of lodging travellers. But, nevertheless, this order has not been complied with nor set right. For there remains to this

day an ancient, persistent custom [συνήθης], little by little advancing into law [νόμος], if it is not prevented from becoming more powerful. I command you also to give attention so that no one should receive a beast of burden, if he does not have my permit. For it is most unjust — whether as a favor to some or a request — to grant permits which it is lawful for no one but me to give; now let nothing happen which will break my orders and which will destroy my most sound inclination towards the cities. For it is just to aid the weak provinces which scarcely have enough for necessities. Do not allow anyone to do them violence against my will. And let no one take a guide if that person does not have my permit. For if farmers are taken forcibly off the fields, then the lands will remain untilled. And you will do best, either using your own beasts of burden or leasing them.

Domitian deals here with an abuse of the imperial post and with traveling dignitaries who illegitimately requisitioned beasts of burden, lodging, and guides at the expense of a province. The post itself was a heavy burden on the provinces, and abuses made it even more burdensome.[81] In dealing with this abuse, Domitian enunciates some of the principles he is following. First, he makes clear that this edict continues the policies of his father, Vespasian, toward the care of cities.[82] He also expresses a concern for all the people in the provinces. In language similar to that of the inscription from Pisidian Antioch he affirms, "It is just to aid the weak provinces which scarcely have enough for necessities." Thirdly, he does not side with the traveling dignitaries, the wealthy and influential. They are abusing the system, and he requires them to stop. Domitian is concerned here with justice for the poor. Finally, Domitian also shows concern for the farmer and the land — the lands must not remain untilled.[83]

The provinces prospered under the Flavians, and Domitian in particular seemed to have a special concern that all the provinces and all the classes of people in a province should be protected so that no one class or province oppressed the others. He hindered governors and upper-class provincials from jointly exploiting others in a province. He sought to be fair and just towards all his subjects. Provincial life in Asia undoubtedly contained tensions between wealthy and poor — and perhaps between Rome and the east — but the Emperor Domitian sought to minimize those tensions. Domitian's care for the provinces is conceded even by Suetonius: "At no time were they [provincial governors] more honest or just, whereas after his [Domitian's] time we have seen many of them charged with all manner of offences" (*Dom.* 8.2).[84] Epigraphy supports Suetonius' assessment and belies Pliny's indictment that Domitian made plundering forays into the provinces and in general treated the provinces poorly (*Pan.* 20.4, cf. 70.5–8). In contrast, Domitian checked abuses by the provincial aristocracy and senatorial governors and treated all classes within the provinces with equity (see Pleket 1961, 312).

From this analysis of economic and political life in the cities of Asia there is no indication of political unrest, widespread class conflict, or economic crisis in the cities of that province. The empire — especially under Domitian — was beneficial to rich and poor provincials; and there were checks against extensive abuse of the poorer provincials by the richer ones. There is little evidence to suggest fundamental conflicts either within the economic structure of the province or between the province and Rome. The writer of the Book of Revelation may urge his readers to see conflicts in their urban setting and to think of Roman society as "the enemy," but those conflicts do not reside in Asian social structures. The urban setting in which Christians worshipped and lived was stable and beneficial to all who participated in its social and economic institutions.

IV

The Play:
The Apocalypse
and the Empire

10

The Book of Revelation
in Asian Society

Broadly conceived, the previous nine chapters have been concerned with religious language and society, specifically with the language of the Book of Revelation and the society of the Roman province of Asia. In actuality the language of Revelation is a part of Roman society: language (chaps. 3–5) and society (chaps. 6–9) form the script and stage, respectively, of the "play"—a larger whole that embraces them both. We are now in a position to begin to look at the play.

Christians and Asian Society

In considering the social order, I have analyzed conditions in the reign of Domitian, local conditions in the province of Asia, and the place of Jews and Christians in those local conditions. A coherent order of society in Asia has emerged from an analysis of relevant social and historiographic elements.

A Review

With regard to Domitian, the standard portrait of him in the Roman sources, written after his reign, does not square with literary, epigraphic, numismatic, and prosopographical evidence from the Domitianic period. From sources written in Domitian's reign, it is clear that Domitian did not modify the imperial cult by demanding greater divine honors than his predecessors (or successors). Nor is there evidence from those sources that Domitian turned into a cruel tyrant, suspicious of those around him and fearful of sharing power. Domitian, like all emperors, no doubt had his informers who were zealous against seditions and insurrections. But Domitian promoted deserving senators, tried to mollify the opposition, and ruled with a fairly broad concern for the whole empire. So far as one can tell, Domitian did not prosecute either Jews or Christians with exceptional vigor. In fact there is a provincial tradition that portrays Domitian as a benevolent emperor towards Jews and Christians alike. All sources recognize how effective Domitian was in administering provincial affairs. He checked the excesses of both greedy senators and the provincial aristocracy.

During his rule the province of Asia prospered and that prosperity most likely benefitted all classes of Asian provincials. Jews in Asia were probably among the most integrated and assimilated of Jews living outside Palestine. They identified socially with their urban environment, and they shared cultural traditions and the prosperity of the cities of Asia Minor. At the same time, they remained faithful participants in their Jewish heritage and religion. Finally, Christians in Asia were located socially in different economic classes, with Christian leaders generally of such wealth that they could travel, have fairly large houses, and act as patrons of small Christian congregations and Christian missionaries. Those Christians—as Asian Jews—also shared in the common life of cities and empire. Imperial officials did not seek out Christians to persecute them; in fact, they preferred not to get involved with the sect. But if local residents opposed Christians and reported them through proper channels, they could force them to trial. For the most part, however, Christians lived peacefully with their neighbors in the Roman political order.

In brief, from this social backdrop one cannot assume widespread oppression and persecution of Asian Christians; nor can we assume that Asian Christians lived in an isolated ghetto as separatists from urban, Greco-Roman life. Christians were on occasion brought to trial, as Pliny indicates, but such accusations did not occur very often in Asia at the end of the first century. It is conceivable that the Book of Revelation was written in response to an otherwise unknown crisis in Asian Christianity, in which Christians were being—or about to be—persecuted in large numbers, but such a crisis does not fit with our other sources for this period. Sources other than the Book of Revelation portray Christians, for the most part, as sharing peacefully in urban Asian life alongside their non-Christian neighbors.

Christians and Asian Society in the Book of Revelation

Little in the Book of Revelation contradicts this portrait of Christianity. First, with regard to the seer himself there is no indication that he suffered banishment or relegation to Patmos by urban or imperial officials. He states, "I John, your brother, who share with you in Jesus the tribulation and the kingdom and the patient endurance, was on the island called Patmos on account of the word of God and the testimony of Jesus" (Rev. 1:9). The relative clause in the first part of the statement ("who share with you in Jesus the tribulation and the kingdom and the patient endurance") refers to what John and other Christians shared with Jesus: Christian existence means, among other things, participating in Jesus' affliction, victory, and steadfastness.[1] The terms *tribulation* and *endurance* could refer to social, political realities just as well as to faithful participation in Christ, but the coordination of those terms with *kingdom* favors the interpretation that all three refer to Christian life with Christ. Social, political realities may have contributed to that description of Christian existence (e.g., the general dislike of Christianity by non-Christian neighbors), but they cannot be seen as the cause or occasion for the seer's statement.

In the main clause of the sentence John refers to his geographical location. He does so in very neutral terms: "I arrived on [*egenomen en*] the island called Patmos."[2] The same verbal construction is repeated in the next sentence: "I came under [*egenomen en*] the Spirit's influence on the Lord's Day" (Rev. 1:10). The bland verb *arrived* intimates nothing about relegation, persecution, or confinement. If anything, the parallel construction (*egenomen en*) between Patmos and the spirit draws Patmos into a sacral, spatial homologue with the sacral state of being "in the Spirit" and the sacred time of the Lord's Day.

Any notion of banishment to Patmos must come from the prepositional phrase "on account of the word of God and the testimony of Jesus."[3] The preposition *on-account-of* (διά) used here, however, designates a very general relationship of cause, occasion, or even purpose. Here, as in the Book of Revelation generally, it signifies either a contributory or necessary cause, in this case, of John's being on Patmos. One possible meaning of the phrase could be that John was on Patmos because he had been preaching the word of God and was banished to the island as a result of the preaching. That was apparently how Eusebius of Caesarea understood the passage in connection with his view of Domitian (see chap. 1). But it could just as well mean that John was on the island because he wanted to preach there. No Roman source designates Patmos a prison settlement or an island for banishment.[4] Nor was it a deserted, barren isle, as is sometimes suggested; it had sufficient population to support a gymnasium two centuries before the Common Era, and around the time of John an inscription refers to the presence of the cult of Artemis (Saffrey 1975, 393–407).

In the message to Pergamum there is a reference to the martyrdom of Antipas: "Antipas my witness, my faithful one, who was killed among you, where Satan dwells" (Rev. 2:13). The agent of death is not indicated: it could have been a mob or an official agency, but most likely it occurred because Antipas was a Christian. His death had occurred sometime in the past.[5] There is no reason to connect the death with demands to worship the emperor. The seer's references to "the throne of Satan" and "where Satan dwells" probably refer to the great altar of Zeus at Pergamum; it could possibly refer to the Asclepios cult and the medical center there. It is unlikely that John had the imperial cult per se in mind, especially since the imperial cult tended to be integrated with indigenous cults (see pp. 162–64). Pergamum received a provincial imperial temple (*neokoros*) in Augustus' reign, but so did Smyrna under Tiberius and Ephesus under Caligula (see Broughton 1938, 709). Pergamum did not receive another provincial temple until 113–114 under Trajan (Broughton 1938, 742).[6]

Conflict with the Jewish synagogue may have been a source of persecution and oppression. To the church at Smyrna is written, "I know your tribulation and your poverty . . . and the slander of those who say that they are Jews and are not, but are a synagogue of Satan" (Rev. 2:9). *Slander* (*blasphemia*) is strong language, for elsewhere in the Book of Revelation the term is reserved for activity of the beast and the Whore (see 13:1, 5, 6, 17:3).[7] In the letter to the Philadelphians the conflict with Jews is seen as a present power structure that will be reversed in the future: "Behold, I will make those of the synagogue of Satan who say that they are Jews and are not, but lie — behold, I will make them come and bow down before your feet, and learn that I have loved you" (Rev. 3:9). Elsewhere in the Book of Revelation, bowing down is done only before superhuman figures such as God, the beast, or the dragon. Even an angel commands John not to bow down before him (Rev. 19:10, 22:8–9). Here, however, the one dictating says that he will cause those false Jews to come and bow down at the feet of the Philadelphians.[8] These indictments against Jews and their "synagogue of Satan" (the basic Jewish institution of the Diaspora) probably stem from the seer's rejection of any Jew or Christian who has a favorable social standing in the cities of Asia. He rejects those Jews just as he rejects prophetic circles in the churches that sanction social intercourse with the non-Christian world (see chap. 7). That his indictments result from Jews charging Christians of antisocial behavior before political officials is, I think, unlikely. The language of the seer indicates his attitude toward Judaism rather than Jewish actions against Christians.

Otherwise, the seer does not refer to actual crisis involving institutions outside the church. From within there are the dangers of false apostles and false teachings; and there is the impending eschatological persecution from Jesus ("I will come to you. . . . "). The seer talks as though he expects tribulation in the near future from the outside world (e.g., Rev. 2:10–11), but such tribulation does not refer to present social distress. So, too, in the visions (Rev. 4:1–22:5) there are sufficient references and allusions to Rome and emperors to conclude that the seer expected tribulation and oppression from political and economic institutions in Asia, but those descriptions do not report past or present hostilities between Christians and any agencies in the Roman and provincial governments.

In sum, John reports surprisingly few hostilities toward Christians by the non-Christian social world. He anticipates conflict, but conflicts stemming from his fundamental position that church and world belong to antithetical forces. In other words John *encourages* his audience to see themselves in conflict with society; such conflict is a part of his vision of the world.

Genre, Crisis, and the Book of Revelation

Although the author of the Book of Revelation reports little in the way of specific incidents involving overt attacks on the church, he is unequivocal in his negative attitude towards Roman urban society and the Christians or Jews who in any way accommodate to it. In contrast to most Christians in Asia, he views urban society and the empire as antithetical to Christian existence and in league with Satan. Jews who succeed socially belong to the "synagogue of Satan." Christian leaders who espouse participation in the life of the empire as harmless and as irrelevant to Christian existence are made homologous to evil, mythic forces such as Babylon, the Great Whore. The peace and prosperity of Roman society is, from his point of view, not to be entered into by faithful Christians. The seven heads of the scarlet beast are the seven hills of Rome (Rev. 17:9); they are also seven Roman emperors (Rev. 17:10). Other petty kings participate in beastly evil through fornication (18:3, 9). Rome is Babylon the Whore who rides the scarlet beast and who falls into, and becomes, a "prison" (Rev. 18:2, cf. 20:7). The political order of Rome is wholly corrupt, belonging to the Satanic realm.[9]

The economic order belongs to the same corrupt realm. Buying and selling require the "stamp" of the beast (Rev. 13:16–17, cf. Deissmann 1978, 341). Merchants are indicted as growing wealthy from the rich wantonness of Babylon (Rev. 18:3, 15, 23). The seer lists in detail all of the goods that merchants can no longer sell because of Babylon's fall (18:11–14); he takes glee in their loss. The sellers, in contrast, weep over Babylon. They were "great men" through her, and without her they are nothing (18:23). Traders and shipmasters are especially singled out as weeping over her fall: "Alas, alas, for the great city / Where all who had ships at sea grew rich by her wealth!" (18:19). Political and economic corruption results in the great city's being empty culturally: neither harpers, minstrels, flute players, trumpeters, the millstone, the light of a lamp, nor the voice of bridegroom or bride shall any longer be found in her (18:22–23). In short, neither craft nor art will be found in her.[10]

In a nutshell, the conflict and crisis in the Book of Revelation between Christian commitment and the social order derive from John's perspective on Roman society rather than from significant hostilities in the social environment. In this regard the Book of Revelation fits the genre to which it belongs. There is a crisis orientation in the Book of Revelation, but it is a characteristic of the genre, not of political circumstances occasioning the genre; that is, the formulation of "crisis situations" is a topos, a commonplace topic, in the genre "apocalypse." As we saw in chapter 2, writers of apocalypses regularly incorporate the theme of crisis as a formal element in the genre (and life is such that there is usually something to which the theme can be attached). Apocalypses are not always occasioned by great political and social crises, nor do political and social crises always result in an apocalyptic response. The presence of the theme tells us nothing about the social and political situation.

Apocalyptic themes such as "exhortations to remain faithful," "assurances of hope," and "messages of comfort" are topoi related to that of crisis. Thus the writer of the Book of Revelation not only develops the themes of comfort, hope, and perseverance; he also constructs a reality such that people see their situation as one in need of comfort, hope, and perseverance. In the process of listening to John's apocalypse, those commonplace topics in the genre may begin to shape the reader or hearer's understanding of his or her situation. Or, from a different angle, one who, for whatever reason, tends to view the world as a state of crisis and conflict may find the social, literary conventions of an apocalypse an attractive expression of that tendency.

Crisis, comfort, hope, exhortation, consolation, and the like are thus themes and formal elements that help to make up the genre "apocalypse." They are a part of the generic elements that constitute the "vantage point" of the apocalypticist. Such formal elements in the Book of Revelation do not point to serious conflicts between Christians and the politics of urban Asia; they point to an apocalyptic point of view towards society. In locating the seer and his work in Asian society, one cannot draw directly on those *formal* elements without first seeing how they contribute to the *meaning* of the genre, that is, how they help to construct an apoca-

lyptic understanding of reality (see chap. 2). That meaning, that apocalyptic understanding of reality, can then be located in a social order.

Thus, in considering the social location of the Book of Revelation, I shall begin
with an element recognized universally as central to *the meaning of the genre*,
namely that an apocalypse is "revelatory literature." Scholars may differ over
whether *revelatory literature* is synonymous with *apocalypses* or whether it is a
broader category of literature, but all agree that "revelation" is essential to the
definition of an apocalypse. "Revelation" is bound up essentially with the meaning
of an apocalypse; by focusing upon that meaning, we shall at the same time make a
move toward the social location of the Book of Revelation.

Revealed Knowledge, Public Knowledge, and Validation

Let us begin by contrasting two sources of knowledge: revealed knowledge, such as
that given in the Book of Revelation, and public knowledge, gained through the
public order. In developing this contrast, I shall draw on a particular brand of the
sociology of knowledge readily accessible in the writings of Peter Berger. First,
consider "public knowledge." We orient to the world—that is, come to know it and
respond to it—through the guidance of society around us. We understand "raw
experience" through language that provides categories for understanding. That
language provided by society (e.g., English) imposes itself on experience and shapes
how we value, understand, and relate to aspects of our environment. Not only
language shapes us; so do social institutions. They guide and structure how we
relate to one another and how we see ourselves in relation to the world around us
(see Berger 1970, 6). Public knowledge, then, involves cognitive structures about
the world that are embodied in, and learned through, public institutions that provide language, roles, identities, norms, myths, rituals, and a cosmic frame of
reference for their members—all of which are integrated into an ordered reality
(see Berger and Luckmann 1967, 63–67, 73–74). All societies depend on that kind
of public knowledge, which is shared, accepted, and taken for granted by their
members. As Berger states, "Every society provides for its members an objectively
available body of 'knowledge.' To participate in the society is to share its 'knowledge'" (1969, 21). That knowledge becomes accessible through public institutions.[11]
In other words, learning public knowledge is socialization.

Validation of Public Knowledge
in the Province of Asia

For example, in the first century of the Common Era, those living in Greek cities in
the province of Asia (including Jews and Christians) gained knowledge about the
world through Greek urban and Roman imperial institutions such as those discussed in chapter 9. Roman stability was established and validated in the public
sphere: in the constitution laid down by Augustus, in the peace and prosperity of
the whole empire, in the very Roman earth so celebrated in the *Aeneid*, and in the
manifest destiny of Roman expansion. In Asia, Roman imperial institutions were
fairly well integrated into city life. Roman administration of political, economic,

and judicial policies was well integrated into various urban, district, and provincial administrative units. Social position, rank, and status were spelled out fairly clearly for all.[12] The imperial cult meshed well with local religious practices.[13] The topography of the cities was such that people came to know the relationships between, for example, religious and political centers, simply by walking the streets.[14] The spatial layout of the city, its public institutions, and its official activities were all expressions of a "civic ideology" or a "public knowledge" (see Price 1984, 111).

The seer wrote in a period of time when imperial Rome offered Asians — by means of the city, the basic social unit — a coherent, ordered structure of reality that unified religious, social, economic, political, and aesthetic aspects of the world. In the language of Peter Berger, cosmic, social, political, and personal dimensions of existence were so integrated that urban Asians shared "knowledge" that allowed them "to move with a measure of confidence through everyday life" (Berger 1970, 6).[15]

In connection with this "public knowledge" there was a certain kind of revealed knowledge, for example, knowledge given to a seer about the divine destiny involved in the birth of a particular king, the founding of Rome, or the afterlife of an emperor (the claim to "finality" and "ultimacy" is always esoteric); but that revealed, esoteric knowledge was authenticated in part by the public institutions to which it was linked and that it founds or supports. For example, the empire itself bears witness to the oracles proclaimed at the founding of Rome, or a person's kingship fulfills publicly the destiny revealed at his birth.

The revealed knowledge about the world that comes from the Book of Revelation does not gain credence from public political and social institutions. If anything, the Book of Revelation rejects those public institutions by giving them a negative force attached to evil, demonic powers. The reader may affirm the negative value of public institutions, but that in itself will not authenticate the apocalyptic view of the world that comes strictly from private, esoteric revelation.

Validation of Revealed Knowledge in the Book of Revelation

The esoteric, nonpublic character of John's knowledge is spelled out at the beginning of the Book of Revelation. Disclosure of the message comes not through public means available for all to see but through special revelation — *apocalypsis* — that God communicates privately to John (Rev. 1:1–2, 22:16).[16] Since this message has no confirmation in public discourse — not even Christians can replicate his seeing and hearing — it requires a special kind of validation. His visions are private revelations, and as such they can be rejected as being nothing but the subjective experience and partisan claims of a particular person or party in the church. They must be authenticated if they are to influence those who read and hear the words (see Aune 1986a, 89). One aspect of any visionary report is, therefore, establishing the authenticity, the authority, the divine truth of what is being reported.

In Revelation, that authentication is done in two complementary ways: (1) a narrative report of the seer's experience with the divine and (2) devices through which the audience is taken into the visions, so that the visions become "internal" to them, rather than remaining "external" messages requiring authentication. In the first instance, the seer's subjective experience is minimized by emphasizing divine initiation in the visions; in the second instance, the seer's subjectivity is downplayed, as links are made directly between God and the reader or hearer, with the seer's role being that of indirect mediator.

The seer introduces his work as a "revelation of Jesus Christ" (Ἀποκάλυψις Ἰησοῦ Χριστοῦ) (1:1), a revelation given to Jesus Christ from God so that he could make known to his servants what must soon come. Jesus had this revelation communicated through his messenger (or angel) to his servant John, who witnessed to the Word of God and the testimony of Jesus. "The Word of God and the testimony of Jesus" refers to the content of the revelation John receives from God through Jesus; thus God and Jesus testify to themselves in the message they send. John simply bears witness to all of that (1:2). This introduction gives a chain of revelation authenticated by means Christians would accept. It is presented to the audience in third-person narrative in such a way that John becomes merely a link in the chain from God and Jesus Christ. John's "seeing" is minimized by reference to it in a subordinate clause at the end of the chain of communication ("which things he saw" [ὅσα εἶδεν], 1:2). Thus, the possibility of considering this revelation to be a partisan, idiosyncratic view of the world is minimized by the narrative style that introduces the work (1:1–2). That introduction concludes with a beatitude—the giver of the blessing remains anonymous—for the one reading and listening to the revelation: "Blessed is the reader and those who listen to the words of this prophecy and keep that which is written in it" (1:3). The hearer completes the chain of communication: God–Jesus–angels–John–reader–hearer. John's subjectivity is buried within that narrated chain.

After an epistolary greeting (1:4–5), a doxology (1:5–6), an eschatological acclamation (1:7), and a lofty, first-person assertion by God—all of which further emphasizes divine initiative in the revelatory process—only then does John give a personal, first-person narration of his visionary experience.[17] John begins his personal account by identifying himself closely with those who read and hear his revelation: "I, John, your brother and fellow participant in the tribulation, kingship, and steadfastness in Jesus" (1:9). After giving the time and place of this initial vision (1:9), John begins in indicative mood: "I heard behind me a great voice" (1:10). The voice speaks to John in the imperative mood: "Write down what you see on a scroll and send it to the seven churches" (1:11). John then continues in first-person indicative mood, telling what he saw. He describes the voice in detail (1:12–16) and his response to it: "I saw [1:12]. . . . I fell at his feet [1:17] . . . and he placed his right hand upon me saying [1:17]. . . . " The voice then issues a command not to fear him, talks about itself in first-person (1:17–18, cf. 1:8), and reissues the command to write (1:19).

In this visionary account, direct communication occurs between John and the voice, who is obviously Jesus Christ. The churches in Asia Minor are, however, brought indirectly into the schema of communication through imperatives—"Write and send to the seven churches" (1:11)—and through an explanation of the seven stars and the seven gold lampstands (1:16, 20).[18] The visionary report is thus so constructed, as it moves from description to command (indicative to imperative moods) and direct and indirect forms and levels of communication, that John's subjectivity is minimized. He becomes merely an instrument for the voice to use to speak to the seven churches. As a result, the vision is self-authenticating. The churches are not being guided and admonished by John but by the Christ whom John saw and heard. The message, that is, that the churches are to comprehend the world the way John does, is the message of Christ. The structure for learning to know what things mean, which is set forth in Revelation, comes from Jesus Christ and needs no further verification or authentication for those who follow Jesus.

Validation, the Seven Letters, and Asian Life

In the seven letters that follow, direct communication continues between Jesus Christ and John, as the former commands John to write (2:1). A slight change occurs in the one to whom John is commanded to write; he writes not to the churches, as before, but to the "messenger of the church" (*angelos*, 2:1). However one is to understand these "messengers" or "angels," the "messages" or "letters" clearly address Christians in Asia Minor. The communication between John and Jesus Christ contains within it communication between Jesus Christ and each of the churches. Messages to each of the seven churches then follow. In the transmission of these letters, John disappears entirely from view.

For each of the seven letters, the one dictating identifies himself by repeating elements of the lengthy description of the voice John heard and saw in his inaugural vision (1:12–16). The one John saw in an esoteric vision now becomes the one communicating directly to the seven churches. He assesses, commands, exhorts, encourages, threatens, and promises. Those at Ephesus toil and endure patiently without growing weary. They hate the works of the Nicolaitans. But he orders them to remember whence they have fallen, to repent, and to do what they did at first (2:5). He knows the tribulation and poverty of those at Smyrna; they are encouraged not to fear what they are about to suffer (2:10). Those at Pergamum hold fast Jesus' name and did not deny him in the days of Antipas. But some hold to the teachings of Balaam; those he warns to repent (2:16). Christians at Thyatira have works, love, faith, service, and patient endurance. Those at Sardis should wake up and strengthen what is about to die (3:2). The Philadelphians have kept Jesus' word and not denied his name. The Laodiceans are neither hot nor cold. He urges them to be zealous and to repent (3:19).

Of greater importance than those exhortations are the symbols, images, and metaphors introduced in the seven letters. That metaphoric language links the church messages to the rest of the Book of Revelation. Those who conquer among the Ephesians will be given to eat from the tree of life in the paradise of God. The *diabolos* is about to throw some of the Smyrnians into prison, but the faithful will receive a crown of life and will not be harmed by the second death. At Pergamum the throne of Satan is situated. Those who conquer, however, will be given hidden manna, a white stone, and a new name that no one knows except him who receives it. Those at Thyatira practicing fornication with Jezebel will be thrown into great tribulation unless they repent. To those who are victorious Jesus will give power, kingship, and the morning star. A few at Sardis have not soiled their garments; they will walk with Jesus in white, and their names will not be blotted out of the book of life but will be confessed before the father and his angels. Those of the synagogue of Satan will bow down before Christians at Philadelphia. Those conquering will become a pillar in the temple of God and will never go out of it. They shall have written on them the name of God and the name of the New Jerusalem coming down from heaven and the new name of Jesus. Those at Laodicea are urged to buy white garments from Jesus so as not to have the shame of nakedness. Jesus stands knocking at the door; to those who open, he will come in.[19] The one from Laodicea who conquers will sit on the throne with Jesus.

The importance of the epistolary elements in the Book of Revelation cannot be overemphasized.[20] John not only addresses his visions to seven churches in Asia (Rev. 1:4), but he also concludes his work with a closing grace (22:21). Further, his first vision involves the messages to the seven churches. Because of that placement, the conversations with the churches are not only given first position sequentially, but also the "messages" to the churches become the initial context – the bass line – for images, symbols, and motifs used later in the transcendent visions. Later usages loop back recursively to the messages given to the seven churches, in which Jesus describes the churches' situations, accuses, warns, and admonishes them, and promises good things to those who persevere in the faith (see chap. 3 above). As an example, terms referring to evil forces such as the *diabolos* and Satan or to demonic activities such as blasphemy, deception, and whoring or to Satan's imprisonment are introduced in the messages to the churches (see chap. 5 above); their later appearances repeat recursively the given elements in the context of the churches of Asia.

In a word, Asia is "home base" in the visions of Revelation. The seer addresses Christians there, and all that he sees and explains is intended for their benefit. He transmits his visions to them so that they may see their situation correctly. The language of the visions, grounded in the seven messages, does not present a symbolic universe alternative to the work-a-day world in which Asian Christians live. Rather, the seer presents a structure through which those Christians may comprehend the proper meaning of objects and relationships in their work-a-day world. As we have seen, proper comprehension occurs when objects and relations of the work-a-day world are transformed into the seer's vision of the world through boundaries that unfold essential structures (see chap. 5). The prophetess Jezebel at Thyatira is understood properly when she becomes a homologue to Babylon the Whore. The martyrdom of Antipas at Pergamum gains proper meaning when related homologously to the Lamb who conquers through his blood. The seer unfolds a sufficient number of those structures so that his audience can know properly everything they encounter in their daily life in urban Asia.

By linking the seven letters on the one hand to the initial vision which John saw and on the other hand to the characters and actions of the visions to follow, the seer links the Christians in Asia integrally to his revelatory visions. They become a part of those visions and do not read or hear them as "external" to themselves. John authenticates not only commandments and exhortations that he addresses to the churches but, more importantly, the essential boundary structures that unfold in the rest of the work. The issue is not simply validation of a fantastic world that John "saw" in his several visions; it is whether John's structure of meaning –

established by boundaries, ratios, homologues, and proportions—is a valid one for Christians. In the messages to the churches, Christians are well on their way to accepting the homologues between Jezebel and Babylon the Whore or between Rome, the city of seven hills, and the seven heads of the scarlet beast. Through the initial vision of the seven messages to the seven churches John integrates exhortations, encouragements, threats, and promises with homologues, ratios, and proportions; and he authenticates that world of meaning and exhortation as true knowledge from God. The exhortations, warnings, and comfortings John offers make no sense apart from how he envisions the world. His particular brand of comforting and exhorting are appropriate only to those who also receive his structures of meaning—his knowledge about the world—transmitted throughout the visions.

Deviant Knowledge and the Critique of Public Discourse

When compared to that "public knowledge" transmitted through institutions, myths, and rituals involving the town fathers and their social order, the Book of Revelation reveals: "deviant knowledge"; that is, its knowledge deviates from the knowledge given and generally taken for granted in the social order (Berger 1970, 6). As we have seen, it is deviant in its source: apocalypses provide knowledge through private, esoteric means apart from larger communal, institutional validation. It is deviant in its assessment of the larger social order: apocalyptic knowledge devalues, rather than supports, the cognitive structures, identities, roles, and norms in the order of society. And it is deviant in its cosmology. In the public order, cosmicizing assures that "the way we do things" reflects "the way that the world really is." Hellenistic kingship mirrors divine government; role acceptance in the social organization of the Roman Empire also fulfills divine obligations. By contrast, in the cosmicizing of apocalyptic, ironic reversals abound. Kings and emperors are disestablished by means of metaphoric links to satanic and evil forces, while the transcendent knowledge transmitted through apocalypses appears in the here and now to be disconfirmed.[21]

As deviant knowledge whose truth does not depend upon public institutions and existing political power, the Book of Revelation serves to censure the public order. If accepted in a thoroughgoing manner, it would even replace the public knowledge by means of which people participate in Roman society. In challenging public knowledge by offering an alternative to the stable empire, the seer of the Apocalypse is not unique. As censure of the public order, the Book of Revelation joins loose affiliation with several other deviants in the Roman world. Ramsey MacMullen (1966) has pointed out that philosophers, magicians, diviners, and prophets of different backgrounds disturbed the "public mind" through private transmission of values and ideals from an older public order or from a philosophical tradition or from a more recent revelation. The genre and rhetoric of the Book of Revelation were only one among several attacks against the empire, and some of the genre's distinctiveness may be brought out by comparing it to other "vehicles of literary attack" (MacMullen 1966, 35).

Critiques of Public Discourse in the Roman Empire

There were families and coteries that censured the empire on the basis of remembered days of old when the republic named the public order. This remembrance mixed with Stoic and other philosophical ideals to make various claims for freedom: free speech, liberty, free access to political power, a senate free from imperial constraint, and freedom from subjection to a higher ruler. *Libertas* was their watchword.[22] Their weapons were strictly literary. They wrote histories filled with exemplary figures who upheld their ideals against past tyrants. Tyrannicide became a genre in narrative, dramatic literature. Dramatic themes from, for example, the mythology of Agamemnon and the house of Atreus coded contemporary protests against the emperor. Satire, explicitly aimed at an emperor, usually came after his death. Relatives and disciples collected and transmitted the writings of earlier martyrs, especially the last words of those who killed themselves—sometimes at the command of a tyrant—in perfect control at a gathering of friends and disciples. Funerals and birthdays of those same martyrs could be the occasion for eulogies, poems, and songs censuring the loss of liberty in the days of the principate.[23] Death and dying were admired widely throughout the latter half of the first century CE as forms of protest against tyranny.[24]

More generally, protest took the form of withdrawal and deliberate inaction—which could evoke formal charges from the government. Upper-class Stoicism was obliged to offer a philosophy of leadership that worked for the common good; but if government was "corrupt beyond cure," reform efforts might be useless, and "the wise man [would] not struggle uselessly." Moreover, one had obligation not only to one's own homeland but also to the whole universe. Sometimes withdrawal from the empire was required in order to be a faithful cosmopolitan.[25]

Protest by this coterie of aristocrats, who had "a philosophic tinge under the Flavians," was so close to the public order of the empire that it can barely be called "deviant" (MacMullen 1966, 46). These protesters came from the same mold as supporters of the throne and the empire. Their hostility was aimed at "persons who were close neighbors in a cultural and social sense" (MacMullen 1966, 94). On the ides of March in 44 BCE Julius Caesar could not tell the difference between friends and enemies. In the Flavian period the opposition could change allegiance and pass over to the other side "without giving up any essential belief." The emperor could support the "literary vehicles" (e.g., rhetorical exercises extolling tyrannicide) that turned against him, for the imperial establishment and this particular form of opposition shared a common background and many of the same ideals.[26]

Philosophy, a prominent form of opposition under the Flavians, included more than ethics and guidance of human conduct. It embraced "powers of the soul" as well as the mind.[27] Philosophers were thus affiliated with ascetics, magicians, miracle workers, astrologers, diviners, and prophets. Teachers presented their philosophical systems as revelations given in direct conversation with Hermes, Isis, or some other deity. Figures such as Peregrinus Proteus (d. 165 CE) or Pachrates (under Hadrian) or Apollonius of Tyana (under Domitian) mixed their teachings with miracles, prophecies, exorcisms, and claims of divinity. They were received among poor and rich, stupid and enlightened, oppressed and oppressors. Any religious system — for example Judaism or Christianity — could be called a "philosophy," for it proclaimed "a way of life that led to God."[28]

Literary vehicles for these philosophical magicians included curse tablets, magical papyri, amulets, reports of conversations with a deity, books on demons, and aretalogies (accounts of individual acts of miraculous powers) developed sometimes into elaborate "lives" such as Philostratus' *Life of Apollonius of Tyana* or Christian "lives" of Jesus. Prophecies were written down in oracles claiming the authority of the Sibyl, a literary convention used by Jews, Christians, and pagans. After Nero's death, his return became a popular theme in these prophecies. In 69, 79, and 88 "Neros" appeared "in the East" winning followers and disturbing the public order. This theme appears in Jewish and Christian writings, including probably the Book of Revelation (e.g., Rev. 17:8). In these latter writings, however, Nero's return took a more negative turn as he was identified with a "lawless" figure who would come in the last days. In Jewish and Christian prophecy, the future of Rome was another prominent theme. Some, like the Book of Revelation, expected the destruction of that great city, while others (e.g., Melito of Sardis) saw Rome's destiny bound to that of the church.[29]

Magic is, by definition, a form of deviance, a threat to the public order requiring social control (see Aune 1986b). Thus, MacMullen makes an insightful tautology when he writes, "There was thus no period in the history of the Empire in which the magician was not considered an enemy of society" (1966, 125).[30] Of greater threat were the astrologers, diviners, and prophets who could foresee and control the future. They disturbed the "public mind" and threatened stability and the peace. Witnesses had to be present during inquiries into the future lest illegal queries be made. Of greatest danger to the emperor were inquiries into the destinies of the royal house or the future of the state. Since emperors might also value and believe in astrology and divination, "they wanted no meddling with their own stars or lifelines." Such inquiry became an act of treason (*majestas*) punishable by expulsion from Rome — MacMullen asserts that "their expulsion from other cities is never mentioned" (1966, 132) — or death.[31]

During the reign of the Flavians both Vespasian and Domitian drove out the philosophers and astrologers (MacMullen 1966, 133). In general Domitian tried to follow the policies of his father in dealing with these people, bearing their "impudence . . . with the greatest patience," but neither he nor his father could tolerate their intervention into politics (Suet. *Vesp.* 13).[32] Domitian's dealings with some of Pliny's philosopher friends have to be considered in light of the general imperial policies sketched out above.[33] Philosophers who told stories about their heroes martyred by emperors or who manipulated maps or who told what had been foreseen from the stars about the emperor's fate could be as great a danger to the emperor as invading Dacians;[34] for they shaped beliefs about imperial power and specific imperial destinies. Most of Pliny's philosophical friends whom Domitian either killed or banished belonged to a "close-knit group who stood in a clearly defined tradition of His Majesty's *dis*loyal opposition."[35]

In dealing with the magical and prophetic dimensions of philosophy we move further from the common, undergirding ideals of the Roman Empire. The ideology and literary vehicles of magicians and prophets do not blend into imperial ideology in quite the same way as the aristocratic themes of republicanism, libertas, and senatorial power. Yet magic and prophecy flourished in all social circles: "The most enlightened people took it seriously" (MacMullen 1966, 126). Perhaps MacMullen is right: there was "black" magic and "white" magic, magic acceptable to good Romans and magic unlawful. But not everyone would have agreed which was which or which side some people were on. The same was true of practitioners of astrology. Emperors were guided in their daily life—especially in relation to state activities—by astrologers, and astrology was a habit of both rich and poor. Yet, as we have seen, astrology and prophecy were seen, as well, to be disturbances in the public order and dangerous activities to pursue.[36]

Critique of Public Discourse in the Book of Revelation

It is also obvious that magic and prophecy bring us closer to circles involving the Book of Revelation. The seer is a Christian prophet who proclaims specially revealed information, including the destiny of the Roman Empire. He records conversation between himself and his God. He seeks to disrupt the peace at least to the extent of dislocating Christians in the cities of Asia and proclaiming in the form of apocalypse the destruction of the public order. He draws freely on the Nero legend that promised danger to the empire. The seer also borrows magical elements from the Hecate cult, vowel chants from the magical papyri, and procedures for summoning a god for purposes of divination (see Aune 1986b). At the same time, John rejects the claims of magic—they are deceptions that lead people astray (see Rev. 13:13–15)—and magicians belong with fornicators, murderers, and idolators who will be thrown into the lake of fire (Rev. 21:8). He rejects magic and its power, but in so doing he shows how close he is to magic socially.

The style of John's writing may also be a critique of public discourse. Yarbro Collins notices that the seer writes in "a peculiar, contemporarily Semitizing Greek" as "a kind of protest against the higher forms of Hellenistic culture" (1984, 47). As we have seen in chapters 3, 4, and 5, his language is highly metaphoric and fluid; through it he constructs soft boundaries, homologies, and ratios. Transformations and metamorphoses abound. This fluidity in his language endangers the public order because it blurs categories that are essential to stability. Language of the public order tends to be univocal with clear conceptual boundaries. In contrast, the Book of Revelation tends to be more like Ovid's *Metamorphoses*, where clear causal connections disappear in the gaps of arbitrary transformations.[37] The playful language of the seer—puns, irony, and other verbal forms that metamorphose into opposites—especially endanger the fixed order. Through such verbal manipulation what appears publicly contrasts with what really is: the wealthy are not really wealthy, those of high rank are not really of high rank, and the glories of Roman peace and prosperity mask satanic forces soon to be defeated. The seer's language does more, however, than that of the *Metamorphoses*. Ovid's language tends to be arbitrary, shapeless, and fickle, subverting without offering an alternative. His constant change becomes too unstable for nature to reflect back a deeper meaning (see Massey 1976, 23). The language of the seer subverts *and* offers an alternative order. The seer descends into a language more like that of private dream than public communication and casts his enemies in the anonymity of mythic beasts. More is involved than lack of daring and courage. By descending into the beastial language of myth he labels the Roman order as demonic; Curran has suggested that "descent" into animal form is also a way of rejecting a world of empty conventions and of reestablishing one's substantiality (see Massey 1976, 32, 63). An unassailability is established independent of human society.

Whether that depth dimension can be established for the language of the seer or not, there is, as we have seen, a consistency and structure in the transformations that occur in the Book of Revelation. In other words, the Book of Revelation contains "stable" metamorphoses that transform according to a consistent spirit of universal ratios. Through that consistency of spirit the seer offers a fresh vision of the world to subvert the public order of Caesar. Moreover, in the process he reclaims all the public order of Rome. He incorporates the whole of everything in his vision. Everything intersects his world. Conflict with the public Roman domain is inevitable.

On this point there is no fluidity or blurring of boundaries in John's language: his revealed knowledge opposes the public knowledge of Roman institutions. John characterized the opposition and his relation to it in ways similar to Pliny's characterization of Domitian and his reign. In order to establish a new era with Trajan, Pliny and his circle describe Domitian (the old era) as extremely sensual and sexually excessive. So John describes Babylon the Great Whore, the beast she "rides," the kings of the earth, and the merchants (14:8; 17:1–4, 9; 18:3). Through such titles as "Our Lord and God" (*dominus et deus noster*), Domitian is said to make false claims to divinity. So, too, the beasts deceive by their false claims, miracles, and demands for worship (Rev. 13). John and his church, like Pliny and his circle, characterize their relation to the opposition as one of oppression, constant danger, and persecution. Finally, the end of Domitian's reign is described as chaos and confusion in both society and nature. So in John all is lost as nature joins in a cacophany of destruction, followed (as in Suetonius) by a new age, a new heaven, and a new earth with those then persecuted now reigning in power and honor. Those hyperbolic, exaggerated surface characterizations appear for the same reason in the writings of Pliny's circle and the Book of Revelation: to manifest the deep structure of binary opposition and boundary formation—to distinguish insiders from outsiders, authentic self-expressions from false ones, old eras from new ones, and true knowledge from deceptive lies. In short, his revealed vision of reality must be kept separate from that vision of reality embodied in public institutions and public discourse. Public knowledge masks lies and even the primordial Liar himself. Only the divinely revealed Word of God given to John can lead to true knowledge.

11

The Social Location
of the Book of Revelation

Before moving directly to the question of the social location of the Book of Revelation, I should clarify further the nature of the relationship between the seer and his opposition, that is, between his revealed knowledge and the public knowledge of the empire. That relationship involves both social and world boundaries. A social boundary exists because the whole of John's social world does not comprehend his vision, even though John's vision comprehends the whole social world; that is, his organizing grid of relationships, power structures, ideals, and values are not shared by all others in Roman society; indeed, his is a minority report. Since the boundaries of John's social location are a dimension of his world vision, we cannot consider the boundary between him and his opposition in simple social terms. The complexity of the boundary will elude any analysis limited to economic status, social position, and political influence. The social boundary is one dimension of the world boundary; it expresses in the region of social experience a difference in the seer's vision of reality and Rome's vision of reality. The two visions involve distinctions between true knowledge and deception, authentic self-expression and false consciousness, service to the true god or idolatry.

Correlations at the Boundary

In order to appreciate the complexity of a social boundary, let us recall earlier comments on the nature of "boundary" (see chap. 5). A boundary designates a situation in which differences touch one another; for example, a social boundary marks the meeting of life inside and life outside the Christian community. A boundary calls attention to the differences, and it serves as the means by which something passes across. There is a dynamism to boundaries as they act as "transformational devices"; they mark the point where anything (an office, a belief, a ritual, an economic transaction) is transformed—in this instance, from one world system to another. Because world systems are by definition totally comprehensive,

186

at boundary points between systems *everything* must flow across, leaving its place in one world system and taking its place in another.

As an item crosses a boundary into the seer's world, its meaning is transformed and it takes its place as homology and/or contrariety with other items in the seer's world. For example, as Jews cross into the seer's world, they become "the synagogue of Satan," contrary to true Christians, who are homologous to God, who of course is contrary to Satan. So when political forces such as the Roman emperor or the Roman Empire cross into the seer's world, they are not simply allegorized, that is, described as a political system in veiled terms. They are redescribed, redefined, and relocated (now in the seer's world) as homologous to the Great Whore or the seven-headed beast or the great dragon from of old. Those homologues do not simply code political forces; they transform their meaning.

When boundary transformations in the seer's world are compared, they disclose essential proportions and fundamental ratios that structure and guide the transformational process (see chap. 5). Those ratios and proportions can be traced through every dimension of the seer's world: God is to Satan as the Lamb is to the beast, as the faithful in the churches are to those who deceive and mislead, as the Christian minority is to the larger Roman world; heaven is to earth as the eschatological future is to the present, as the temple is to space around it, as cultic activity of worship is to everyday activity, as being in the Spirit is to normal consciousness. Proportions are thus formed among social, political, religious, theological, and psychological aspects of the seer's vision; for every boundary situation unfolds an essential proportion that structures the seer's world. A boundary in any one of those dimensions can illumine another, for a boundary situation is nothing more or less than a fundamental ratio unfolding in one dimension an order implicated in all others. Like Einstein's notion of time and space, boundaries in the seer's world are coordinates of a multidimensional reality.

These different dimensions are not like provinces on a map, each separated from the other. Rather they are laminated in overlays so as to coincide with one another. They have neither center nor peripheral edges. No one province can thus claim to be prior to and determinative of other domains. This is a crucial point in understanding social boundaries. There is no social-historical context, psychological context, or any other context prior to and occasion for other elements in the vision. All "contexts" are transformed as they flow into the seer's multidimensional world. And all unfold the essential structures of that world: temporally in eschatological expectations, spatially in heaven-earth connections, liturgically in sacral action and sacred speech, and behaviorally in social practices and social relations.

The processes of cognition involved here may seem awkward and foreign precisely because no dimension—in particular, no social, historical dimension—is given privileged status. Rather than conceiving of a social boundary as a point where outside forces (especially social, historical forces) cause inside effects (linguistic, liturgical, visionary responses), I consider social boundaries to be part of a larger transformational process by which "formally homologous structures, built out of different materials at different levels of life—organic processes, unconscious mind, rational thought—are related to one another" (Lévi-Strauss 1967, 197). In transformational exchanges everything is affected. We often think of the symbolic as malleable to the "real" social, political situation; and conversely, we see social experience (political policy, economic exchange) as impenetrable, essentially unaffected by religious, mythic, and linguistic symbols. It then follows naturally (1) that there are two separate spheres, the social and the symbolic; and (2) that the social *causes* the symbolic. Both points reflect a faulty understanding of cognition. Within the process of world building, John's social boundaries—the place where John and his audience meet the rest of the world—operate in the same way as liturgical boundaries, spatial boundaries, and temporal boundaries: they unfold an order of essential proportions that are implicated in all other boundaries. John does not encounter Roman society at the edge of his vision; that social boundary, like all others, forms coordinates in John's multidimensional structure of overlayed laminations without center or periphery.[1] In other words, the social boundary unfolds an order implicated in John's world vision: the whole is contained in every part.

Tribulation at the Boundary

Complexity leads to not-very-tidy discussions. In order to minimize the untidiness, let us limit the following inquiry into social boundaries to the subject of tribulation.

The theme of tribulation dominates many of John's visions. At the opening of the fifth seal, the seer "saw under the altar the souls of those who had been slain for the word of God and for the witness they had borne; they cried out with a loud voice, 'O Sovereign Lord, holy and true, how long before thou wilt judge and avenge our blood on those who dwell upon the earth?' Then they were each given a white robe and told to rest a little longer, until the number of their brethren should be complete, who were to be killed as they themselves had been" (Rev. 6:9–11). In the second vision of chapter 7 an innumerable crowd of people wearing white garments with palm branches in their hands appear before the throne and the Lamb, whom they acclaim in the form of a doxology (Rev. 7:9–10). The crowd is identified as those "who have come out of the great tribulation; they have washed their robes and made them white in the blood of the Lamb" (Rev. 7:14). Several visions portray tribulation in connection with conflicting or warring situations. In various guises demonic forces distress those who follow the true God: the beast and the two witnesses (Rev. 11:1–13); the dragon and the woman with her offspring (Rev. 12:1–17); the beasts from the sea or earth and the saints (Rev. 13:1–14:12); or Babylon the Harlot and the witnesses of Jesus (Rev. 17:1–19:5). The seer does not here report social, political realities. He constructs visions with their own spatio-temporal frame that includes Asian society.

The theme of tribulation is also present in the messages to the seven churches (Rev. 2:1–3:22). For the most part distress is created by matters internal to the church (e.g., false apostles, false teachings) or by eschatological threats ("I will come to you. . . . "). There are, however, a few references to crises stemming from relationships outside the community. At Pergamum, there is reference to the martyrdom of Antipas (Rev.2:13).[2] The seer expects tribulation to come soon to those at Smyrna (Rev. 2:10–11). Also at Smyrna—and at Philadelphia—the seer alludes to conflict with the Jews (Rev. 2:9, 3:9). Involving himself, the seer makes the statement, "I John, your brother, who share with you in Jesus the tribulation and the kingdom and the patient endurance. . . . " (Rev. 1:9). But, as we have seen, there is remarkably little indication anywhere in the Book of Revelation of sustained attack on the church from those outside. There are, however, plenty of indications that the seer views the Roman Empire antagonistically and that his evaluation of the empire affects the social boundary.

Tribulation in the Book of Revelation has an explicitly religious connection. References to tribulation consistently occur in connection with the imitation of Jesus. As indicated in Revelation 1:9, "tribulation" is part of what Christians share with Jesus. For John, "suffering" is one of the most essential ingredients in Christian proclamation and Christian living. The gospel message is linked to Jesus' suffering and crucifixion, just as the eschatological Lord is portrayed throughout as the slain Lamb or the faithful martyr. Images of suffering, death, and sacrifice carry through the introductory vision of Jesus to his final victory.

That message becomes embodied in Christians when they "imitate Christ." In different ways the seer compares the situations of Christ and Christians. Those conquering at Laodicea are promised they will sit with the Son of Man upon his throne just as he conquered and sat with his Father upon his throne (Rev. 3:21, cf. 2:26–28). Moreover, both Christ and faithful Christians conquer by being slain; at Revelation 12:10–12, for example, they imitate Christ, conquering the dragon "by the blood of the Lamb" and "by the word of their testimony, for they loved not their lives even unto death." At Revelation 6:9 victorious ones in white "had been slain," just like the worthy Lamb (Rev. 5:9). And at Revelation 7:14 those from the great tribulation wore robes made white through the blood of the Lamb. As martyrs they imitate Jesus Christ, the prototype of all faithful martyrs to come (Rev. 1:5, 2:13, 11:3, 17:6). Christians witness to the message by both holding to and reenacting the testimony made by Jesus.

Imitative witness combines, in the seer's verbal symbolics, with another aspect of tribulation that is both linguistic and religious. According to the seer's formulation of the Christian message, victory and kingship are disclosed ironically through suffering and crucifixion. The seer actually uses irony in several different contexts (see chap. 3). Most germane to tribulation, however, is what may be called kerygmatic irony, that is, irony involved essentially in the formulation, proclamation, and embodiment of the Christian message.[3] In the greeting of Revelation 1:4–5 Jesus Christ is given epithets drawn from the royal ideology of Israel that in their original context express and enhance the kingly power and royal authority promised to the Davidic line (cf. the Septuagint version of Ps. 88:38, Isa. 55:4). In the verbal symbolics of the Apocalypse, however, those epithets of power and authority are transformed through irony. The witness is faithful through crucifixion; and the "firstborn" is numbered among the dead. In short, the seer celebrates the enthronement of the king who reigns in blood, crucified on the cross. In the continuing liturgical piece of Revelation 1:7, John even identifies the eschatological Lord— who comes on the clouds in the sight of all—as the "pierced one" whom all will mourn.[4] That theme of the king who reigns in blood is carried through the entire Apocalypse by such combined images as the Lion/Lamb (Rev. 5:5–6), the slain Lamb, and the Word of God clothed in blood (Rev. 19:13).

Christian existence shares in the same ironic structures of royalty and kingship (Rev. 1:6). Those at Smyrna are exhorted, "Be faithful unto death, and I will give you the crown of life" (Rev. 2:10). Life comes through death for the Christian, just as it did for Jesus. As life appears ironically in the guise of death, so power appears as powerlessness. To the Philadelphians is given the message, "I have placed before you an open door which no one has the power to close because you have little power and you kept my word and you did not deny my name" (Rev. 3:8). The Laodiceans who think that they are rich but are really poor are urged to buy from the one who bought them by his slain blood (Rev. 3:18, cf. 5:9). Color of clothing becomes one means of expressing irony here, for the "white clothing" that they are urged to buy is made white in the blood of the Lamb (Rev. 7:14). In the message to the Ephesians the "tree of life" carries a similar ironic message: those who conquer are promised they will eat from the tree of life (Rev. 2:7), which in the verbal symbolics of the Apocalypse is linked to the "healing of the nations" and "no more accursedness" and thereby to the cross (Rev. 22:2–3).[5] Kerygmatic irony also offers the clue to the compatibility of *tribulation, kingship*, and *steadfastness* in Revelation 1:9: these three terms characterize both Christ and Christian existence, for tribulation in itself reveals victory, conquest, and kingship.[6] If the oxymoronic foolishness of this overt irony—Paul calls it the folly of the cross—is blunted by temporalizing (present cruelty, future glory) or by compartmentalizing (bodily cruelty, spiritual glory), the seer's ironic statement about Christian existence is lost. These linguistic and religious themes obviously make a social impact by anticipating persecution, but irony also invites collusion with the author. An ironic understanding is not an obvious one. Only those "in the know" may discern its presence, and that contributes to the sense of community John shares with his audience.[7]

Tribulation and suffering in the life of Jesus and in the imitative witness of faithful Christians express something more than momentary events in history. Tribulation correlates in John's world with true knowledge, authentic self-expression, and service to the true God. The slain Lamb appears not only on earth but also in heaven, close to the throne (Rev. 5:6). Further, the Lamb is slain from the foundation of the world and reigns in that form (Rev. 13:8, cf. 1 Pet. 1:19).[8] The crucifixion lies in the deep structure of reality that enfolds all historical disclosures. It is, in turn, unfolded in Christ's rule and Christian existence so as to form a homology between Christ and Christians. A life of tribulation and social oppression expresses on the social boundary *how* Christians should reign with their crucified king and *how* they should participate in the power and glory of God. Disclosure of that reality constitutes a central ingredient in true knowledge and authentic selfhood.

Non-Christian attitudes towards Christians flow into the social boundary, from the outside in. Whatever the official basis for the interrogation and sentencing of Christians (the legal grounds for this activity are still debated), it is clear that the government had popular support for prosecuting Christians. From the viewpoint of the Roman world, Christians had "lost their shelter of tradition" with Judaism, were recognized as atheists who did not worship the communal gods, were nonconformists adhering to a recently formed religion from the East, and possibly participated in cannibalism. Indeed, if Pliny's correspondence is typical, the populace was more adamant than were Roman officials about bringing Christians to trial (*Ep.*

10.96). Trajan (probably following guidelines from the time of Domitian) directs Pliny not to listen to anonymous accusations and not to initiate prosecution by seeking Christians out. At the same time Christianity is viewed clearly as a social ill to be dealt with if Christians are brought before a tribunal; in that case they would probably be killed if, after due opportunity was given to them, they did not confess the religious dimension of the common, public Roman life. In John's world, Christians should seek out clashes with the state, for that would unfold the essential structures of reality into history.

Finally, the way the seer opposes his revealed knowledge to public knowledge affects the social boundary. As we have seen, the seer sets up a thoroughgoing contrast between the two. In language similar to that used by Pliny and his circle (see chap. 10), the opposition is characterized in terms of false divinity, sexual excess, and oppression.[9] This language is used to contrast revealed and public knowledge as binary oppositions. It includes conflict at the social boundary, but once again it is far more inclusive: it distinguishes insiders from outsiders, authentic self-expressions from false ones, old eras from new ones, and true knowledge from deceptive lies.

All of the above are dimensions of the social boundary. Within the seer's world all of the elements related to tribulation form a feedback loop that grows larger with each cycle: religious identification with the crucified king shapes psychosocial identity, which leads to patterns of behavior that support social-political prosecution, which loop back to create a more intense religious identification with the crucified king. In the process, mutual adjustments are made so that tribulation – so prominent a theme in the seer's knowledge of the world – enters fully into all provinces of meaning, including the social dimension. A myriad of qualities; behavioral traits; religious commitments; psychosocial understandings; and social, political interactions coalesce into a term like *tribulation*. At times it may gain impetus from the social boundary (for example, if a Christian is brought before the authorities), but the theme is more prominent in the religious commitments and psychosocial understanding of the seer. Thus, in order to understand the dynamics of tribulation on the social boundary, much more is involved than an analysis of social, economic oppression or political persecution.

Cognitive Minority and Social Location

Let us do a quick review. We are trying to locate the Book of Revelation socially in the Roman Empire, more precisely, in the cities of the province of Asia. We can rule out any portrait of Asian Christians as a beleaguered, oppressed minority living as separatists in an isolated ghetto. Christians, for the most part, lived alongside their non-Christian neighbors, sharing peacefully in urban Asian life. There is not even much evidence in the Book of Revelation itself for persistent hostilities towards Christians by Roman officials or non-Christian neighbors. At the same time John is unequivocal in his negative attitude toward Asian society and the empire. This negative attitude is expressed through topics commonplace to the apocalyptic genre such as conflict, crisis, assurances of hope, and exhortations to steadfastness. As generic topoi they do not necessarily indicate anything about the circumstances in which the Book of Revelation arose. They are not, therefore, appropriate clues to the social location of the Book of Revelation in the empire.

Instead of beginning with those topoi, I have argued that we should begin with a more basic element of the genre, namely, that it is "revelatory literature." The Book of Revelation offers to its audience revealed knowledge that may be contrasted with the public knowledge available through such sources as civic institutions, public roles, and spatial arrangements in the city. As revealer of esoteric knowledge the Book of Revelation has its own means of establishing the authenticity and authority of the divine truth it reports. When compared to that public knowledge generally taken for granted in the Roman social order, the revealed knowledge of the Book of Revelation is a deviation from, and a censure of, that public order.

As censure of the public order its author can be compared to philosophical aristocrats and to magicians, diviners, astrologers, and prophets who disturbed "the public mind" through private transmission of values and ideals that went against the order of the empire. In contrast to the coterie of aristocrats with a philosophical bent, the seer of Revelation works out of an ideology far separated from the public realm. Caesar would have had no difficulty differentiating his friends from the seer, had he been on the capital steps in 44 BCE. He is not quite so easy to distinguish from other prophets, exorcists, and magicians in Domitian's empire, yet they, too, are closer to the public order. John operates from a Christian framework that will be integrated fully into imperial structures only centuries later. In contrast to the writings of Paul or to 1 Peter, the seer of Revelation rejects any recognition of the empire as a godly order. In both style and content the writer of the Book of Revelation sets his work against the public order. More accurately, he reclaims all the public order for himself, incorporating the whole of the empire and everything else in his vision of what the world is really like and how one should live in it. His revealed knowledge, not public knowledge, integrates religious, social, economic, political, and aesthetic aspects of the world properly. Here John is unambiguous. Within his vision of reality, he and all those who wear the white garments are pitted against the evil empire.

Finally, the location of social boundaries — by means of which the seer is given a place in the Roman world — cannot be separated from an investigation into *all* of the boundaries operating in the seer's world. An examination of boundary formation offers a way to understand how the seer comprehends his environment. It involves an inquiry into those "laborious *mutual* adjustments" whereby symbolic expressions and social relations reflect a *common* structure that arises out of the process of John's adapting to his specific environment (see Lévi-Strauss 1966, 214, my emphasis). What we are here resisting is the notion that the social location of the seer precedes and offers the clue for understanding how the seer views his world. Social location is an important factor in the boundary formation process, but it is not determinative. Social boundaries are only one dimension of the seer's world boundaries. In diminishing the causal power of the social-historical, I am not returning to a phenomenological position like Schmithals' (see chap. 2 above) or to a theologically driven model that allows little or no force to social-historical realities. The social boundary has its place alongside all other boundaries in the seer's world.

Cognitive Minority/Cognitive Majority

In comparison to the public knowledge embodied in the empire, John reveals a deviant knowledge, that is, one that deviates from public knowledge taken for granted in everyday Roman life (see chap. 10). Peter Berger locates such deviant knowledge socially in a "cognitive minority," which is "a group formed around a body of deviant 'knowledge'" (1970, 6). The cognitive conflict between Roman public knowledge and John's deviant knowledge is not between institutions and roles in society on the one hand and religious commitments on the other: it is a conflict over what the world is really like. The cognitive minority formed around the Book of Revelation refuses to accept the majority's definitions of reality as knowledge, just as (of course) the majority refuses the minority's "knowledge" of what the world is like (see Berger 1970, 7). Each, however, offers an all-inclusive knowledge that incorporates even problematic features such as death, dreams, and other anomic experiences (see Berger 1969, 23, 43). No social space is left out, not even that occupied by the opposition; each explains the other as ill-informed, ignorant, mad, or fundamentally evil; but each has a place in its cognitive system for the other.

In explaining the other and its knowledge, the "cognitive majority" has the edge. Since the majority has the greater number of visible, social institutions and opportunities to socialize members, the "status of a cognitive minority is . . . invariably an uncomfortable one" (Berger 1970, 7). The majority will be condescending to the minority; their deviant knowledge will be considered with "tolerant amusement" or viewed as an "ethnological specimen" (see Berger 1970, 7). The account in Acts 17 of Paul at Athens illustrates this. Epicurean and Stoic philosophers ask, "What is this chatterer [*spermologos*] trying to say?" Others suggest that Paul is promoting foreign gods (Acts 17:18). They find his teaching a curiosity (Acts 17:19–22). After Paul's little speech some scoff, and others say that sometime later they will listen again (17:32–33). So far as we know, Paul never went back to discuss his "deviant knowledge" with Athenians. If, on the other hand, the minority becomes too irritating, it can evoke minor hostilities and temporary attacks against it.

The cognitive minority will find it more difficult to view majority opinions so benignly. A member of a deviant community is uncomfortable with his or her placement in the social order simply because the majority "refuses to accept the minority's definitions of reality as knowledge" (Berger 1970, 7). Negative and hostile expressions toward the majority — and the public social institutions representing the majority — derive in large part from that fact. The rhetoric of the minority may become intense and excessively hostile; for in order to find a place for the majority in their "deviant knowledge," they must explain the majority's refusal to "believe the truth." If the majority tends to assimilate the cognitive minority by considering them ignorant or mad, the minority assimilates the majority by viewing them as evil and satanic. The countercommunity of the cognitive minority becomes "a 'fellowship of the saints' in a world rampant with devils" (Berger 1970, 17–18). Because they are socially located as deviants, members of the countercommunity tend to "know" the dominant social order negatively, so that it becomes identified with chaos, anomy, and evil that constantly endanger their universe of meaning (see Berger 1969, 39).

Ill will towards the majority and extreme statements by the minority about the majority need not presuppose oppression or persecution by those dominant social forces. Their refusal to believe is sufficient cause; that is, social location as cognitive minority is itself a powerful cause for distress, for encouraging steadfastness in the faith, and for comforting the faithful. I suppose that one could describe a cognitive minority as perpetually in a state of crisis because of its social/cognitive location. If so, it should be noted that the crisis stems from the deviant knowledge, rather than vice-versa; that is, because of the character of revealed knowledge, those committed to that knowledge are located socially in a cognitive minority and, therefore, in crisis. Deviant knowledge, status of being a cognitive minority, and a state of "crisis" are all elements in the social location of an apocalypse.[10] Whether or not one finds *crisis* an appropriate term to characterize that location, it is clear that the cognitive structures of the majority continually threaten and endanger the plausibility of deviant knowledge.

Cosmopolitan Sectarianism

In order for cognitive deviance to exist, it must have a "social base." In Berger's terms, "Worlds are socially constructed and socially maintained" (1969, 45). The revealed knowledge needs to be plausible to real, live human beings. If it is not, it is imperiled.[11] According to Berger, members of a cognitive minority tend to organize into a sectarian countercommunity with high boundaries and a strong sense of solidarity among its members (see Berger 1970, 18). In the typology of religious organization, the countercommunity is sectarian: "The sect, in its classical sociology-of-religion conception, serves as the model for organizing a cognitive minority *against* a hostile or at least non-believing milieu" (Berger 1969, 164). So in apocalypses there is often a sharp distinction between the chosen and the rest of the world or the true remnant and the rest of the so-called chosen (e.g., 1 Enoch 93:10, Rev. 1–3). According to Yarbro Collins the Book of Revelation elicits the response of "withdrawing from Greco-Roman society into an exclusive group with rigorous rules and an intense expectation of imminent judgment against their enemies and of their own salvation" (Yarbro Collins 1984, 137). That sect-type countercommunity fits well with the Book of Revelation as deviant knowledge in a cognitive minority; as a revelatory book with its own authority separate from public structures its social base may be seen as a ghetto shielded from the rest of the world by high social boundaries.

The sect-type countercommunity does not, however, fit so well with the all-encompassing scope of knowledge in the Book of Revelation. As a book that voraciously engages all other "knowledge" and assimilates all world structures — public and private — to itself, it calls for a bold cosmopolitan existence in the empire. The key factor in locating the Book of Revelation socially in the Roman Empire involves this somewhat paradoxical relation with the larger, public order stemming from the book's claim to a revealed, but all-encompassing, knowledge. Its source in self-authenticated revelation leads to separation from the larger society, but its all-embracing content calls for a cosmopolitan existence.

That exclusive, yet comprehensive, knowledge transmitted through the Book of Revelation creates flexibility in the types of institutions that can support it; it belongs to a genre not limited to one kind of social base. Some apocalypses may have arisen in conventicles, ghettos, and other sect-type countercommunities, as at Qumran; but others may not have a setting "in a movement or community at all" (Collins 1984b, 21); that is, the social base of an apocalyptic cognitive minority may take the form of separate, sectarian institutions; or it may be more parasitic, attached to "host institutions" found throughout Jewish and Christian life, for example.[12] The writer of the Book of Revelation does not call for a separate institutional formation. He operates more as a "parasite" on existing host churches in the cities of the province of Asia. He calls for reform along specific lines, but he does not quite treat his opposition in the churches as "outsiders." Jezebel, Balaamites, and others named pejoratively by the seer are recognized as members in the churches. For whatever reason, John accepts the existing churches in the cities as his social base.

Perhaps some house churches in the cities were formed along the lines called for by the seer, and other house churches in the same cities followed different guidelines on how to relate to the larger, social order; but of that division we have no evidence. Even if it were true, "John's houses" would be part of the Christian community in those cities. It is true that John, like Ignatius, tries to polarize the churches in Asia (see Schoedel 1980, 32); that is, John draws sharp boundaries within the churches by claiming that there is only one proper attitude to take towards the world. Those who accept the necessity (and pleasure) of living quietly with their neighbors are in the wrong; good Christians assimilate the Roman world as demonic. In fact, one could argue that John's major challenge comes not from outside the church but from Christians who are open to living in the world.

Social Location of the Book of Revelation

By now it should be clear that the Book of Revelation, its author, and its audience cannot be located in the social world of Asian cities through one or two social variables such as economic status or political oppression. It is obviously written within and to Christian communities of Asia, but, as we have seen, Christians included people from various social and economic statuses. It is always tempting to locate apocalyptic spirituality and millenarian movements among those who are oppressed, downtrodden, persecuted, or impoverished; but the appeal of the Book of Revelation has never been limited to Christians who find it difficult to get on with their lives because they are socially oppressed. Wealthy Christians and poor ones may find it attractive; so, also both clever and stupid Christians may be attracted to the Book of Revelation.

John and his audience can, however, be located in Roman society as a group of people who understand themselves as a minority that continuously encounters and attacks the larger Christian community and the even larger Roman social order.

That communal self-understanding leads us back to the paradox of a "cosmopolitan sectarianism." The universal, cosmic vision of the Book of Revelation is grounded in first-century Asian life and necessarily entangles itself in all power structures in all dimensions of human society. But it entangles itself as opposition. It opposes the public order and enters the fray as other "deviant" groups in the empire, not by joining rioters in the streets but by a literary vehicle, a written genre—in John's case, a genre offering revealed knowledge as an alternative to the knowledge derived from the public order. The Book of Revelation struggles to speak for the whole world; yet if it lost its minority status, it would lose its raison d'être. The book could not remain an apocalypse yet speak for the public order. Given this ambivalent relationship with the larger order, the Book of Revelation could be accepted by a person without even relocating into a sect-type social base. It only requires engaging the larger order as the enemy from a particular Christian perspective. The social base of the Book of Revelation must be conceived in such a manner as to allow this essentially ambivalent, symbiotic relationship with the larger society.

The ambivalent relationship of the Book of Revelation to a larger public order helps to explain the diversity of the audience attracted to it. Its audience is not limited to those who are persecuted or oppressed by the larger society. Nor need we limit the book's communicative power only to those alienated by rapid social change involving cross-cultural contact. When one people is conquered by another (as in the exilic and postexilic periods of Judaism) a cognitive majority may be transformed into a cognitive minority, but that condition is neither sufficient nor necessary for the emergence of an apocalypse. The notion that rapid social change dissolves social control, brings distress, tension, and lack of norms, as well as disorganization and disorder may not even hold in very many situations (see Tilly 1984, 53–56). Without doubt, such a notion can be used too facilely to explain why the postexilic or early Christian situations were ripe for apocalypses.[13]

To be sure, the book does communicate an alienation from the larger society that may be attractive to those who are persecuted and oppressed, but it may be just as attractive to those who are momentarily frustrated or dissatisfied with the public order or to those who are disengaged from some aspect of public knowledge. McGinn has shown that the revealed knowledge of an apocalypse could even be called upon when public knowledge and public structures were endangered or implausible (McGinn 1979b, 9); that is, the revealed knowledge of the Book of Revelation can even support a public (Christian) order if "public knowledge" has lost its credibility.

The Book of Revelation offers knowledge—a certain perspective on life—that is not class-specific or status-specific. It belongs to a genre not limited to one kind of social base. It could flourish in a ghetto; in a separate, sectarian institution; or in other, often parasitic social bases. John operates within the existing churches of the Asian cities. Burridge once introduced a disquieting notion to those who seek a clearly delineated social location for the Book of Revelation. He suggested that "apocalyptic messages may be natural to human groups, culturally prescribed, or an intrinsic part of the evolutionary process" (1982, 100). If considered in that light, our attempt to locate John and his audience in a particular social class or social, economic status is wrongheaded. Certainly, the attempt to link the Book of Revelation with upheaval and crisis is wrongheaded.

Down through the ages the book has held appeal for a great variety of people in different social locations and in different kinds of historical situations (see Cogley 1987, 393). Its language and vision hold great explanatory power; for the language is supple and versatile, and the vision grounds human existence in a meaningful, expanded universe. Moreover, the cognitive certainty of the seer is attractive — especially since he lards it with guarantees of significance to human life. Further, the Book of Revelation has been a literary vehicle for providing a "cognitive distance" from the public, social order and thereby providing space for critiques of the public order, for creating a satisfying dissonance in human activity (a bulwark against boredom) between public and revealed knowledge. Various combinations of elements in the Book of Revelation have made it attractive to a wide variety of people in every age. Its original audience need not have been any less diverse.

12

Postscript: Religious Language and the Social Order

Where are the Christian-filled lions? Where is crazed Domitian who sends out his legions against hapless Christians? Where are the oppressed slaves with crosses proudly worn inside their tunics? Where is the crazed seer caught in a state of spiritual madness?

It all seems so ordinary: an emperor who negotiates political decisions with senators and provincials so as to assure stability in the empire; economic prosperity and social integration of cities in the Asian province where John and his audience live; for the most part a coincidence between life-styles and attitudes of Christians and non-Christians in those cities. The Book of Revelation does not contradict this portrait: there are few indications of crises involving institutions outside the Christian communities, every indication that the seer is writing in a particular generic tradition that influences his choice of images, his construction of scenes, and his portrait of the world. His audience is made up of ordinary people committed to Christianity: people sharing in the provincial prosperity, albeit some more than others; a diverse group from different social locations who find an alternative vision to the public order attractive.

Can religion, especially apocalyptic spirituality, come to expression in such ordinariness? Does an approach to the social and psychological dimensions of apocalypticism (at least one that claims scholarly credibility) not require some kind of abnormal, out-of-the-ordinary social, historical, political, economic, or psychological situation as explanation for the outbreak of apocalyptic spirituality? In short, can religious expressions, especially of this apocalyptic variety, be understood as part of those normal, ordinary processes through which humans accommodate to everyday realities of existence?

Perhaps not all apocalypses can be so understood. Some—for example, the Book of Daniel—may be the product of a crisis in the political history of Judaism. Others may reflect deprivation (relative or otherwise) on the part of their audience.

198

Jung may be right about some writers of apocalypses—that they express long-pent-up negative feelings, chronic virtuousness, such brutal impacts of contrarieties as a sea of grace met by a seething lake of fire (see Jung 1973, 75–76, 87, 89)—but those comments are probably not appropriate to the seer of Revelation. The Book of Revelation arises from normal processes of adaptation to ordinary human conditions. Social or psychological deviance is not a necessary—though it may be a sufficient—cause for the writing of an apocalypse; powerful religious expressions, even of the apocalyptic type, may arise in other social and psychological contexts.

This hypothesis of ordinariness may be seen by many as radical. It goes against opinion about the Book of Revelation specifically, more generally about the Roman world into which Christianity was born, and even more generally, about most social and historical inquiry into millenarianism. The notion that in each of those instances the social historian may be encountering ordinary, normal life that does not involve a state of spiritual bankruptcy or social chaos or political oppression may be unpalatable, for it does not distinguish in any qualitative manner early Christian life and experience from our own life and experience, and it brings the seer's language and vision contiguous to all other kinds of language and visions. As human expression it becomes available to human experience. That contiguity does not filter out the mythic power of John's language or the awesomeness of his vision; rather, it enlarges human expression and the possibilities of human experience.

Broadly speaking, I place the seer's product in the context of those "laborious adjustments" by means of which humans adapt to their environment. As with all such adjustments, there is a local context: a particular place; a particular time; a particular social, historical situation; a particular psychological and biological makeup; and a particular set of communal, social institutions. The seer refracts those particulars through his apocalyptic language so as to create a particular whole vision of what the world is like and how he and his audience fit into it. Among the various media the seer could use to contribute toward the "laborious adjustment" (e.g., institutional development, flag waving, education), a literary vehicle takes pride of place. He offers to his audience a linguistic vision.

Language, the substance of that vision, is a powerful force. Words, for the seer, are not simply spoken and written thoughts; they are worn, held, and eaten. They can cause an upset stomach. They control destinies. They initiate action. Words and social experience are not discrete entities striking each other like billiard balls. To speak of his language as merely a response to a destructive and oppressive society separates verbal symbolisms and social experience too much. To describe his rhetoric as simply encouraging others, strengthening their resolve, and exhorting them in the face of death externalizes too much the relationship between verbal expression and social intercourse. Each is essentially affected by the other, as both participate in the same cultural system.

The reciprocal relations between language and experience can be seen in John's treatment of the Christian proclamation (kerygma). The kerygma is a verbal proclamation that enfolds a mode of Christian existence in the world. Similarly, Christian existence enfolds the verbal proclamation. The repeated formula "the Word of God and the witness of Jesus Christ" testifies to both verbal and social proclamation and to an adaptive strategy by means of which Christians engage with their world.

There is, of course, a surplus to experience that cannot be refracted through language; yet John's language (like all language) catches and mirrors our social experience while it shapes and structures that experience. His language works its "poetic," creative function as it creates and molds both human and environment into a livable, workable relationship. Working in that powerful medium of mythic language, the seer relates his local conditions to what is most inclusive, most cosmic, most total.

Must inclusive visions necessarily be uncompromising? Must John's inclusive vision require hard boundaries between esoteric, Christian knowledge and public, Roman knowledge? His vision is attractive in part because he offers a "rock-bottom" fundamental orientation to the world that is at the same time totally inclusive. And his total inclusiveness seems to require cognitive boundaries between true and false knowledge. The cognitive majority must be wrong if his cognitive minority is right. Yet it is this total inclusiveness—that God is all in all—that is belied in those hard cognitive boundaries.

Would we belie his vision if we asked him to relativize slightly his rock-bottom orientation so that his knowledge was not absolutely identical to God's? Does the all-inclusiveness of his vision not call for at least the possibility that God may be at work in both his and his opponents' cognitions? We are dealing here in part with the place of revenge and retribution in the Book of Revelation—with the confidence that those outside the circle of true knowledge cannot succeed, that they must be outside the city gate and beyond the temple wall. The cognitive boundary becomes harder and harder as the seer moves from Nicolaitans and Jezebel to Jews to denizens of the Asian cities. Is there hope for those outside? Is the city gate of the New Jerusalem closed forever on such people? Can God be all in all if the gate is closed for eternity?

The seer would be more faithful to his vision of an unbroken wholeness if he did subvert his cognitive exclusiveness. John's radical monotheism should leave a place for Roman knowledge. Then the God would embrace a higher dimensional ground than the dimensions of John's visions. There would be a divine surplus not captured by John's language. At times we get glimmers of this possibility, most especially in connection with the irony of the crucified king. Here power, royalty, and maybe even true knowledge are susceptible to ironic turns. The slain Lamb may provide a metaphor for subverting revenge and retribution while assuring cognitive stability. Could rulers and ruled, cognitive confidence and cognitive humility, rock-bottom certainty and openness to something more fundamental be allowed their creative tension under that metaphor without this becoming a justification for either the status quo or silent weakness?

In that neverending "laborious process" of adaptation, rock-bottom certainty has its necessary place. We need to be able to express in words what we know, to know with certainty what we encounter, and to encounter absolutely the true "God." Such certainty is important for psychological well-being and social location. In that process, hard boundaries emerge separating us from others — for example, from those who claim a different formulation of rock-bottom knowledge in a different social location. Such certainty leads to the closing of the millenarian mind. In the Book of Revelation, countering that process, there is a continual softening of boundaries that leads to the opening of the millenarian mind: the boundaries between contrarieties are not as hard as they first appear; ironic and eschatological reversals separate appearances — including hard, inpenetrable boundaries — from reality. In this process vengeance, exclusiveness, and certainty about the adequacy of particular human insight (even revelation) fade.

This dialectic between hard and soft or closed and open is a dimension of the seer's language that millenarians (as well as scholars) should find of interest. Boundaries are not done away with; we (like John) need them to orient ourselves to the world. But the boundary is no longer fundamentally between us and them but on another plane — that of closed and open. No boundary, not even the cognitive boundary between majority and minority, becomes the ultimate divide. Further, there is a different relationship to the boundary. Those with vengeance, moral exclusiveness, and certainty about the absolute adequacy of their insight are masters of the boundary. The seer's language of transformations and reversals makes mastery of boundaries very difficult. In fact, he is time and again taken across boundaries over which he has no power. Entering fully into the seer's transformations, homologies, and ratios, the reader experiences along with the seer something like Menelaus' ride on Proteus. Menelaus holds tight while Proteus, the shapeshifter, passes from stability to transformation to new stability. Together, they experience surprise and discover novelty.

Any serious reader of the Book of Revelation will want to reckon with that protean aspect of the work. Judgments and claims will still be made; convictions will still be held. But if attention is given to the transformational character of the Book of Revelation — if attention is given to powerful ironic images such as the slain Lamb — judgments and convictions — even those on which we make decisions and take actions — may be subverted by the seer's language. Ultimately the boundaries of the seer's world are soft; they are points of transaction whereby religious promises, social encounters, political commitments, biological givens, and cultural demands undergo mutual adjustments and form homologous relations. In short, the seer would draw the reader into a protean world where God alone is master of all boundary transactions.

APPENDIX

Recent Theories about the Social Setting of the Book of Revelation

Although most scholars may conclude that the latter part of Domitian's reign offers the best social context for the Book of Revelation, the text of Revelation is not sufficiently explicit about time and date to provide ready-made connections between Revelation and a specific social, historical context. As we have seen (chap. 1), even when the seer's language seems to make very specific references, as in the description of the seven kings in Revelation 17:9-10, the political allusions are not clear. So even though the language of Revelation must be the beginning point and primary source for all study, it does not in itself make connections with a social situation. Those connections can be made only with the help of an underlying theory about how the language of Revelation is shaped by its originating setting and how the language functions in its social situation.[1] The following review of some recent approaches to the social setting of the Book of Revelation illustrates some of the different connections that have been made between the language of Revelation and its social situation.

Colin Hemer

Colin Hemer's book (1986), although limited to the letters to the seven churches (Rev. 2-3), should nonetheless be mentioned if for no other reason than that it carries on the tradition of Sir William Mitchell Ramsay, the late-nineteenth-century historian of Asia Minor. Ramsay traveled extensively in Asia Minor, collected inscriptions, observed terrain, and linked the language of the New Testament to the situation in which it arose. His work is sometimes appealed to to prove that the New Testament is "true," that is, that it refers to actual situations at the time in which it was written. Hemer began his work as a "reassessment of Ramsay" and has obviously remained close to Ramsay's approach and conclusions (1986, x).

Hemer plays the historian rather than the literary critic or sociologist with the Book of Revelation. The visionary, symbolic language of Revelation is referential language; symbols and images find explanation in the historical, topographical, and social setting of the seven churches (Hemer 1986, 16-20). For example, those at Ephesus who conquer will "eat of the tree of life" (Rev. 2:7). This image refers to the Genesis passage, but for Hemer such an image must also have "peculiar applicability" to those at Ephesus. In this case he thinks that John uses this image in the message to those at Ephesus because there it "had an analogue in the Artemis cult"(p. 42). The connection with the setting has, for Hemer, more explanatory power than any other connection.

In explaining verbal and imagistic similarities and repetitions in the Book of Revelation or between the Book of Revelation and some other work — for example, Ignatius' letters—Hemer plays down the importance of literary or generic influence and looks instead to common situations.[2] Ignatius uses in his letter to the Ephesians the phrase "that we may be his *naoi* [*temples*]," an image somewhat similar to that in Revelation 21:3. The similarity arises not from John and Ignatius using a common stockpile of images or from drawing upon a common body of literature but from "features of the city" of Ephesus that both John and Ignatius knew (p. 54). As with Ramsay's work, connections between the language of the Book of Revelation and the situation of the seven cities are often tenuous and of little individual significance (pp. 7, 26), but Hemer argues that the tenuous, insignificant examples accumulate into a defensible case for the situational significance of the specific language used in Revelation (e.g., p. 224).

According to Hemer the letter to the Laodiceans (Rev. 3:14–22) demonstrates his approach most convincingly (see p. 224). Here, in brief, is his treatment of several phrases from that letter:

1. "The words of the Amen, the faithful and true witness, the beginning of God's creation" (3:14). Hemer reviews allusions to the Old Testament, something he criticizes Ramsay for not doing. The most striking parallel, however, is to the New Testament letter to the Colossians (Col. 1:15)—Christians living near Laodicea in the Lycus Valley. The similarities are explained as replies "to similar tendencies of thought persisting in the district" (p. 185).
2. "You are lukewarm." The special appropriateness of this language relates to the water supply at Laodicea. The water in that area is tepid and bad, functioning as an emetic (see Rev. 3:16). Details are somewhat obscure.
3. "I am rich" (Rev. 3:17). Laodicea was a wealthy, self-sufficient city with woolen industries and a district bank. Residents reconstructed the city after an earthquake in Nero's reign. Revelation 3:17 should be read "against the background of the boasted affluence of Laodicea, notoriously exemplified in her refusal of Roman aid and her carrying through a great programme of reconstruction in a spirit of proud independence and ostentatious individual benefaction" (p. 195).
4. "Salve to anoint your eyes" (Rev. 3:18). Though there is no solid evidence for ophthalmology at Laodicea, there is sufficient evidence to consider this passage in connection with a medical center nearby and with the manufacturing of eye salve.
5. "White garments to clothe you" (Rev. 3:18). This is "an allusion to the clothing industry of Laodicea and in particular a contrast with the glossy black wool of its sheep" (p. 199).
6. "He who conquers, I will grant him to sit with me on my throne, as I myself conquered and sat down with my Father on his throne" (Rev. 3:21). Local allusions here include references to a branch of the Zenonid dynasty, a dynasty important throughout Asia Minor.

Hemer's work represents one way of trying to relate the language of the Book of Revelation to a social setting. He assumes that the language is transparently referential and that a major task of the social historian is to find the references in the social situation. To do that, Hemer must sometimes search through a wide range of literary, epigraphic, and numismatic sources; and after such a search, the connections often remain very tenuous. He and others who relate language and social setting in this manner must rely on the accumulative effect of a number of uncertain connections. Even if one leaves aside some of the issues about language and society explored in chapters 1 and 2, Hemer's approach remains unconvincing for two reasons: (1) several tenuous connections do not accumulate into a strong group of connections; and (2) the casting of the net over literary, epigraphic, and numismatic sources is overcontrolled by a determination to find situational parallels to the text in the Book of Revelation. In principle everyone engages in net casting, in finding support for a hypothesis about text and social context; but Hemer's approach does not have any built-in check: complete disconfirmation would be rare, given how he proceeds. Nevertheless, Hemer's concern for "the specifics of place and time" should be included in all theories about social setting.

John Court

John Court approaches Revelation as a work formed by a "creative literary artist" who has combined in a deliberate and skillful manner traditional mythological ideas and allusions to historical situations contemporary with the artist (1979, 164). Interpreters of Revelation always need to discern both historical allusions and traditional mythology; otherwise, they will lose control of the material and fail to grasp the form in which the seer has presented his work.[3] In combining history and traditional mythology, the author of Revelation follows the lead given in the Old Testament, where myth is historicized and history is mythologized (p. 165). Moreover, the combination of history and myth draws Revelation firmly into the circle of apocalyptic, where "intensive and esoteric use of imagery" (i.e., myth) is coupled with an interpretation of history (p. 165). In that way history is given meaning through "the apocalyptists' vision of what — or, more accurately, who — transcends history" (p. 165). Koch's observation about apocalyptic — "An apocalypse is . . . designed to be 'the revelation of the divine revelation' as this takes place in the individual acts of a coherent historical pattern" — is nowhere better demonstrated than in the Book of Revelation (Court 1979, 165). The historical situation controls the use and selection of mythological ideas, and the mythological ideas interpret and provide a framework for understanding the historical situation (pp. 10, 19).

Court's method of relating traditional mythic themes and historical situations contemporary with the seer is especially well exemplified in his interpretation of the opening of the first six seals (Rev. 6:1-17; Court 1979, 50-70). According to Court the seer here draws from two traditional sources: the prophet Zechariah (Zech. 1:8-15, 6:1-8) and the Synoptic apocalypse (Matt. 24, Mark 13, Luke 21). The four horsemen reflect a traditional theme traced back to Zechariah — the divine control over all history. By means of this traditional theme, the writer of Revelation offers "a prophetic reinterpretation of the contemporary situation and recent events" (Court 1979, 59). Court relates the opening of the third seal (Rev. 6:5-6) to famine in Asia Minor and to Domitian's attempt to limit the growing of vines and olive trees and to encourage the growth of wheat and barley. The white horse of the first seal is depicted to reflect features of Mithras, "the warrior-god whose cult was well known in Asia Minor" (p. 62). The saber of the second seal (Rev. 6:4) may allude to the rumours of war that were regularly reported on the eastern border of the empire or, more specifically, to Domitian's despotic cruelty of putting Civica Cerialis, the proconsul of Asia, to death (pp. 63-64).[4] The sick, yellowish green color of the fourth horseman (Rev. 6:7-8) may reflect "an unrecorded outbreak in the time of Domitian" of some pestilence often associated with famine. Together, the four horsemen represent "God's 'four sore acts of judgement' in Ezekiel 14:21 (sword, famine, evil beasts, and pestilence)" (p. 65) and relate those judgments to current concerns in the immediate situation in Asia Minor. The combination of myth and history reinforces the point that in the contemporary difficult situation God is in control. Even martyrdom — reflected in the opening of the fifth seal (Rev. 6:9-

11), embracing martyrs from Steven to Antipas and those soon to come—is placed in "the working out of God's plan" (p. 67): "The author's own experience is sufficient evidence for a limited persecution, if only of individual Christians, at the time of writing, and the expectation of much more. . . . It is likely that John saw the threat of much worse to come, and composed this work to meet that threat" (pp. 66–67). The opening of the sixth seal (Rev. 6:12–17), which reinterprets material from Luke 23:28–31, anticipates the great cosmic cataclysm in the end days. Altogether, the opening of the six seals reappropriates earlier prophetic and apocalyptic traditions to the situation in Asia Minor so that Christians who have suffered in various ways during Domitian's reign can read the "signs of the time" and be encouraged and reassured that in all the disasters God is in control (p. 70).

Court's approach to Revelation is illuminative for what it does not do, as well as for what it does do. Court—like Hemer—usually begins his analysis of a block of material by looking for possible allusions to actual historical situations, for he assumes that traditional mythic and eschatological motifs are drawn upon to illumine actual situations in the lives of the seer and those he is addressing (see p. 19). In that regard he often seeks to make a one-to-one connection between a passage in Revelation and some aspect of an historical situation, for example, the associations of the four horsemen to situations and events contemporary with the seer. This approach becomes especially evident in his analysis of the letters to the seven churches. There Court draws on the research of Ramsay and Hemer to show connections such as the "tree of life" in the Ephesus letter and tree shrines associated with Artemis of Ephesus (p. 29) or the "white stone" in the Pergamum letter with white marble at Pergamum (p. 33). By linking text and history in this way, Court never considers the *whole* text of Revelation as a symbolic universe or a constructed literary world, to be related to the social, historical world of Christians in Asia Minor. So Schüssler Fiorenza refers to his work as a "somewhat unsophisticated discussion of the imaginative, mythopoeic language of Revelation" (1985, 22).

Schüssler Fiorenza

In contrast to Court, Schüssler Fiorenza considers separate themes or topics in the text as parts of a "total configuration." An analysis of smaller units must be complemented by considering their place in the work as a whole, for "small units . . . change their formal characteristics and function differently when they are incorporated into the new framework of a complex literary type or genre (*Gattung*)" (1985, 164). She views Revelation as "a unique fusion of content and form" (p. 159) that can best be analyzed and understood through the tools of literary criticism.

For Schüssler Fiorenza the first step in interpreting Revelation involves an examination of its poetic, evocative language and symbols (p. 183). In this examination the specific poetic language of Revelation must not be left behind in favor of theological or philosophical abstractions, for form and content are inseparable (p. 184, 159). She rejects both archetypal analysis, in which specific symbols represent "transhistorical realities" and theological analysis, which restates the content of Revelation in nonpoetic language (pp. 23, 26, 183–86). When approached as poetic, evocative language, the symbolic language of Revelation—like all poetry—opens up, rather than limits, interpretations, "evoking rather than defining meanings." "It becomes necessary for interpreters to acknowledge the ambiguity, openness, and indeterminacy of all literature" (p. 186). To understand fully this multiplicity, a lexicon—or perhaps better a grammar and syntax—of the imagery and symbols of Revelation would have to be written.

Schüssler Fiorenza, however, does not exactly pursue that solution in trying to control the ambiguous language of Revelation. Instead of developing a complex network of meaning relationships in the poem, she calls for the following two criteria to assess a particular interpretation: (1) it "must make 'sense' with regard to the overall structure of the book" (p. 187), and (2) it must "fit" the historical-rhetorical situation to which it is a response (p. 183). The first criterion is appropriate for any poetic analysis, but the second criterion in a sense belies the emphasis on Revelation as poetry. Revelation should perhaps be read as poetry, but finally she states that there is a "difference between a work of literature and the NT writings. Whereas a work of art is considered as a system or structure of signs serving a specific esthetic purpose, . . . the NT books are theological and historical writings." She then concludes that "one must not only analyze the literary patterns and structure of a writing but also their relation to its theological perspective and historical setting" (p. 159). In other words, the language of Revelation is not "just linguistic-semantic but also always social-communicative" (p. 183), and that "social-communicative" must involve not only the reader or hearer of a text but also the originating situation of the text: "In other words, we are never able to read a text without explicitly or implicitly reconstructing its historical subtext within the process of our reading" (p. 183). Literary analysis of Revelation is ultimately controlled by social, historical analysis or, in Schüssler Fiorenza's terms, by rhetorical analysis.

Since "John did not write art for art's sake" (p. 23), interpreters must do more than a "purely formalistic literary" analysis of the work (p. 23); they must consider "the communicative situation and literary-social function of Revelation." (p. 23).[5] Here the epistolary framework of Revelation is crucial; John's purpose in writing is similar to that of Paul in his letters; but John strengthens, encourages, and corrects "Christians in Asia Minor who were persecuted and still must expect more suffering and harassment . . . not simply by writing a hortatory treatise, but by creating a new 'plausibility structure' and 'symbolic universe' which he communicates in the form of an 'apostolic'-prophetic letter" (pp. 23–24);[6] that is, "John creates a 'literary vision' instead of a sermon and tractate" (p. 6), but that vision functions rhetorically to motivate and persuade Christians in Asia Minor to remain faithful "in the face of harassment and victimization" (p. 6). For Schüssler Fiorenza, the situation of "tribulation and persecution" under Domitian is fundamental for understanding the rhetorical function of the work as one that strengthens endurance and hope (pp. 8, 114). Revelation as a symbolic-poetic work combines with its visionary rhetoric so as to invite imaginative participation in the symbolic world created in the text: "The strength of its persuasion for action lies not in the theological reasoning or historical argument of Revelation but in the 'evocative' power of its symbols as well as in its hortatory, imaginative, emotional language, and dramatic movement, which engage the hearer (reader) by eliciting reactions, emotions, convictions, and identifications" (p. 187).

The priority that Schüssler Fiorenza gives to the "rhetorical function" thus leads her to a commonly held understanding of apocalypses: the Book of Revelation arises in response to harassment and persecution, and its rhetorical function among its readers and hearers mirrors its originating situation (distress/comfort, persecution/perseverence, despair/hope). In contrast to the rich variety and multivocal meanings of Revelation as a symbolic-poetic work, as a rhetorical work it has a fixed and precisely limited meaning.[7] Criticism on the symbolic-poetic level is finally not crucial, for the "true" interpretation reduces that rich symbolism to a social functionalism: "The symbolic universe and world of vision . . . is a 'fitting response' to its socio-political 'rhetorical situation'" (p. 6).[8] Social experience determines the rhetorical elements in the work.[9]

John Gager   •

John Gager has not written extensively on Revelation, but his provocative remarks in *Kingdom and Community* opened up a new brand of sociological analysis of the early church that has had great importance for the study of Revelation. Gager also interprets Revelation as a response to oppression, more specifically to persecution and martyrdom, and as a work that functions to console its readers and hearers (1975, 50). He discusses Revelation as one way in which "the millennium has in some sense come to life in the experience of the community as a whole" (p. 49).[10] Drawing on the work of Claude Lévi-Strauss, Gager states that Revelation brings the millennium through "mythological enactment" (p. 50). Following Lévi-Strauss's structural analysis of myth, Gager segments Revelation into blocks of text (mythemes) that are then organized into two contrasting categories: victory/hope on the one hand and oppression/despair on the other. Revelation 4:1–5:14, 7:1–8:4, 10:1–11:1, 11:15–19, 14:1–7, 15:2–8, 19:1–16, and 21:1–22:5 belong to the first category; Revelation 6:1–17, 8:5–9:21, 11:2–14, 12:1–17, 13:1–18, 14:8–15:1, 16:1–20, 17:1–18:24, and 19:17–20:15 belong to the second, contrasting category (p. 53). No middle ground can reconcile those contrasting categories because they reflect an irreconcilable tension in the life of Christians: "On the one hand was the belief that, as Christians, they were the chosen people of God, protected by him and assured of eternal life in his kingdom. On the other hand was the overwhelming experience of suffering, deprivation, and death at the hands of those whom they most despised" (p. 51). Or the conflict may be seen as a tension between hope and reality, what ought to be and what is, an ideal past or future and a flawed present (p. 51). In brief, Christians experienced an unbearable conflict between their religious convictions and their social, political experience.

The language and symbols of myth function to soften that experience of conflict. In the words of Lévi-Strauss, "mythic thought always progresses from the awareness of oppositions toward their resolution" (1967, 221). The Book of Revelation resolves those oppositions by two means. First, the text oscillates between symbols of victory/hope and oppression/despair so that the reader gradually assumes that oppression and despair are penultimate. So, for example, "the opening of the sixth seal [oppression/despair] is followed not by the seventh seal but rather by a vision of the one hundred and forty-four thousand who bear the seal of God [victory/hope]" (Gager 1975, 54). The oscillation from oppression to victory serves "to undermine any tendency among the audience to treat" the conflicts in their experience as "permanent, unbearable contradiction" (p. 54). Secondly, as myth Revelation is structured so as to function as a suppressor of time. Through "mythic enactment" the faithful readers and hearers experience future bliss in the present moment (p. 50). Thus Gager argues that hearing or reading Revelation becomes a form of therapy, "much like the technique of psychoanalysis, whose ultimate goal is to transcend the time between a real present and a mythical future" (p. 51). Drawing heavily from Lévi-Strauss, Gager argues that like the process of psychoanalysis, apocalyptic myths "manipulate symbols" so as to change the reality of the patient/reader (p. 51).

Unfortunately, however, the suppression of time is only illusory. In contrast to psychoanalysis, the manipulation of symbols in myth does not finally change the reality of Christian existence. Listening to the Apocalypse read aloud in worship may provide "a fleeting experience of the millennium," but the "real world" of persecution and deprivation reasserts "itself with dogged persistence for Christian communities" (p. 56). Nothing fundamental is changed by the myth. The world of social, political realities is too real for the alternative symbolic world of the myth. At best, the transitory, alternative mythic world provides "energy needed to withstand the wrath of the beast" (p. 56). The oscsillating text establishes the penultimacy of oppression and despair, but victory and hope elude the machinery of myth over the long haul.[11]

Schüssler Fiorenza and Gager agree that the apocalypse arose in response to the felt contradiction between faith and experience. In language similar to Gager, Schüssler Fiorenza writes, "Like John, Christians of Asia Minor suffered a deep tension between their faith and their experience. They believed in the ultimate power of God and Christ, but at the same time they experienced daily their powerlessness in the face of harassment, oppression, and persecution. Their everyday experiences ran counter to their belief in God's power and undermined their hope in God's empire, glory, and life-giving power" (1981, 28). Schüssler Fiorenza and Gager disagree, however, on how Revelation responds to that conflict between religious belief and social experience. For Gager, Revelation takes the Christian temporarily into a world of millennial bliss; it functions as a brief reprieve from the hard political realities. For Schüssler Fiorenza Revelation mobilizes, persuades, and provokes Christians to stake their lives on faithful obedience. She would agree with Gager that Revelation assures Christians of the penultimacy of oppression, but she questions Gager's notion of mythic suppression of time. In fact, she argues for the reverse. The author of Revelation "stresses that Christians do not yet actively exercise their kingship. Eschatological salvation is near but not yet present. . . . He speaks of future salvation for the sake of exhortation" (1985, 168). The rhetorical strategies of Revelation require the immediacy of the rhetorical situation out of which they arise; but for Schüssler Fiorenza the seer never offers a *present* alternative to the actual world of inescapable political realities through manipulation of symbols. Symbolic manipulation cannot change those hard realities; it can only provide reassurance and strengthen the faith of those who hear: "Language cannot remove or correct 'the brute realities' of the social-political exigency and of religious 'tensions,' but it can help us to control their destructive effects" (1985, 198). In sum, both Schüssler Fiorenza and Gager agree that the Apocalypse was a response to actual social, political persecution and oppression and that Christians felt a conflict between their political situation and their religious faith. They disagree fundamentally on how the work comforts and encourages those Christians in crisis.

Yarbro Collins

Yarbro Collins presents a subtle sociological analysis of the Apocalypse (1984). In contrast to Gager and Schüssler Fiorenza, Yarbro Collins does not view Revelation as a response to political oppression or persecution. There were elements of crisis in the social environment of Christians in Asia Minor at the end of Domitian's reign that are probably reflected in the Apocalypse—conflict with the Jews, mutual antipathy toward neighboring gentiles, conflict over wealth both within the society of Asia Minor and between Rome and the eastern empire, and the somewhat precarious relations between Christian organizations and Roman officials (1984, 84–99).[12] But such elements of crisis cannot be seen as the sole "or even primary" explanation for the apocalyptic message (p. 105). Factors such as personal background, temperament, and theological perspective "are at least as important as aspects of the sociohistorical situation" (p. 105). Yarbro Collins formulates the relationship between the Apocalypse and its social environment in the following way: "The book of Revelation is not simply a product of a certain social situation, not even a simple response to circumstances. At root is a particular religious view of reality, inherited in large part, which is the framework within which John interpreted his environment" (pp. 106–7). As a result, she argues that the Book of

Revelation is a product of the interaction between John's preunderstanding of his situation and the sociohistorical situation itself (p. 106).

As a way of recognizing and including those subjective factors of background, temperament, and theological perspective, Yarbro Collins introduces the terms *perceived crisis* (see chap. 2 above) and *relative deprivation*. The term *perceived crisis* makes it possible to connect apocalypses to social crises while recognizing that no *evident* social crises are necessarily present. So Yarbro Collins writes that Revelation "was indeed written in response to a crisis, but one that resulted from the clash between the expectations of John and like-minded Christians and the social reality within which they had to live" (p. 165). Crisis dimensions of the social situation are evident only through the seer's angle of vision, and other people discover those crisis dimensions only by reading Revelation. That approach contrasts with Schüssler Fiorenza's, which requires a "rhetorical situation" that "is not just a product of the mind and psyche of the author" (1985, 8). *Relative deprivation* refers to any situation where there is "marked disparity between expectations and their satisfaction" (Yarbro Collins 1984, 106).[13] In John's case relative deprivation occurs because of the "conflict between the Christian faith itself, as John understood it, and the social situation as he perceived it. A new set of expectations had arisen as a result of faith in Jesus as the Messiah and of belief that the kingdom of God and Christ had been established. It was the tension between John's vision of the kingdom of God and his environment that moved him to write his Apocalypse" (p. 106).

Thus Yarbro Collins shares with Schüssler Fiorenza and Gager the assumption that disparities between religious belief and social, political experience create the crisis to which Revelation is a response: for the seer this conflict was virtually irreconcilable because his "hopeful faith" included "the conviction that God's rule must be manifest in concrete political ways and that acknowledgment of God's rule is incompatible with submission to Rome" (p. 143).[14] That conflict between the two kingdoms is reflected in the Apocalypse in the contrasting symbols and sets of symbols portraying God, Jesus, and his followers on the one hand and Satan, the beasts, and their devotees on the other (p. 141).[15] The seer wrote Revelation in order "to overcome the unbearable tension perceived by the author between what was and what ought to have been. . . . Its task is to overcome the intolerable tension between reality and hopeful faith" (p. 141). The "dualistic structure" of Revelation serves two purposes: it points the conflict out to those unaware of it, and it overcomes or meditates the tension through manipulation of symbols along the same lines discussed in connection with Gager (Yarbro Collins 1984, 141–42). Yarbro Collins also underscores a certain rhetorical function to the book: it calls for commitment on the part of the readers and hearers "by the use of effective symbols and a narrative plot that invites imaginative participation. This combination of effective symbols and artful plot is the key to the power of apocalyptic rhetoric" (p. 145).

Revelation creates its effect primarily by means of expressive, not referential language (p. 144). It both displays and evokes attitudes and feelings that lead the reader or hearer to participate in the imaginative world constructed in Revelation. The seer claims the authority of divine revelation and true interpretation of Scripture that draws analogies between scriptural and contemporary situations. He also draws analogies between the contemporary situation and "traditional myths of combat and creation," by means of which the struggle with Rome is presented in terms of "the old conflict between order and chaos." That analogy clarified the situation of the hearers "and gave it meaning" (pp. 148–150). By such techniques the seer manipulates the feelings of his hearers and readers so that they experience revealed truth and assurance that their powerlessness is only temporary, for their victory is as certain as was creation over chaos (p. 152). Fear, suffering, and death are not denied, but they are penultimate (p. 152).

Finally, however, for Yarbro Collins the expressive language of Revelation deals with the crisis of faith by providing a catharsis for its hearers. Just as tragedy provides a catharsis of fear and pity for those who share in it, so the Apocalypse offers a catharsis of fear and resentment (p. 153).[16] Hearers of the Apocalypse are cleansed of fear and resentment in part simply by the intensification and "objective expression" of those emotions in the motifs of conflict, oppression, destruction, judgment, and salvation in the Apocalypse: "The feelings are thus brought to consciousness and become less threatening" (p. 153). Here Yarbro Collins depends heavily on psychoanalytic theory as applied to biblical material by Gerd Theissen (1977). By projecting the perceived conflict with its accompanying emotions onto a larger cosmic screen (that of God and Satan, for example), the Book of Revelation clarifies and objectifies negative feelings and displaces them upon Jesus or God (Yarbro Collins 1984, 153, 161). Finally, Christians are called on to turn their emotions of resentment and aggression back upon themselves so as to create greater moral demands, for example, sexual abstinence, poverty, more intense isolation from the social order, or even martyrdom (p. 157). In such ways the symbolic world of Revelation affects the behavior and action of Christians in Asia Minor: "It is a text that enables hearers or readers to cope in extreme circumstances" (p. 156). It does so, however, through "an act of creative imagination" (p. 155). As an imaginative act, the apocalyptic process is analogous to the creative imagination of a schizophrenic who "feels the pain of the human existential dilemma more acutely than others"; "by means of elaborate fantasies, the schizophrenic is able to live with the terror of reality" (p. 155). For Asian Christians Revelation offered an escape from reality by enabling them to experience the present as a time when religious hopes were realized: "What ought to be was experienced as a present reality by the hearers in the linguistic and imaginative event of hearing the book read" (p. 154). Yarbro Collins thus offers a solution similar to Gager's in that both resolve the religious crisis of faith in a temporary imaginative experience that does not affect the hard social and political realities of Asisn life (see Schüssler Fiorenza 1985, 8).

ABBREVIATIONS

Loeb refers to editions in the Loeb Library series of Greek and Latin texts with facing translations. *Ante-Nicene* refers to translations in the Ante-Nicene Fathers series. All other editions are specified by date and/or editor's name.

Ap. Abraham	Apocalypse of Abraham (Charlesworth, 1983)
Apol. Tyan. *Ep.*	Apollonius of Tyana, *Letters* (Loeb, see Philostr. *VA*)
Apul. *Met.*	Apuleius, *Metamorphoses (Golden Ass)* (Loeb)
Aug. *Conf.*	Augustine, *Confessions* (Loeb)
Barn.	*Epistle of Barnabas* (Ante-Nicene)
Cic. *Flac.*	Cicero, *Pro Flacco* (Loeb)
Cic. *Leg. Man.*	Cicero, *Pro Lege Manilia* (Loeb)
CIG	*Corpus Inscriptionum Graecarum* (1828–77)
CIJ	*Corpus Inscriptionum Judaicarum* (1952)
CIL	*Corpus Inscriptionum Latinarum* (1863–)
1 *Cl.*	*First Epistle of Clement to the Corinthians* (Ante-Nicene)
Corp. Jur. Civ.	*Civil Law* (S. Scott, 1932)
Dessau, *ILS*	H. Dessau, *Inscriptions Latinae Selectai* (1892–1916)
Did.	*Didache (Teaching of the Twelve Apostles)* (Ante-Nicene)
Dig. Just.	*Justinian Digest* (see *Corp. Jur. Civ.*)
Dio Cass.	Dio Cassius, *Roman History* (Loeb)
Dio Chrys. *Or.*	Dio Chrysostom, *Orations (Discourses)* (Loeb)
Dittenberg. *SIG*	W. Dittenberger, *Sylloge Inscriptionum Graecarum* (3rd ed., 1915–24)
1 Enoch	1 Enoch (Charlesworth, 1983)
2 Enoch	2 Enoch (Charlesworth, 1983)
3 Enoch	3 Enoch (Charlesworth, 1983)
Eus. *Eccl. Hist.*	Eusebius, *Ecclesiatical History* (Loeb)
4 Ezra	4 Ezra (Charlesworth, 1983)
Frontin. *Aq.*	Frontinus, *Aqueducts of Rome* (Loeb)
Frontin. *Str.*	Frontinus, *Strategems* (Loeb)
IG	*Inscriptiones Graecae* (1873–)
IG Rom.	*Inscriptiones Graecae ad Res Romanas Pertinentes* (1906–)
I *Magn.*	Ignatius, *Epistle to the Magnesians* (Ante-Nicene)
I *Philad.*	Ignatius, *Epistle to the Philadelphians* (Ante-Nicene)
I *Polyc.*	Ignatius, *Epistle to Polycarp* (Ante-Nicene)
Iren. *Haer.*	Irenaeus, *Against heresies* (Ante-Nicene)
I *Rom.*	Ignatius, *Epistle to the Romans* (Ante-Nicene)

I *Smyrn.*	Ignatius, *Epistle to the Smyrnians* (Ante-Nicene)
Joseph. *AJ*	Josephus, *Jewish Antiquities* (Loeb)
Joseph. *Ap.*	Josephus, *Against Apion* (Loeb)
Joseph. *Vit.*	Josephus, *Life* (Loeb)
Just. 1 *Apol.*	Justin Martyr, *First Apology* (Ante-Nicene)
Just. *Dia. Tryph.*	Justin Martyr, *Dialogue with Trypho* (Ante-Nicene)
Juv.	Juvenal, *Satires* (Loeb)
MAMA	*Monumenta Asiae Minoris Antiquae* (1928–39)
Mart.	Martial, *Epigrams* (Loeb)
Mart. Polyc.	*Martyrdom of Polycarp* (Ante-Nicene)
Mart. *Spect.*	Martial, *Spectacles* (Loeb)
OGI	*Orientis Graecae Inscriptiones Selectae* (1903–5)
Philo *Leg.*	Philo, *Embassy to Gaius* (Loeb)
Philostr. *VA*	Philostratus, *Life of Apollonius of Tyana* (Loeb)
Philostr. *VS*	Philostratus, *Lives of the Sophists* (Loeb)
Plin. *Ep.*	Pliny the Younger, *Letters* (Loeb)
Plin. *HN*	Pliny the Elder, *Natural History* (Loeb)
Plin. *Pan.*	Pliny the Younger, *Panegyricus* (Loeb)
Polyc. *Phil.*	Polycarp, *Epistle to the Philippians* (Ante-Nicene)
POxy.	*Oxyrhynchus Papyri* (Grenfell & Hunt, 1898–)
Quint. *Inst.*	Quintillian, *Institutio Oratoria* (Loeb)
SEG	*Supplementum Epigraphicum Graecum* (1923–)
Sib Or.	*Sibylline Oracles* (Charlesworth, 1983)
Sil. *Pun.*	Silius Italicus, *Punica* (Loeb)
Stat. *Achil.*	Statius, *Achilleid* (Loeb)
Stat. *Silv.*	Statius, *Silvae* (Loeb)
Suet. *Dom.*	Suetonius, *Domitian* (Loeb)
Suet. *Galb.*	Suetonius, *Galba* (Loeb)
Suet. *Ner.*	Suetonius, *Nero* (Loeb)
Suet. *Tit.*	Suetonius, *Deified Titus* (Loeb)
Suet. *Vesp.*	Suetonius, *Deified Vespasian* (Loeb)
Tac. *Agr.*	Tacitus, *Agricola* (Loeb)
Tac. *Ann.*	Tacitus, *Annals* (Loeb)
Tac. *Germ.*	Tacitus, *Germania* (Loeb)
Tac. *Hist.*	Tacitus, *Histories* (Loeb)
Tert. *Ad Scap.*	Tertullian, *To Scapula* (Ante-Nicene)
Tert. *Apol.*	Tertullian, *Apology* (Ante-Nicene, Loeb)
Carn.	Tertullian, *Flesh of Christ* (Ante-Nicene)
Test. Sol.	*Testament of Solomon* (in Charlesworth, 1983)
Ulp.	Ulpian (see *Corp. Jur. Civ.*)

NOTES

Chapter 1

1. On other external evidence for dating Revelation to the reigns of Trajan, Claudius, Nero, or Domitian, see Charles 1920, 1:xci–xciii. Charles notes, "The earliest authorities are practically unanimous in assigning the Apocalypse to the last years of Domitian" (1:xcii).

2. Irenaeus does not mention any kind of persecution at this time, but many scholars have taken his witness as fitting with evidences of persecution in the Book of Revelation (e.g., Kümmel 1975, 467).

3. See also Eus. *Hist. Eccl.* 3.39; 4.18; 5.8, 18; 6.25; 7.25.

4. For a full discussion of the evidence see Canfield 1913, 74–76.

5. In the following paragraphs I present the point of view of most scholars toward Domitian. This view does not take into account several difficult historiographic problems. For a detailed discussion and reassessment of Domitian's reign, see chapter 6.

6. On the use of "signs" in the imperial cult see Scherrer 1984, 599–610.

7. Contrast Rissi 1966, 80: "It is striking that in the whole context no word is spoken of the emperor cult, which shows that it did not yet play any role for Christians in Asia Minor at the time of the book's writing." Rissi argues for a dating of the original apocalypse under Vespasian.

8. See Charles: "Our author from his ascetic standpoint had sympathized with Domitian's decree, which according to its own claims was directed against luxury, and was accordingly the more indignant when it was recalled" (1920, 1:167–68).

9. See Collins 1984a, 2–8 and Aune 1986, 66–76 for some of the most important issues involved in the question of genre in general and specifically the genre "apocalypse." Hellholm (1983) discusses genre issues in apocalypticism in the whole of the Mediterranean world as well as the Near East. Sometimes generic classifications take some odd twists. Walter Schmithals, for example, argues that John's Revelation does not reflect "the apocalyptic understanding of existence at all" (Schmithals 1975, 169–71). For him, then, the Apocalypse — the source for delineating apocalyptic literature — does not embody apocalyptic thinking. The Apocalypse is not apocalyptic! We should also be wary of identifying the genre "apocalypse" with the meaning of the word as it has entered into popular U.S. culture, that is, as a way of describing totally destructive and horrible occurrences. Ancient apocalypses may include descriptions of horror in the future, but that is not their most salient feature.

10. There are also apocalyptic elements in parts of some books, e.g., Mark 13; 2 Thess. 2; and sections of some Hebrew prophets.

11. Vielhauer treats the social setting of apocalypses only briefly and tentatively. He

reviews the possibility of Iranian influence, the continuity between apocalypses and prophecy, connections with the wisdom tradition, and possible links to Qumran. He assumes cautiously that apocalypses arose "in those eschatologically-excited circles which were forced more and more by the theocracy into a kind of conventicle existence" (1965, 598).

12. For more detailed reviews see Hartman 1983, Olsson 1983, and Sanders 1983 (from the 1979 Uppsala Conference published in Hellholm 1983); Collins 1984a, 1–32; and Collins 1984b, 2–5.

13. An introduction and English translation of this whole work is readily available in Isaac 1983.

14. For a recent commentary see Collins 1984b.

15. On the transmission of the text of 4 Ezra and its relation to other writings associated with Ezra, see Metzger 1983, 516.

16. See Hanson 1976, 28–30; Collins 1984a, 1–11; Collins 1979, 3–4.

17. Schmithals 1975, 42–43, 77, 81. Schmithals thus contrasts apocalyptic eschatology with both prophetic theology in the Old Testament where the god acts in history and Christian salvation where the savior once again takes a place in history (pp. 159–160, 171). Few scholars would agree with this radical differentiation of apocalyptic eschatology from prophetic and Christian eschatologies. See Rowland 1982, 29; Bloch 1952.

Chapter 2

1. On the similarities between rabbinical writings and apocalypses see Bloch 1952, 73–82, 89–111.

2. For a critique of this notion of social change see Tilly 1984, 33–56.

3. On the appeal to contemporary millenarian movements see, e.g., Wilson 1982.

4. I adapt material from Collins 1984a, 142–154.

5. For example, so far as we know the frightful conditions surrounding the war with Rome did not immediately produce any apocalypses even though Rabban ben Zakkai certainly had interest in a speculative mysticism that is related closely to apocalyptic thought (see Rowland 1982, 282).

6. So an apocalyptic attitude toward the world cannot be explained as a product of an essential development within Judaic prophetism or wisdom, as a conglomerate of Jewish and Iranian conceptions, or as an outgrowth of either a particular social reality or a new social, political, or psychological situation (Schmithals 1975, 118–20, 127, 130, 145, 148).

7. Schmithal quotes with approval the following explanation of apocalyptic from Rudolf Otto: "This removal of the world from the direct sphere of divine control has been traced back to the political conditions of late Judaism. There are no proofs. Rather it seems to me that the operative factor was an idea necessary to religion, and necessarily pressing its way more definitely into consciousness, viz. the idea of the transcendence of the divine over all that is of this world" (1975, 150).

8. The "idea of the transcendence of the divine" is detachable not only from every "social reality" but also from the language that expresses it. For Schmithals language objectifies an understanding of existence that in some essential way remains independent of its linguistic objectification. So he asserts that in spite of the "fundamentally different understanding of the world and of existence" in Iranian and apocalyptic pieties, they are objectified in "widely identical conceptual material" (1975, 122). Searches for the origins of a religious movement only yield information about the

language used to objectify piety, not piety itself (see pp. 118–19, 130). So neither prophetism, wisdom, nor Iranian religion "represent[s] the sum of" apocalyptic piety, "even though apocalyptic is presented *objectively* as a combination of Old Testament-prophetic and Iranian elements with additions provided by the Wisdom movement" (pp. 138–39).

9. "What counts is not a neutral observer's view of whether things are good or bad," writes Nickelsburg, "but the apocalyptist's *perception* and *experience* that the times are critical" (1983, 646); and Collins states, "We must also reckon with the fact that what is perceived as a crisis by an apocalyptic author cannot always be accepted as objective reality" (1984a, 30). The combination in one phrase of one term referring to a social, historical situation (*crisis*) and one referring to an attitude of mind (*perceived*) is tricky. "Crisis" appears only through perception. Without perception there is no crisis. The issue, then, is not perception over against a social, historical reality but perception over against perception, i.e., "perceived crisis" over against "perceived noncrisis." Further, in most cases, the perception is not a private, individual matter but an element within a social, symbolic construction, so that the issue becomes how one group (however defined) perceives over against how another group perceives. Finally, as we shall seek to show later, "perceptions" do not exist "outside," but are part of the social, historical network.

10. If "crisis" or "perceived crisis" continues to be important in studies of apocalypticism, "crisis" should be more precisely defined than is now the case. Billings and colleagues (1980), for example, consider "perceived value of possible loss," "perceived probability of loss," and "perceived time pressure" as three key elements in defining the extent of perceived crisis; they contrast these elements with those which Hermann had proposed earlier—"threat," "decision time," and "surprise." Billings's shift to perceptual elements in crisis is typical of recent research, but it assumes the curious ontology that there is always out there an untouched "thing in itself": "Crisis is defined by a set of variables as perceived by the decision maker. These perceptions may differ from the objective situation or from the perceptions of others" (1980, 306).

11. Speech need not take the form of propositions and assertions. It may express affective dimensions of a relationship, e.g., comforting, lamenting, or assuring. It may express injunctions—"Remain faithful," "Keep steadfast." It may create self-contained aesthetic objects, as in certain kinds of poetry (see Guiraud 1975, 5–15). In every case, however, language shapes our perception of reality, which always includes a social component.

12. Cf. Austin 1962, 114; Skinner 1969, 45–48; Hartman 1983, 334–35.

13. In the language of Quentin Skinner, they must draw on a socially conventional intention, one "within a given and established range of acts which can be conventionally grasped as being [a case] of that intention" (1970, 133). Searle also emphasizes that "in our analysis of illocutionary acts, we must capture both the intentional and the conventional aspects and especially the relationship between them" (Giglioli 1982, 145). In the Daniel example, Lacocque's observation that the Hebrew preposition *I* is used rather than *qdm* may be seen as a clue to the social convention there used (1979, 116).

14. Illocutionary elements are located publicly in written or spoken language available to all and are distinct from the question of authorial intention if such is seen as an intentionality located in the mind of the speaker or writer. Presumably, the shared social covention expresses an authorial intention, but we can never have absolute certainty about the exact relation of the typical and the specific. "Social conventions" express typicalities, but situations are both specific and typical. In the case of Daniel, I do not assume that we have a report of historical occurrence; I am interested only in the

dynamics of the conversational situation; they are the same for "fiction" and "history."

15. That is, at the level of genre a sentence, scene, or vision becomes a formal unit, not a unit of meaning. So Benveniste: "The *form* of a linguistic unit is defined as its capacity for being broken down into constituents of a lower level" (1971, 107).

16. The prominence of a theology of creation also assures this interplay between transcendent realities and the social order, social institutions, and social norms.

17. Hellholm returns "form critical analysis" to its beginnings when formal characteristics, common thoughts and moods, and similar life situations all contributed to the definition of genre (1986, 26). He develops genre analysis from a linguistics approach that can consider sequential aspects of a text as well as a hierarchical listing of generic characteristics. That approach opens new possibilities for describing and defining a genre, but see Yarbro Collins' cautionary remarks (1986, 3–4). Except for the reference to a group in crisis, Hellholm's description of intention is similar to that of John J. Collins: "The intention of an apocalypse then is to provide a view of the world that will be a source of consolation in the face of distress and a support and authorization for whatever course of action is recommended, and to invest this worldview with the status of supernatural revelation. The worldview may or may not serve as the ideology of a movement or group" (1984b, 22).

18. See Hartman's reflections on the *Semeia 14* definition of genre: "It seems to me, though, that the time has come to deepen the analysis and to take into account as exactly as possible the hierarchic structure and literary function of the propositional elements, the illocution of the texts, and their sociolinguistic functions" (1983, 339).

19. It refers, for example, to shared beliefs and new assertions about the world; to injunctive, affective, and aesthetic elements; to self-authenticating intentions; to devices of reactualization and participation; or to its social conventions and illocutionary force—all of which are to be considered in relation to genre, not as assertion on the sentence level.

Chapter 3

1. See Thompson 1978, 182–88, 203–5. I do not assume a sharp distinction between *prophetic* and *apocalyptic*.

2. See Feuillet 1965, 27. Source critics have used those doublets as evidence for various sources behind the Book of Revelation.

3. Note also how the seven angels with the seven bowls reappear at Rev. 17:1 and 21:9 as another device unifying segments of Revelation.

4. For the importance of heavenly worship see chapter 4.

5. See Beasley-Murray 1974, 238–39 for a synoptic tabulation of the two series. The sixth in the series of seals, trumpets, and bowls are also interrelated through structural reversals and cumulative images; see below.

6. The "time, and times, and half a time" in Rev. 12:14, a parallel passage, probably refers to the 3½ years of Dan. 7:25, which also, of course, equals 42 months or 1,260 days. Compare also the time of the exposure of the bodies of the witnesses (Rev. 11:9, 11).

7. For a discussion of what these "times" may mean in a larger context, see Sweet 1979, 182–83.

8. Three and four, factors of twelve, appear regularly in the Book of Revelation, e.g., 8:7, 13; 9:18; 16:13, 19; 21:13 (three); 4:6; 6:6; 7:1; 9:13, 14; 14:20; 20:8 (four).

9. At Rev. 16:12 "the rising of the sun" and "Euphrates" appear together as refer-

ences to the East. That is the only other passage in which those terms occur. On the altar as a microcosm see, e.g., Eliade 1963, 371–74.

10. The content of the open scroll and the message of the seven thunders are somehow related in this scene.

11. So Yarbro Collins 1979, 66; Caird 1966, 126; see also Schüssler Fiorenza 1985, 53–54.

12. καὶ εἶδον; see Thompson 1969, 332.

13. Rev. 19:11–16, 19:17–18, 19:19–21, 20:1–3, 20:4–10, 20:11–15, 21:1–8.

14. Later at Rev. 20:7–10 there is the "warless" battle between Satan and the saints.

15. With sufficient cleverness one is supposed to be able to figure out the number of the name of the beast, for it is human (Rev. 13:18); for the woman, Mystery is only one of several names written on her forehead (17:5). Those victorious at Pergamum are promised a share in Christ's mystery (2:17).

16. If notated as an elaborate score, those Christ images could be seen in the structuralist language of Lévi-Strauss as synchronic "mythemes." Like music, a piece of language can be viewed both statically as a score or dynamically as performance. Most literary critics—whatever their school—assume that.

17. *Ring composition* and *inclusion* are other terms used for this phenomenon.

18. See also οὐκ ἔτι in Rev. 18:11, 14; οἱ πλουτήσαντες and πλοῦτος in 18:1ϛ, 17; and γράψον in 1:11, 19.

19. The new city also shares characteristics with the throne in chapter 4; cf Rev. 4:1; 21:11, 18–20.

20. That cluster of similarities is only one of several links between the letters and the visions indicating that the former cannot be isolated from the latter. For example, at Rev. 4:1 the "first voice" speaks to John again; or note similarities between the throne scene in chaps. 4–5 and the letters (throne, white clothes, gold crowns). See also Yarbro Collins 1984, 73–76; Schüssler Fiorenza 1981, 47.

21. Rev. 22:16 repeats several elements from the introduction (1:1–2).

22. Abrams (1984, 343) uses *recursive* in a way that seems almost synonymous with *typology*. Bruner defines it more precisely as "the process whereby the mind or a computer program loops back on the output of a prior computation and treats it as a given that can be the input for the next operation" (1986, 97).

23. Cf. Rev. 5:13, 7:1, 2; 12:12; 14:7. With the new heaven and new earth, the sea passes away and is no more (21:1).

24. Cf. Rev. 8:8, 9:17, 11:5, 14:10, 16:8, 17:16, 19:20, 20:9.

25. The other instances of *lion* are metamorphic: 9:8, 17; 10:3; 13:2.

26. Contrast the "horses" of the sixth trumpet, who have heads like lions; from whose mouth comes forth fire, smoke, and sulfur; and whose tails are like serpents with heads (Rev. 9:17–19). Only evil forces seem to have tails (9:10, 19; 12:4).

27. The abandonment suggested in that combination differs markedly from the enforced drinking of blood that takes the form of a *lex talionis* in Rev. 16:6.

28. In a sense all language is ironic, for all conceals (see Frank Kermode 1979). Sometimes, however, the author calls attention to the "concealing" in an explicit manner and invites a reader or hearer to join in collusion with the author. Irony thus contributes to the community John shares with his readers and hearers. The irony in Revelation is stable, to use a term of Wayne Booth (1974); that is, it operates from a definite standpoint. For example, reigning through suffering is itself not susceptible to ironic manipulation.

29. Note especially verse 3, where the motivation for the fall is given: "drunk the

wine of her impure passion," "committed fornication," and "wantonness." This is not the language of lament. See also Yarbro Collins 1980, 138.

30. See Thompson 1978, 224–26.

31. See also Rev. 19:13, where the Word of God is soaked in blood and 1:7, where the "pierced one" comes on the clouds.

32. The ὅτι clause could be linked to οἶδά σου τὰ ἔργα, with the intervening words a parenthetical comment; but as it stands, the play δύναται/δύναμιν cannot go unnoticed. So also the καὶ after δύναμιν may be taken as adversative ("and yet"), but καὶ connects ἔχεις, ἐτήρησάς, and ἠρνήσω in a series. To read the sentence with a parenthetical phrase and an adversative καί complicates the syntax unnecessarily.

33. On the white garments, see Rev. 7:14, where they are made white in the blood of the Lamb.

34. Cf. Gal. 3:13; Acts 5:30, 10:39; Polyc. *Phil.* 8. At Rev. 13:3 there is a parody of the crucifixion of Jesus in the "healing" of the wound of the beast.

35. Lohmeyer 1953, 42. It is used as a connector, for example, in Rev. 7:1, 9; 15:5; 18:1; and with a specific temporal meaning in 9:12, 11:11, 20:3.

36. ὑπάγω with εἰς is a common idiom for purpose.

37. Compare to the pouring of the second bowl, where the sea becomes "like the blood of a dead man" (Rev. 16:3) and to 6:12, 8:7, 16:4.

38. For example, Tracy's comment that "all authentic limit-language seems to be initially and irretrievably a symbolic and a metaphorical one" (1974, 295).

39. See Yarbro Collins 1984, 42, 48; Schüssler Fiorenza 1985, 102, 135.

40. See also Rev. 12:14 and Exod. 19. In Rev. 12 patriarchal birth stories and Exodus elements are fused to describe the formation of a new community of God, a "new seed," with whom the dragon makes war (12:17).

41. For a detailed discussion of this scene see Court 1979, 82–105.

42. See Schüssler Fiorenza's objection to Gager's segmentation as dividing "the text in a topical rather than formal manner" (1985, 167).

Chapter 4

1. For acclamations see Peterson 1926, 141–45.

2. Norden calls this form of praise an *essentielle Prädikationsart* (1956, 221–22). It is a hellenistic form of divine predication; see Deichgräber 1967, 101.

3. The use of precious stones in describing the throne scene is not part of the traditional throne scenes in apocalypses. They are used to describe the New Jerusalem, the breastplate of the priest, and the garden of Eden (cf. Rev. 21:11, 18, 19, 20; Isa. 54:12; Exod. 28:17–20, 39:11; Ezek. 28:10–13).

4. Various terms are used to designate "worship," e.g., λατρεύω (7:15, 22:3), προσκυνέω (13:4), and αἰνέω (19:5).

5. The thrice-holy (*trisagion*) also occurs in Jewish liturgy, alluding to creatures singing in heaven; see Werner 1939, 141.

6. κύριος ὁ θεός ὁ παντοκράτωρ translates Isaiah's *Yahweh Sᵉbaoth*.

7. For "Lord God Almighty" cf. Rev. 1:8, 11:17, 15:3, 16:7, 19:6, 21:22. See also Deichgräber 1967, 49.

8. See E. Peterson 1926, 176–80; also Mowry 1952, 79.

9. *New* always has eschatological overtones in the Book of Revelation; cf 2:17; 3:12; 21:1–2, 5.

10. See Rev. 7:12, where seven terms honor God; six are the same. In 7:12 εὐχαριστία

occurs instead of πλοῦτος as in 5:12. ἀχαριστία is used only in connection with God in Revelation.

11. ναός was introduced earlier in the letter to the Philadelphians (Rev. 3:12).

12. σκηνόω ("dwells") refers to God's presence in the temple; see Rev. 15:5.

13. Cf. Rev. 7:10, 12:10-12. That cry probably derives from the enthronement psalms; see Deichgräber 1967, 54.

14. The hierophany of stars, lightning, thunder, and hail (Rev. 16:18-21) repeats 11:19. In the series of the seven bowls, however, the revealing of the ark of the covenant (15:5) is separated from the sky hierophany (16:18-21) by the pouring out of the bowls.

15. These last two lines may not be part of the hymn.

16. Or one could say that ontology undergirds eschatology; or that the "temporal" flows from the "eternal."

17. Rev. 1:6; 4:11; 5:12, 13; 7:12; 19:1, 7.

18. At Rev. 15:8 glory is like smoke (καπνός), which prevents entrance into the temple, not shining like the sun.

19. ἴασπις Rev. 4:3 (throne), 21:11, 18, 19 (New Jerusalem); σάρδιον 4:3, 21:20; σμαράγδινος 4:3, 21:19. Cf. the priest's breastplate (Exod. 28:9, 17; 35:9, 27; 39:6, 10); Jerusalem (Tob. 13:16); and the garden of Eden (Ezek. 28:13).

20. φοῖνιξ occurs only here in the Book of Revelation.

21. Cf. 1 Chron. 24:25, Mishnah, tractates Taanith 2.6, Sukkah 5.6-8. Cf. Yoma 1.5 in the Mishnah, where these royal priests are even called elders.

22. Yarbro Collins suggests that the courtyard-temple boundary parallels the earthly-heavenly boundary in the Apocalypse (1984, 68).

23. Cf. Eph. 5:18-19; Odes of Solomon 6:1, 14:7.

24. Van der Leeuw, in his phenomenological study of religion, notes that "crying aloud and singing set [sacral] power in motion" (Leeuw 1963, 430). In the more traditional language of Christian theology, hymns express the *fides quae creditur*, and they are a means of evoking the *fides qua creditur*.

25. See *Hekhalot Rabbati* 1.1, quoted in Gruenwald 1980, 103. Gruenwald discusses in detail this literature about the "heavenly palaces."

26. Cf. Acts 20:7, *I Magn* 9:1. The "Day of the Lord" also echoes the eschatological day.

Chapter 5

1. Structural analysis of the Lévi-Strauss variety usually presupposes conflict at the "deep" level; for the purpose of mythic language is to resolve fundamental contradictions in our experience and understanding of the world that otherwise cannot be resolved. According to structural analysis if such conflicts did not exist, mythic language would have no purpose: "Mythic thought always progresses from the awareness of oppositions toward their resolution" (Lévi-Strauss 1967, 221).

2. Those differences depend, of course, on the particular world construction.

3. Generally speaking, trees and plant life are classified with redeemed humans under divine care. See Rev. 7:1, 9:4; cf. 8:7, where judgments from heaven do harm the plants. At 14:17-20 vine and the grape are used as symbols of judgment.

4. E.g., ἐκ πάσης φυλῆς καὶ γλώσσης καὶ λαοῦ καὶ ἔθνους (Rev. 5:9).

5. They are thus offered as sacrifices to the sacrificial Lamb as well as to God.

6. Rev. 1:6. For Christ as priest, note the garb in 1:13 where Daniel 10 has been

modified. Schüssler Fiorenza comments, "These explicit changes of the Daniel text seem to stress Christ's royal/priestly character" (1981, 53).

7. The "brightness" (λαμπρόν) of the linen is also a quality of the river of life (22:1) and the morning star (22:16).

8. See also the "golden girdles," which only the seven angels with the seven last plagues and the one like a Son of Man wear (Rev. 15:6, 1:13, cf. 5:8, 8:3).

9. Here there is a play on the verb μολύνω which means both "soiled" and "defiled," like the English word *stained* (see Rev. 14:4, "those not stained [i.e., defiled] with women"). The soiled clothing contrasts with the clean clothing worn by the Bride of the Lamb, the seven angels, and the warring army mentioned above.

10. Contrast Babylon the Whore who wears fine linen, but in purple and scarlet (Rev. 17:3–4, 18:16).

11. The connection between Rev. 9:11 and 17:8 is made even firmer by the fact that Abaddon (9:11) is translated in the Septuagint as ἀπώλεια; cf. Job 26:6, 28:22; Ps. 88:11; Prov. 15:11. Abaddon parallels death, the grave, and Sheol.

12. That throne is given to the beast by the dragon (Rev. 13:2).

13. Cf. Rev. 13:14, 19:20; also the sequencing of dragon and beasts at 12:1–13:18, 16:13, and 20:10.

14. φυλακή (wilderness, or desert) and *abyss* function homologously in various religious structures; see Thompson 1978, 95.

15. On the surface level those conflicting oppositions support the structuralist notion that opposition and conflict are fundamental to the seer's world; for clearly demarcated boundaries separate the faithful from evil forces.

16. ὡς ἐσφαγμένην or ἐσφαγμένον.

17. In each case the verb is ἔζησεν. There are two renditions of the beast's wound: that he had a mortal wound that was healed (Rev. 13:3, 12) and that he had a wound from a sword and lived (13:14).

18. πνευματικῶς and other terms based on the root meaning of "spirit" function throughout the Apocalypse as transformational agents.

19. Jeremiah warns that Jerusalem will become a "desert" (Jer. 22:6, cf. Rev. 18:19). Thus the transformation of Jerusalem into wilderness, evil, and chaos is also a part of the prophetic tradition. In Rev. 18:2 the great Babylon becomes a φυλακή, a haunt of every unclean demon, spirit, bird, and beast, reminiscent of descriptions of the desert (see Thompson 1978, 95–96).

20. It does not indicate that from the seer's point of view the city as such "is the social and political embodiment of human self-sufficiency and rebellion against God" (Sweet 1979, 187).

21. Wilderness—like chaos and Egypt—thus reflects in the Apocalypse its place in prophetic tradition: to return to the wilderness is to be judged and punished (Rev. 17:16), but wilderness is also the setting for divine deliverance and nourishment from God (12:6, cf. Thompson 1978, 193).

22. Πνεῦμα is also associated with transformations connected with cultic time (Rev. 1:10), resurrection (11:11), and prophecy (19:10). At 11:8 nominal transformation is designated by the adverb πνευματικῶς. Typological and allegorical exegesis may thus be seen as a form of spiritual transformation.

23. Note also at Rev. 15:5–8 the polysemantic character of smoke as wrath (θυμός) and as glory (δόξα).

24. Ἀναβαῖνον is used to indicate the beasts coming up from the sea (Rev. 13:1) and the earth (13:11); κλεῖν, ἔβαλεν, ἔκλεισεν, ἐσφράγισεν, λυθῆναι, λυθήσεται, ἐξελεύσε-

ται, refer to Satan's movement in and out of the bottomless pit (20:1–3, 7).

25. The "firstborn from the dead," for example, has the keys to Death and Hades (1:18) and can thereby control death and life.

26. On the insertion of another myth at Rev. 12:7–12, see Court 1979, 112–15; Schüssler Fiorenza 1981, 124–25.

27. That judicial dimension of the divine needs to be taken seriously in any study of the important theme of justice in Revelation.

28. Generally speaking, "descent," in the Apocalypse, results in loss — destruction, judgment, moral decline, death — whereas "ascent" signifies deliverance, moral perfection, and life.

29. Proper identification is further complicated by eschatological reversals whereby greatness is brought low — even to be no more — and lowliness is transformed into greatness (e.g., Rev. 7:9–17, 18:21, 19:2). Such transformational reversals occur in the text through cultic liturgies as well as symbols of clothing, color, and ascent.

30. See Schüssler Fiorenza: "Only when we acknowledge that Revelation hopes for the conversion of the nations, in response to the Christian witness and preaching, will we be able to see that it does not advocate a 'theology of resentment' but a theology of justice" (1981, 119); Contrast Yarbro Collins 1984, 170.

31. On "first" see Rev. 20:5 (the "first resurrection") or 22:13, where God declares himself to be "the first and the last." *Disappear* or *pass away* ($\dot{\alpha}\pi\acute{\epsilon}\rho\chi o\mu\alpha\iota$) is used to describe — among other things — "woes" (9:12, 11:14), the fruit for which Babylon longs (18:14), and the dragon (12:17). Rev. 16:2 ($\dot{\alpha}\pi\tilde{\eta}\lambda\theta\epsilon\nu\,\dot{o}\,\pi\rho\tilde{\omega}\tau o\varsigma$), referring to the first of the seven angels, reads strikingly similar to 21:4 ($\tau\dot{\alpha}\,\pi\rho\tilde{\omega}\tau\alpha\dot{\alpha}\pi\tilde{\eta}\lambda\theta\alpha\nu$).

32. In the Apocalypse "newness" is written (Rev. 2:17, 3:12), sung (5:9, 14:3), seen (21:1, 2) and done (21:5).

33. This blurring of old age and new age has its counterpart in the syntax of the book: as we have seen, the seer has not organized his material along a strong narrative line with an appropriate climax at the end. Rather, there is repetition, concentric movement, and ever-widening circularity.

34. For example, there are two clearly defined types of "servant parables" as well as the obvious generic differences between the narrative of the Passion and apocalytic sections; see Thompson 1978, 256–61.

35. Organic change is an essential element in all world building. We are constantly adapting to different environmental conditions, roles, religious, and philosophical understandings. In spite of this change, a person's world does not usually crumble: modifications occur, contours of the world change; but boundaries and distinctions are observed.

36. In both Rev. 11:1 and 21:15 $\kappa\acute{\alpha}\lambda\alpha\mu o\varsigma$ is used for "measuring rod."

37. See also Zech. 2:1–5, where measuring Jerusalem reflects its sacredness. On other possible meanings of measuring see Court 1979, 86.

38. Cf. the derivation of such terms as *medicine, moderation,* and *meditation* from the root "to measure." In every instance, an inner measure and harmony are assumed; see Bohm 1983, 20. The association of measuring with rebuilding, restoring, judgment, destruction, and preservation is secondary to this inner/outer correspondence.

39. Schüssler Fiorenza suggests that the difference in proportion "indicates that the universal cosmic salvation by far exceeds that prefigured in the Church" (1981, 205).

40. "Fall" and "repentance" are contrasted at Rev. 2:5.

41. The Bride of the Lamb forms a homology with faithful Christians; probably,

therefore, the seer is comfortable with faithful Christians' participating in "proper" sexual relations. The virgins on Mount Zion in chapter 14 would then be viewed as a special group, not representative of all Christians.

42. Cf. the two lists of vices: δειλοῖς, ἀπίστοις, ἐβδελυγμένοις, φονεῦσιν, πόρνοις, φαρμάκοις, εἰδωλολάτραις, and ψευδέσιν (Rev. 21:8); κύνες, φαρμακοί, πόρνοι, φονεῖς, εἰδωλολάτραι, πᾶς φιλῶν καὶ ποιῶν ψεῦδος (22:15).

43. As pointed out above, all Christians share royal and priestly attributes with Christ.

44. The social boundary that keeps the Christian pure from acculturation and assimilation reveals through its various homologues and contrarieties a fundamental structure in John's world. Social forces external to the church do not *occasion* that boundary or any other; rather, they unfold one dimension of the essential order in the seer's world.

Chapter 6

1. Book 10 of the *Letters* was published posthumously.

2. See also Plin. *Ep.* 2.1; 4.13, 15; 6.9; 9.14, 23.

3. Trajan accedes reluctantly to the request (Plin. *Ep.* 10.95).

4. Other sources from this same period include Martial, whom Pliny helped to support; Juvenal—never mentioned by Pliny—who published satires from 115 to 127; and Dio Chrysostom, born before 50 at Prusa in Bithynia, who went to Italy as an orator and was relegated from Italy and Bithynia early in Domitian's reign, apparently because of his friendship with Flavius Sabinus (C. Jones 1978, 46–47). He later had some difficulty at Prusa; cf. Plin. *Ep.* 10.81–82; Dio Chrys. *Or.* 40.11–12; and C. Jones 1978, 54, 103.

5. Philostratus, who wrote at about the same time as Dio Cassius, also mentions Domitian in his *Life of Apollonius of Tyana*. Both Philostratus and Dio probably came into contact with the circle around the Empress Julia Domna and at least met each other.

6. See Plin. *Pan.* 50.5

7. Cf. Plin. *Pan.* 90.5–7, *Ep.* 1.12.6–8.

8. Dio Cassius writes that Domitian actually murdered Agricola (Dio Cass. 66.20.3).

9. See also Dio Cass. 67.3.2, Juv. 2.28–38, Plin. *Ep.* 4.11.6.

10. Suetonius has a different anecdote: Vespasian "was surprised that he [Domitian] did not appoint the emperor's successor with the rest" (*Dom.* 1.3).

11. He did absurd things like impaling flies on a stylus. Cf. Dio Cass. 65.3.4, which seems to be a doublet of 65.9.3; or perhaps both are literary topoi.

12. Titus's last words were supposedly "I have made but one mistake." Some conjecture that he referred to taking his brother's wife, Domitia. Dio Cassius, however, is inclined to think that Titus meant that he was mistaken in not killing Domitian (66.26.3–4).

13. Given these supposed relationships with brother and father, some scholars appeal to Freudian theory to explain Domitian's mad character, e.g., Caird 1966, 20, 23.

14. Tac. *Agr.* 2–3; Plin. *Pan.* 33.3–4, 42.1; Suet. *Dom.* 10, 12, 14–15; Dio Cass. 67.13.2–14.5, cf. 68.1.1–2.

15. Suet *Dom.* 15.3, 16.1–2; Dio Cass. 67.16.2.

16. This impression could be made, according to Tacitus, because Domitian's character was as yet unknown.

17. For the account in this paragraph see Waters 1964, 52–65.

18. On Domitian's poetry, cf. Plin. *HN*, pref. 5, Quint. *Inst.* 10.1.91, Sil. *Pun.* 3.618-21, Stat. *Achil.* 1.15, Mart. 5.5, Tac. *Hist.* 4.86.2, Suet. *Dom.* 2.2, see also McDermott and Orentzel 1977, 29. For the influence of the standard sources on modern scholarship, see Butler in the Loeb of Quint. *Inst.* 10.1.91 and n.

19. On a coin from Lugdunum dated 70-71, Titus and Domitian appear on the reverse, with Vespasian on the front; Titus honors his brother in 72 as COS DES II (designated second consulship), featuring Domitian's head on the reverse of his own; in 79-80, during Titus's reign, there is no falling off of Domitianic material, as Domitian is now DIVI F (deified Flavian) (see Waters 1964, 62).

20. An inscription from Galatia (79 CE) refers to Domitian in his fifth consulship, designated sixth (IG R*om.* 3.223); and in 80 (under Titus) an inscription commemorating the paving of a road from Derbe to Lystra refers to Domitian as consul 7, designate 8 (*CIL* 3.12218). See also the milestones between Ancyra and Forylaeum (Dessau, *ILS* 263) and in Almazcara, Spain (McCrum and Woodhead 1961, 417).

21. See Newton 1901, no. 168=*CIL* 10.5405.

22. See B. Jones 1973, 8-12; Jones points out that only in military training was Domitian's education incomplete, but even here "his practical inexperience . . . was not necessarily a serious drawback" (p. 12).

23. On *divus* as gods who had been men see Hopkins 1978, 202; Weinstock 1971, 391-92. K. Scott there comments, "The opinions of Pliny, Suetonius, and Dio . . . are altogether too prejudiced against Domitian" (1975.62).

24. Cf. Stat. *Silv.* 1.1.94-98, Mart. 9.34.

25. One is from the council and assembly of Pinara in Lycia, and one is from Celei, Noricum, dated sometime between 90-96 (see McCrum and Woodhead 1961, 111-12).

26. See also Waters 1964, 60: "The story savours too much of Periander and Nero to be anything but the stock-in-trade of anti-tyrannical invective."

27. On Frontinus see Plin. *Ep.* 4.8. For other references to Domitian's military successes see Mart. 5.3 (cf. Mart. 6.10, Dio Cass. 67.6.5), 6.76; Juv. 4.3; McCrum and Woodhead 1961, 140; Stat. *Silv.* 4.2.66-67.

28. For an appreciation of Domitian's military successes, cf. Henderson 1927, 98; Syme 1936, 162-64; McDermott and Orentzel 1977, 28.

29. E.g., Benko and O'Rourke 1977, 67-68.

30. See, e.g., John Elliott: "The last years of Domitian's reign (93-96 C.E.) were marked by the opposition of the senatorial aristocracy, philosophers and religious groups alike. The reign of terror which the suspicious emperor instigated against all of his enemies, real and imagined, was punctuated with his own assassination in 96 C.E. . . . The oppressive measures undertaken by Domitian from 93 C.E. onward would suggest a likely *terminus ad quem* for 1 Peter" (1981, 86-87); see also Caird 1966, 20-21.

31. For more on the imperial cult, see pp. 158-64.

32. See *POxy.* 1143.4 and discussion by Deissman (1978, 349-62). At Pergamum an inscription names the living Octavius as "Imperator Caesar, son of god, divine [Θεός] Augustus" (*IG Rom.* 4.309; *OGI* 458).

33. Cf. Taylor [1931] n.d., 35-57, 181-204; Hopkins 1978, 205-8.

34. Waters 1969, 395-98. See also references to the arch at Beneventum in Ferguson 1970, 96-97.

35. He also gives his niece the title *Augusta*; cf. Waters 1969, 397-98.

36. On the fluidity of the meaning of *dominus* see Sherwin-White 1966, 557-58. It can be a simple title of respect (like "Sir"), but with reference to the Emperor, some sense of the inperial cult is implied.

37. τῶν ὅλων κύριος. On the various Greek terms for "master," "lord," and "tyrant," see the interesting passage in Philostr. *VA* 7.42.

38. Plin. *Pan.* 2, 33.4, 52.7; Dio Chrys. *Or.* 1.22–24, 45.1, 50.8; cf. Mart. 10.72.

39. See Dio Cass. 67.14.1. Toward the end of Statius' poem the Sybil does give a lofty title to Domitian: *Hic est deus, hunc jubet beatis pro se Juppiter imperare terris.*

40. Dilke notes that A. W. Verrall used the absence of divine epithets as evidence "to try to show why Dante thought Statius a convert to Christianity" (1954, 81).

41. The meaning of *DNImperator* in *CIL* 2.4722 is uncertain. The *dominus et deus* formula does occur later in reference to Antoninus Pius in an inscription from Tauric Chersonesus (Crimea) (Latyschev 1965, no. 71).

42. Statius' comment on Domitian's reticence about being called *dominus* is probably an accurate reflection of Domitian's view throughout his reign (Thompson 1984, 469–75).

43. Tac. *Agr.* 2–3; Plin. *Pan.* 33.3–4; Suet. *Dom.* 10, 12, 14–15; Dio Cass. 67.13.2–14.5, cf. 68.1.1–2.

44. On Josephus' relations to the Flavians see chap. 8.

45. See *Dom.* 10.1; K. Scott 1975, 126–32.

46. This explanation, says Waters, "is quite inadequate, though it may well refer to symptoms of treachery" (1964, 72).

47. Cf. Plin. *Ep.* 7.19.5, Tac. *Agr.* 2, Dio Cass. 67.13.2, Suet. *Dom.* 10.3–4. Rogers has shown that those motives, if operative at all in the trials, were "trivial incidentals"; those three belonged to a "close-knit group who stood in a clearly defined tradition of His Majesty's *dis*loyal opposition" (1960, 22); see also Syme 1958, 1:76; MacMullen 1966, 37. Pliny's facile presentation of Domitian's punishment of the vestal virgin also shows clearly his anti-Domitian bias; cf. *Ep.* 4.11.

48. On school exercises see MacMullen 1966, 35. Waters gives a complete list of "proved cases" (1964, 76 but read *Ascletarion* for *Ascleparion* and add, probably, *Arrecinius Clemens*).

49. See also 1.13, where Martial refers favorably to *casta Arria*, Paetus's wife.

50. Szelest 1974, 107–9. Recall that Martial received patronage from that circle when he went to Rome in 64; see also Macmullen 1966, 1–45.

51. Cf. Mart. 2.60; 5.75; 6.2, 22, 45, 91; 9.6, 8; 11.7; Szelest 1974, 111–12. On the Julian law cf. Stat. *Silv.* 3.4.74–77, 4.3.8–19; Suet. *Dom.* 8.3; Dio Cass. 67.12.1; Philostr. *VA* 6.42; it continued to be enforced in the reign of Trajan, cf. Plin. *Ep.* 6.31.

52. Szelest offers the standard temporal explanation for this — seventeen of the twenty-three epigrams were written before 90 — with the additional explanation that the seventeen are often laced with praise of Caesar (1974, 113).

53. See B. Jones 1979, 30–45. Suetonius may be using a fairly common literary device of starting with the good and ending with the bad — what Brunt calls "a kind of principle of chiaroscuro in ancient literary portraiture" (1961, 221); see also K. Scott 1975, 104.

54. See Pliny: "I am not asking you [Trajan] to model yourself on him [Domitian] whose successive consulships dragged the long year out without a break. . . . And so ordinary people enjoyed the honour of opening the year and heading the official calendar [under Trajan], and this too was proof of liberty restored: the consul now need not be Caesar" (*Pan.* 58).

55. See McDermott and Orentzel 1977; David Vessey 1974, 113–15. Probably it is significant that Italicus was not in Trajan's coterie during the time of Pliny's writing; see *Ep.* 3.7.6–7. Others, including Frontinus, also continued to have a positive attitude

toward Domitian after his death (see McDermott and Orentzel 1977, 67–68), though Frontinus' reference to Domitian in *Aq.* 2.118 is somewhat compromising.

56. B. Jones 1979, 4. The first of a new dynasty is not wont to deify his predecessor; *damnatio* is always more likely.

57. Octavius remained *ab epistulis* from Domitian to Trajan, cf. Dessau *ILS* 1448 and Syme 1958, 1:38; on Veiento see Dessau *ILS* 1010 and Syme 1958, 1:6; on Corellius Rufus see Plin. *Ep.* 4.17.4 and B. Jones 1979, *album senatorium*, no. 83. No doubt many others could be added to this list (see Magie 1950, 579–80; Viscusi 1973, 209–16).

58. Contrast that to Nerva, who chose several top men having in common "a lack of conspicuous favour from Domitian," old men passed over in the recent past (Syme 1958, 1:3–4).

59. Trajan's legends contrast with Nerva's but follow Domitian's; see Waters 1969, 393–97.

60. See Waters 1969; Pleket 1961, 310; Keresztes 1973, 22.

61. For the mutiny see Plin. *Pan.* 14.5; for the consulship see McCrum and Woodhead 1961, 9. There is probably a connection between his activity in 89 and his consulship in 91; see B. Jones 1979, 35.

62. Compare to what Pliny says about himself in *Pan.* 90.5. Earlier, Pliny had explained how Domitian had trusted in Trajan during the Saturninus revolt (14.5).

63. In the literature of Trajan's day, Domitian's avarice, along with his sensuality and vaingloriousness, became a topos; see C. Jones 1978, 121.

64. *Dig. Just.* 48.22.1 of *Corp. Jur. Civ.* (trans. S. P. Scott). Trajan sold off—but did not return—properties claimed by Domitian for the *fiscus*; see Plin. *Pan.* 50; also Millar 1964, 110.

65. For example, the Flavians had earlier invoked a "new age" ideology; cf. K. Scott 1933, 255–56. Statius illustrates this nicely in *Silv.* 4.1.17–32.

66. Dessau *ILS* 2927. See Sherwin-White 1966, 732–33 for other relevant inscriptions.

67. See Sherwin-White 1966, 75; Syme 1958, 2:657; Radice 1975, 125.

68. Pliny later revised it, see *Ep.* 3.18.2.

69. See also Plin. *Ep.* 8.14.3. In *Hist.* 1.1 Tacitus also contrasts Domitian's reign with the "new era" after Domitian, but there he does admit his debt to the Flavians.

70. Benario 1975, 145–46.

71. Ogilvie and Richmond 1967, 140.

72. Dio Chrysotom says that during his exilic travels, on the way from Heraea to Pisa, he met an old woman at a sacred grove of Heracles who divined "that the period of my wanderings and tribulation would not be long, nay, nor that of mankind at large [i.e., the rule of Domitian]. . . . 'Some day,' she said, 'you will meet a mighty man the ruler of very many lands and peoples. Do not hesitate to tell him this tale of mine even if there be those who will ridicule you for a prating vagabond'" (*Or.* 1.55–56). As C. P. Jones observes, "this prophecy fits 'too neatly'" in "a speech delivered before an emperor who had a special devotion to Heracles" (1978, 51). Dio Chrysostom also claimed to bear up "under the hatred . . . of the most powerful, most stern man, who was called by all Greeks and barbarians both master and god . . . without fawning upon him or trying to avert his hatred by entreaty but challenging him openly, and not putting off until now, God knows, to speak or write about the evils which afflicted us, but having done both already, and that too in speeches and writings broadcast to the world" (*Or.* 45.1). Jones assesses Dio's claim as follows: "Either this claim is grossly exaggerated or almost all of these works have perished" (1978, 50).

73. See also Plin. *Ep.* 10.2. Sometimes Pliny has to manipulate words at great length to make the contrast; cf. his comments about mimes (*Pan.* 46).

74. Waters observes that both terms, along with the related words *princeps* and *dominus*, were in use as early as the reign of Tiberius and as late as Constantine. There was no "sudden retroversion from the Dominate of Domitian to the Principate of Trajan" (1969, 399, quoting Beranger); see e.g., Suet. *Galb.* 9.2; Jupiter's priest at Clunia predicts that "there would come forth from Spain the ruler and lord of the world [princeps dominusque rerum]."

75. So Nerva's famous dinner party: "Where would Catullus Messalinus be if he were alive today?" Answer: "Here, dining with us" (Plin. *Ep.* 4.22). Though Pliny drips an acid tongue on Fabricius Veiento, he was there in Nerva's closest circle, just as he had been in the closest circles of the Flavians and of Nero before them.

Chapter 7

1. See also 1 Cor. 15:32, where Paul refers to "fighting wild beasts" in Ephesus, probably also a metaphoric reference to conflict in his missionary activities.

2. Acts 18:18. Irenaeus states that Paul founded the church at Ephesus (Iren. *Haer.* 3.3.4). This is not clear from Acts.

3. This assumes the integrity of the letter to Rome. One could then reconstruct their movement as from Rome to Corinth, then to Ephesus, and then eventually back to Rome.

4. Their equal status is noteworthy; in fact Priscilla is often mentioned first in Acts, as though she was of greater importance than her husband. She has a good Roman name.

5. In 1 Corinthians Paul mentions Apollos in a section condemning factionalism. Paul says that some of the Corinthians claim to "belong to Apollos" just as others "belong to Paul" or "Cephas" (1 Cor. 1:12, cf. 1 *Cl.* 47.3). Perhaps Paul's attention to "human wisdom" and "effective speaking" in 1 Cor. 1–4 is related to Apollos's rhetorical powers (Meeks 1983a, 61, 117).

6. See *Test. Sol.* 8.11 and perhaps 7.5 where Lix Tetrax claims to be "the direct offspring of the Great One." The *Testament of Solomon* may come from Ephesus; see Duling 1983. On Artemis' ability to overcome fate, see Oster 1976, 40–41.

7. That is an especially appropriate theme in Acts at this point, as Paul travels toward Jerusalem. When Paul is in Jerusalem, Asian Jews stir up the people there over Trophimus, a Gentile Ephesian who traveled with Paul (Acts 21:27–36, cf. Acts 20:4, 2 Tim. 4:20).

8. On manual labor see Malherbe 1983, 22–28; Hock 1980, 39–42, 56–59; Meeks 1983a, 64–65.

9. On Cerinthus see Iren. *Haer.* 1.21, 3.3.4, which tell how John the disciple rushed out of a bath house in Ephesus when he saw that Cerinthus was in it, and Eus. *Hist. Eccl.* 7.25, where Eusebius gives Dionysius of Alexandria's views on Cerinthus as author of the Book of Revelation.

10. Strictly speaking, Rev. 2:6 suggests that the Nicolaitans could not get a foothold at Ephesus.

11. Regarding churches in Asia, there is also a brief reference to Paul's missionary activity up in the northwestern corner of the province at Alexandria Troas; cf. Cor. 2:12 and Acts 20:5–12, in which the infamous Eutychus falls out the window. Perhaps there was a church also at Assos (Acts 20:13–14). By the end of the first century, Christian writings assume that there is a church at Troas (cf. 2 Tim. 4:13; I *Philad.* 11.2, *Smyrn.*

12.1). Paul refers to Asian "churches" (plural) that send greetings to the church at Corinth, but we do not know where those churches were (cf. 1 Cor. 16:19).

12. Even if Paul is not the author of Colossians, it was most likely written before Paul's death or soon thereafter, no later than c. 65.

13. Paul says that the Colossians learned the grace of God in truth from Epaphras and calls him a beloved fellow servant and a "faithful minister" of Christ on behalf of the Colossians. Paul apparently had not ministered in that area; cf. Col. 1:4, 2:1, which also mentions the church at Laodicea.

14. It is not clear exactly where Paul is; he may be at Ephesus, Caesarea, or Rome; cf. Kümmel 1975, 324–32.

15. According to Acts he and Trophimus are identified as Asians who accompanied Paul to Jerusalem to deliver the collection (Acts 20:4). Tychicus is referred to in the letter to Ephesus (6:21–22) in language almost identical to that in Colossians. Tychicus is also mentioned in the pastorals; cf. 2 Tim. 4:12, Titus 3:12.

16. See Philem. 1, where he is referred to as a fellow soldier ($\sigma\upsilon\sigma\tau\rho\alpha\tau\epsilon\acute{\omega}\tau\eta\varsigma$). Archippus was a fairly common name in Asia.

17. There is no compelling reason to identify him as the bishop of Ephesus referred to by Ignatius (Eph. 1.3, 2.1, 6.2). The name is common, e.g., *IG Rom.* 806, 921, 1441, 1732.

18. These names seem to be common to the area. For Philemon see *IG Rom.* 4.864 (from Laodicea) and 4.1435 (from Smyrna); for Apphia see 4.868 (a priestess at Colossae) and 4.796 (a woman from Apamaea); for Archippus see *CIG* 3143, 3224 (from Smyrna). See Arndt and Gingrich 1979 under each name.

19. For excavations there see Yamauchi 1980, 149–54.

20. On the identification of the followers of Balaam with the Nicolaitans, see Caird 1966, 38–39.

21. Rev. 2:19. That she and her followers were a part of the church is made clear in 2:24, where the seer refers to $\tauο\hat{\iota}\varsigma$ $\lambda o\iota\pio\hat{\iota}\varsigma$ $\tauο\hat{\iota}\varsigma$ $\grave{\epsilon}\nu$ $\Theta\upsilon\alpha\tau\acute{\iota}\rhoο\iota\varsigma$ ("to the rest of those among the Thyatiran Christians").

22. This idolatry is linked to Balaam in Num. 31:13–20.

23. Given the practices of some of the religions in Asia—as in Moab—both meanings could of course be applicable; but I incline, with Caird, to think that the seer would not have been so "tolerant" of sexual deviation (Caird 1966, 44, 39).

24. 1 Cor. 8:1–9:23, 10:14–22, 10:23–11:1; and Theissen 1982, 121–43 against Johnson 1975, 93.

25. The *Didache* notes that traveling prophets generally have a trade (*Did.* 12).

26. Rev. 2:2, 6; cf. Schüssler Fiorenza: "He praises the church of Ephesus for hating the works of the Nicolaitans, to whom the people 'who call themselves the apostles' (2:2) probably belong. These apostles appear to be itinerant missionaries" (1985, 115).

27. See Schüssler Fiorenza 1985, 117–25 for one way of relating the "opponents" of John and the situation of the churches in Paul's day.

28. For a more theological treatment of the same groups and same issues see Schüssler Fiorenza 1985, 114–32.

29. There were probably two separate situations with which Ignatius came into conflict: Judaizers and docetists (cf. Schoedel 1980, 32, 221; Johnson 1975, 111). Docetists were in Smyrna and Tralles. Judaizers were in the churches of Philadelphia and Magnesia, in the Hermus Valley, on the same roadway as Sardis. Sardis is conspicuous for its absence in the Ignation corpus. He would have passed through there from Philadelphia to Smyrna, but there is no indication that he stopped to visit the church.

30. Dittenberg. *SIG* 985, with a translation in F. Grant 1953, 28–30.

31. On the sabbath issue see Schoedel 1980, 34–35.

32. See Meeks 1983a, 54. Meeks is especially interested in "persons of low status crystallization, that is, those who are ranked high in some important dimensions but low in others" (p. 55). People with such "ambiguous status" may have been especially attracted to Christianity (pp. 72–73).

33. Theissen's article on social stratification in Corinth is an important one for understanding what classes and statuses were represented in the early Christian church (1982, 69–119); contrast R. Grant 1980, 16–17.

34. E.g., 1 Cor. 7:21; Philem.; Eph. 6:5–9; 1 Tim. 6:1–2; Tit. 2:9–10; 1 Pet. 2:18–20 (cf. Malherbe 1983, 52); Col. 3.22; 1 Tim. 6.2; 1 *Cl.* 34.1; James 5.4; I *Polyc.* 1.3, 4.3; *Barn.* 10.4, 19.7; *Did.* 4.10, 12.3–5; Eus. *Hist. Eccl.* 3.20.2–3.

35. *omnis aetatis, omnis ordinis, utriusque sexus*; R. M. Grant (1980, 17) on the one hand accepts the variety indicated by Pliny but also points out that Pliny "had the key case of the Bacchanalia in mind and knew that among the Bacchants of Italy, suppressed in 186 BC, there had been nobles, men and women, and persons of various ages."

36. See, e.g., R. M. Grant 1980, 21–23; Schoedel 1980, 43; Malherbe 1983, 62–68, 95–96; Meeks 1983a, 16–19, 57.

37. Colwell 1939, 70. See 1 Cor. esp. 5:9–11; 9:22; 10:14–21, 27–28; 14:23–24; also Phil. 2:15.

38. See also Schoedel 1980, 44–46, 52–54 on Ignatius.

39. Tacitus wrote the *Annals* after his stint as proconsul of Asia in 112–113. Thus, that experience could have helped shape his understanding of the Neronian persecution.

40. That is, the legal issue here is one not of religion but of specific associations (*collegia*, cf. Applebaum 1974a, 460).

41. See Millar 1972, 145–146. Except, possibly, for the reign of Nero "there is no authentic and concrete evidence of Imperial pronouncements about the Christians *except* in the form of letters" until the Decian persecution: Trajan to Pliny; Hadrian to Minicius Fundanus; Antoninus Pius to Larissa, Thessalonica, Athens and "all the Greeks"; Antoninus Pius or Marcus Aurelius to the *koinon* of Asia; Marcus Aurelius to the *legatus* of Lugdunensis. All of these letters were "almost certainly" responses to inquiries (Millar 1972, 158).

42. That equals what the seer calls "steadfastness in the faith."

43. On *hetaeria* see Wilken 1984, 31–47.

44. Pliny supports the "vertical" stratification of Christians. He says that they were drawn from all classes (*omnis ordinis, Ep.* 10.96.9) including slaves (*ancillae*, 10.96.8).

45. Cf. Johnson 1975, 93; R. M. Grant 1977, 21–25.

Chapter 8

1. On the *fiscus* and its administration see Stern 1980, 129; Smallwood 1981, 371–76.

2. On the economic state of the empire at the end of Domitian's reign see Syme 1930, 55–70 and the response by Sutherland 1935, 150–62; also Viscusi 1973.

3. From the legend on the reverse side, the coins were stamped in 96 CE.

4. For Flavia Domitilla as a Christian and a niece of Flavius Clemens see footnote 9 below and discussions by Smallwood (1956, 7–8) and Keresztes (1973, 7–15).

5. "Josephus is not an objective writer; but the Palestinian prejudices . . . have a deeper effect on his writing than the Roman bias which tends to be automatically ascribed to him" (Rajak 1984, 185). Josephus never received the official title of *amicus*

Caesaris and did not penetrate into the circle, e.g., of Statius. Yavetz suggests that Josephus "must have been a member of the lower entourage, in the same category as doctors and magicians, philosophers and buffoons" (quoted in Rajak 1984, 196).

6. Josephus is another example of a person who praised Vespasian and Titus profusely and remained honored to the end of Domitian's reign. Domitian was a Flavian who honored those whom his brother and father had honored earlier.

7. Cf. Geffcken 1902, 183–85, 188–89; also Pleket 1961, 303. This assessment of Domitian is all the more striking, because *Sib. Or.* 12 at this point is patterned upon *Sib. Or.* 5. Both oracles give a brief resume of the Roman emperors and offer similar assessments — except for Domitian. In *Sib. Or.* 5.40 Domitian is referred to as a "cursed man" (see Collins 1983, 394, 448), the reverse of *Sib. Or.* 12.

8. See Syme 1930, 63; Pleket comments that the oracle "reflects the current opinion of the average provincial about the emperor" (1961, 303).

9. Eusebius cites as example that Flavia Domitilla — a Christian niece of Flavius Clemens — was banished along with many others to the island of Pontia. This is another version of the story in Dio Cassius about the Jewish convert Flavia Domitilla, wife of Flavius Clemens, condemned to the island Pandateria (Dio Cass. 67.14.2).

10. The other two were Vespasian and Trajan; see Eus. *Hist. Eccl.* 3.12, 19–20, 32.

11. According to Eusebius those interviewed by Domitian lived on into the reign of Trajan and his persecution of Christians, and Trajan's persecution stopped when Pliny expressed alarm over the number of martyrs in Asia: Trajan decreed "that members of the Christian community were not to be hunted, but if met with were to be punished" (*Hist. Eccl.* 3.33). This reference is to the Pliny correspondence with Trajan (Plin. *Ep.* 10.96–97), but Eusebius draws the information from Tertullian (Eus. *Hist. Eccl.* 3.33).

12. On Clement of Rome see Barnard 1963–64, 255–57.

13. Cf. Kraabel 1968, 5–6, 199–200; Applebaum 1974a, 432; Broughton 1938, 632–33. Applebaum interprets the second phrase to mean "state-support for the upkeep of the religious functionaries of the newcomers" so that they were "accorded an organization in which the religious functionaries . . . were the responsible officials" (1974a, 468–70, 472). For the inscription discovered at Sardis in 1960 referring to this resettling, see also Robert 1964, 9–21 and Kraabel 1978, 15–18.

14. *Religio licita* is not a Roman legal expression but one from Tertullian; cf. Applebaum 1974a, 460. For a discussion of the authenticity and dating of documents in Josephus see Millar 1966, 160–62; Applebaum 1974a, 440–44.

15. This letter is probably from about the same time as the Flaccus incident, though Magie (1950, 1586) lists it with undatable names from the republican period.

16. Jews possessed their own archives; cf. *CIJ* 775, 776, 778; Applebaum 1974b, 483; on the meaning of *katoikia* in *CIJ* 775 see Schürer 1986, 27–28, 89.

17. *CIJ* 770. On the side of the curse he identifies himself as having served as βουλεύσας and ἄρξας; on other side he is identified as Εἰρηναρχία, Σειτωνία, Βουλαρχία, Ἀγορανομία, and Στρατηγία.

18. On wealth see Joseph. *AJ* 14.112–13.

19. Prymnessus, east and north of Acmonia, also is the source of "children's children" curses (on this curse formula see Kraabel 1968, 82–86) and are probably Jewish. On the possibility that some of the inscriptions referring to "the highest god" are Jewish, see Kraabel 1968, 93–108.

20. On the reference in Obadiah 20 to Sepharad and the possible identification of Sepharad with Sardis by means of a bilingual inscription from Sardis, see Schürer 1986, 20–21.

21. For this argument I am grateful to Marianne P. Bonz for sharing her unpublished paper.

22. Bonz cites from the *Justinian Digest*: "The deified Severus and Antoninus [Caracalla] allowed those who profess the Jewish superstition to hold office, but also imposed on them only those obligations which would not damage their superstition" (4.50.2; 3.3). It is not clear, however, whether Jews had never before been allowed to hold office or whether some limitation had been put on them in the recent past.

23. See Kraabel 1968, 181–90. For possible meanings of the last sentence see Schoedel 1980, 34.

24. This assumes that σαμβατεῖον in *IG Rom* 4.1281 equals σαββατεῖον (sabbath house, or synagogue), as most scholars are inclined to think; cf. Schürer 1986, 19. On Thyatira in general see A. Jones 1983, 83 and n. 93; *IG Rom.* 4.1205; Applebaum 1974b, 480; Broughton 1938, 763.

25. Edicts from Dolabella and others (Joseph. *AJ* 14:223–27, 228–29, 234, 236–40, 262–64); one edict from Marcus Vipsanius Agrippa, Augustus' close associate (*AJ* 16.167–68, cf. 12.125–28, 16.27, 60); and one from Julius Antonius (*AJ* 16.172).

26. Cf. *CIJ* 745, 746, reedited by Louis Robert; see Schürer 1986, 23.

27. Regarding Polycarp and the Jews see chapter 7.

28. Cf. Theopempte, another woman, *archisynagogos* from Myndus in Caria, south of Melitus (*CIJ* 756). On the office of *archisynagogos* see Kraabel 1968, 71–73. Further north of Smyrna, at Phocaea (Kyme), Tation is honored for her contribution to the construction of a synagogue there (see *CIJ* 738).

29. Other collection centers of the tax that Flaccus confiscated were at Adramyttium, Apamea, and Laodicea; cf. Cic. *Flac.* 68.

30. Josephus states that the Pergamenes had been friends of the Jews since the time of Abraham. This may indicate long-time Jewish influence in that region (see Schürer 1986, 18).

31. A "house synagogue" has also been excavated at Priene (second or third century CE), similar in structure to the Dura synagogue (see Kraabel 1968, 20–26). At Teos a funerary inscription has been found referring to a Jewish Roman citizen who built a synagogue there. Robert rejects the proper name "Proutioses" (cf. *CIJ* 744) for "P. Routilios Ioses" (1940, 27–28). Robert assumes that this Jewish *archisynagogos* is a Roman citizen. Inland at Colophon there is also some indication of a Jewish presence (cf. Stern 1974, 152; Schürer 1986, 22). Evidence of a Jewish community at the port town of Elaea is uncertain; it "may be the place from which the 'synagogue of Elaea' in Rome takes its name" (see Kraabel 1968, 179).

32. On the latter see Applebaum 1974a, 458–60; Applebaum 1976, 719. The assurance of these rights does not necessarily indicate the need for protection. It could also indicate the prestige of the Jewish community (see Kraabel 1978, 18).

33. On the complex problem of legal status of Jews in the cities, see Applebaum 1974a, 434–61. Early citizenship in the towns of Asia is probable: "It is entirely probable that the processes which enabled individual Jewish families to achieve Greek citizenship in such towns as Acmonia had commenced before the period of Augustus" (pp. 443–44).

34. On connections of Jews and guilds see Applebaum 1974b, 476–82 and references.

Chapter 9

1. See the description of an estate of one Mnesimachus in the Sardis inscription quoted in full in Broughton 1938, 631–32.

2. Cf. Dio Chrys. *Or.* 31.54; Broughton 1938, 645; Bogaert 1976, no. 36; Broughton 1951, 245.

3. In Broughton's list of estates of various sizes, note the following, in cities mentioned in John's Revelation: at Smyrna Marcus Antonius Polemo owned the best house (Philostr. *VS* 1.25; Broughton 1938, 667); in the third century Heracleides the sophist bought a small suburban estate at Smyrna; a woman at Philadelphia offered land to members of the council, the income to be divided among them yearly on her brother's birthday; a boundary stone at Sardis divided the land of Hermeias, a consular magistrate, and Marcellinus, a municipal magistrate (see Broughton 1938, 671).

4. Broughton 1938, 697–98. Under the Flavians several new cities were established, especially in the eastern plateau; see Magie 1950, 570. At that time the Moccadeni in the Hermus River Valley were also organized around Silandus and Temenothyrae; see A. Jones 1983, 93.

5. Cities bore much of the expense for road building; see A. Jones 1940, 140.

6. Cf. Magie 1950, 571, 574–75; French 1980, 709. After the Flavians, French has discovered no evidence for the building of any new road system in Asia Minor (1980, 711). The extensive development of roads under the Flavians made possible Trajan's rapid advances into Armenia and Mesopotamia (see B. Jones 1984, 149).

7. On the number of cities in Asia, cf. Joseph. *BJ* 2.16.4, Apol. Tyan. *Ep* 58, Philostr. *VS* 548; but see Habicht 1975, 67, who suggests that three hundred is a more accurate number.

8. Pergamum also received the right to build a temple to Augustus; and under Tiberius Smyrna secured the right to build a temple to him. Hypaepa, Tralles, Laodicea, and Magnesia were passed over in the competition; cf. Tac. *Ann.* 4.55.3; Broughton 1938, 709.

9. Local courts still existed, but "jurisdiction tended more and more to be concentrated in the hands of the governor" (see A. Jones 1940, 134).

10. Macro 1980, 671 summarizes evidence for assize centers. Habicht 1975, 90 offers evidence that assize districts were used for purposes other than judicial, such as for working out costs of the cult of Roma and Augustus.

11. See Broughton 1938, 710.

12. Broughton lists building, foundations, and gifts in Pergamum, Sardis, Smyrna, Ephesus, Thyatira, Philadelphia, and Laodicea, among others (1938, 716–26).

13. If vacancies in the council were filled by popular election, nominations for those vacancies were made by the council members (see A. Jones 1940, 183). So, for example, when Hadrian supported his friend Erastus for a seat on the city council of Ephesus, he addressed the magistrates and the council (A. Jones 1940, 183; Dittenberg. *SIG* 838). Each city council usually had a fixed number of seats, but popular athletes and actors often became honorary members of councils in several cities; see Magie 1950, 641.

14. The name of the municipal head varied from city to city (see A. Jones 1940, 163). In the imperial period this office was diminished by the city priesthood of Rome (and the emperor) who was also viewed as chief official of the city (see p. 174).

15. In the larger cities two or more people might carry out the duties of one office.

16. A peace officer arrested Polycarp (see *Mart. Poly.* 6; Magie 1950, 647).

17. A famous decree by the council at Hierapolis ad Lycum forbids the sheriff from taking advantage of surrounding villages (cf. *OGI* 527; Abbott and Johnson 1926, 117; Magie 1950, 988, n. 25 and 1515, n. 47).

18. Magistrates often had as their chief qualification "the possession of wealth and a readiness to spend" (Magie 1950, 649). Designated officers had to be connected with

money. For example, in one instance, a dead person was appointed so that his money could be used by the city (see pp. 649–50).

19. That shift could be a factor in the rise of prominence of Jewish locals in the third century.

20. See Magie 1950, 63, 653 and 854–860, nn. 37–38.

21. On city revenues generally, see Broughton 1938, 797–803.

22. This request is apparently being made on behalf of all Asia; cf. Broughton 1938, 840.

23. For this paragraph see Broughton 1938, 838–41; cf. Hock 1980, 35.

24. See Broughton 1938, 841–49; Burford 1972.

25. See *IG Rom*. 4.1419, 1432, where Septimus Publius is a citizen of Pergamum, Smyrna, Athens, and Ephesus; and Apollinarius is a citizen of Thyatira, Smyrna, Philadelphia, Byzantium, and councillor of Smyrna. Cf. Broughton 1938, 855.

26. For this paragraph see Broughton 1938, 849–57.

27. Rostovtzeff left Russia in 1917, lived a few years at Oxford, and then accepted an appointment at the University of Wisconsin, Madison in 1920. Although hostile to Bolshevism, he belongs to the generation of Russian historians who used socioeconomic analyses to explain historical change.

28. See, e.g., Macro 1980, 689 for proof that the social distinction between *splendidiores*, or *honestiores* (the senators, *equites*, provincial aristocrats, and legionaries), and *humiliores* (the lower classes) did not emerge until the second century.

29. Oliver 1953, 957. Compare the case of the archons, council and demos of the city of Ephesus against Vibius Salutaris, who tried to avoid liturgical obligations (p. 958).

30. Dio Chrys. *Or*. 34.21. See Broughton 1938, 809–12 for this and other examples; also Rostovtzeff, 1957, 2:621 for other secondary references.

31. This occurred late in the reign of Vespasian, before Dio Chrysostom's banishment in 82; cf. Magie 1950, 1443.

32. Titus to Munigua in Baetica, Domitian to Lappius Maximus and other proconsuls in Bithynia, and later Trajan to the proconsul of Achaea; cf. Millar 1966, 164. Methods for appointing a proconsular legatus, however, were different from those of an imperial legatus, and the two offices had different lengths of tenure (see Millar 1966, 165–66).

33. See Millar 1966, 164. For communications involving Christians see, of course, Pliny's correspondence with Trajan (*Ep*. 10) and also Eus. *Hist. Eccl*. 4.8.6, 9.1–3; Just. 1 Apol. 68–69.

34. On the prosperity of the cities mentioned in Revelation—Ephesus, Smyrna, Pergamum, Sardis, Philadelphia, Thyatira, and Laodicia, see Magie 1950, 583–86 and nn. On continuities within the Flavian dynasty see B. Jones 1984, 122–52.

35. See Broughton 1938, 740–73 for a detailed discussion of this prosperity, which continued into the second century.

36. For more information about Marcellus, including his attempted conspiracy and death, see B. Jones 1984, 87–93.

37. Magie, in his list of proconsuls of Asia (1950, 1582), includes the name Sextus Vettulenus Civica Cerialis, who, so far as I can tell, is different from Gaius Vettulenus Civica Cerialis; see B. Jones 1979, *album senatorium*, nos. 303, 599.

38. Marcus Ulpius Trajanus, father of the later emperor, was governor when a stadium was built in Laodicea and an aqueduct at Smyrna and a temple was rebuilt at Ephesus, among others.

39. See B. Jones 1984, 130 for appointment of senators from the Greek East as

assistants to consular governors and as military commanders. Appointments to higher offices in Trajan's reign followed, as did promotions of earlier Flavian appointments.

40. For the issuing of coinage in Asia by the Flavians see Broughton 1938, 883.

41. None of the seven cities of Revelation was among the several free cities in Asia. At this time free cities were not necessarily immune from tribute (Broughton 1938, 740).

42. Magie cites Dessau *ILS* 1517 for reference to Fortunatus, son of an imperial freedman, becoming under Domitian the *procurator fisci Asiatici* (1950, 1425).

43. Broughton cites evidence that in the reign of Hadrian, Hadrian's procurator and the proconsul of Asia were both involved in the new foundation of Stratoniceia Hadrianopolis (1938, 653).

44. See Pleket 1961, 310. There is some debate whether this office came in under Domitian or Trajan's reign. Magie states that it "may have been created under Domitian," but he favors Trajan (1950, 597, cf. 1454, n. 13). Broughton, on the other hand assumes Domitian (1938, 744, 810). Oliver also assumes a Domitian origin (1953, 974).

45. See Magie 1950, 471. For example, the Cyzicenes suffered under Tiberius for failure to finish their temple to Augustus (see Price 1984, 66).

46. Contrast Brian Jones: "Participation in the cult in the provinces was essentially a political and social matter" (1984, 153); or Bowersock: "Provincial priesthoods were viewed as civic duties suitable for the wealthy and ambitious but in no sense a display of piety" (1973, 183). See also Pleket 1965, 347.

47. According to Price (1984, 59), thirteen imperial temples were built from 50 BCE to 0, ten from 0 to 50 CE, seven from 50 to 100, fifteen from 100 to 150. Note the small number during the Flavian dynasty.

48. This intensity of imperial cultic activity involving the emperor dropped off after Augustus. There was what Price calls a "predictable routinization" of Augustus' charismatic authority.

49. See Magie 1950, 572, 594, 613–14; the stadium dedicated in Laodicea (Magie 1950, 1431); *IG Rom* 4.636, 845 (79 CE), 846, 861 and *SEG* 2.696.

50. For this paragraph see Price 1984, 55–58.

51. That is simply an indication of the importance of those cities in the province of Asia; it need not suggest that John selected those cities because of their relation to the cult.

52. Associations such as the *augustales* or *flamines*, distinctively western and Roman, are found only in Roman colonies in the eastern empire (Price 1984, 88). Gladiators and animal fights originated in Rome; and although they were present in imperial festivals, they were not central to the imperial cult in Asia Minor (pp. 89, 124, 170). Provincial elites would sometimes adopt Roman practices in order to display their cosmopolitan sophistication and their knowledge of how things were done in Rome, but this attempt of provincials to win prestige could make sense only "because the community as a whole expressed its identity through the maintenance of Greek traditions" (p. 91).

53. The provincial cult (for which see below) was probably closer to Roman practices than were the city cults for the emperor; see Price 1984, 76.

54. Price notes that the imperial cult did not draw on heroic cults, but rather the cult of the gods, in Asia. That is probably relevant to Christian response to the imperial cult. It was not so much the imperial cult per se but celebrations affiliated with Greek religion that Christian objected to.

55. For coins with images of this temple at Pergamum, see Magie 1950, 1293. Cf. also the monument erected at Ephesus to Julius Caesar, "the descendant of Ares and Aphrodite, a god made manifest, and the common saviour of all human life," erected in

response to Caesar's reduction of taxes required from the province of Asia; see Magie 1950, 407.

56. *OGI* 458=Ehrenberg and Jones 1955, no. 98. See also Magie 1950, 1342, n. 39 for other fragments.

57. During the first three centuries six of the seven cities mentioned in Revelation had claim to the title *neokoros*: (1) Pergamum, *neokoros* II (113-14 CE), III (215); (2) Smyrna, rival with Ephesus, *neokoros* II (under Hadrian), III (after 209); (3) Ephesus, *neokoros* II (under Hadrian), III (under Septimus Serverus); (4) Sardis, *neokoros* I (under Hadrian), II, (under Albinus), III (under ?); (5) Philadelphia, *neokoros* (under Caracallus); (6) Laodicea, *neokoros* (under Caracallus) (Broughton 1938, 742).

58. When the provincial assembly would call the emperor's attention to outstanding citizens in the province (see Millar 1977, 389), some self-serving recommendations were probably made.

59. Dio Chrysostom is here (*Or.* 35.15) referring to assizes, but the same point could be made for the imperial festivals (cf. Price 1984, 107; Broughton 1938, 871).

60. All citizens were recipients of these distributions and banquets, sometimes non-citizens as well; cf. Price 1984, 113.

61. Price proposes that those not participating by sacrificing and having wreaths outside their homes would be noticed (1984, 123).

62. Price makes this point in reaction to those who see the imperial cult as primarily an elite phenomenon (e.g., 1984, 108).

63. For this paragraph see Price 1984, 107-14.

64. Cf. also the difference in verbs used to address the deities and the emperor: Aristides indicates that people "pray" (εὐχόμεθα) to the gods but "petition" (δεόμεθα) the rulers; see Bowersock 1973, 199-200.

65. On the distinction between "honorific" and "sacred" significance of statues including the terms *eikon, andrias,* and *agalma,* see Price 1984, 177-79; Bowersock 1973, 185.

66. Price 1984, 183. See the coins from Cilicia where Trajan is in his temple "enthroned as Zeus holding thunderbolt and sceptre."

67. So Price concludes: "It was difficult for the assumption of divine attributes by a man of flesh and blood to be successful. The tensions between mortality and immortality, visibility and invisibility could best be solved by the subtle collocation of attribute and image" (1984, 184).

68. Price 1984, 211.

69. Price 1984, 211; *MAMA* 8.492b. Price argues that προθύτης, a term designating an imperial cultic official, indicates that sacrifices were made "on behalf of" (προ) the emperor. See Price's discussion of sacrifices on behalf of the emperor on his accession, on his arrival in a provincial city, after imperial victories, on a member of the imperial family's coming of age, or by annual vows (1984, 210-20).

70. See Pleket 1965 for a discussion of evidence for imperial mysteries.

71. Pleket 1965, 341; see also Magie 1950, 448 and 1297, n. 58.

72. Pleket 1965, 344. Compare to the illumination and revelation of Isis in Apuleius' *Golden Ass.*

73. Price 1984, 192 from Ulp. 21.1, 19.1.

74. For a brief critique of images in Christian and Greek circles see Price 1984, 199-204. Bowersock (1973) overstates the case with intellectuals and the imperial cult.

75. Recall that the imperial cult occurs in connection with the traditional Greek religious pantheon that had been adapted in the Asian cities.

76. Among the seven cities of Revelation, in Smyrna one Apollonius contributed a

statue, altar, and tables in marble to the cult of Apollo Cisauloddenus (Broughton 1938, 751; *SIG* 996); in Laodicea towers and gates with adornments were built in the time of Domitian from private funds (see Broughton 1938, 769); and in Ephesus Domitian supported extensive building: an addition to the stage building, an entrance to the market, decorations in the stoa, and baths (see p. 753). For the most part, however, the peak of prosperity for the seven cities — as for Asia more generally — came in the second century under Marcus Aurelius (see p. 794).

77. Some, e.g., Rostovtzeff (1957, 599), would connect this scarcity at Antioch with a wider famine in Asia at this time.

78. Note how saving and rationing is more viable than hauling in more grain, which was expensive because of the cost of transportation; see Broughton 1938, 868.

79. See Suet. *Dom.* 7.2, 14.2; Philostr. *VS* 520; Stat. *Silv.* 4.3.11–12; Magie 1950, 580.

80. Cf. Mouterde and Mondesert 1957 and Lewis 1968; also Pleket 1961, 304–5.

81. Under Hadrian the post was for a time restructured to make it less burdensome, but that reform was apparently short-lived (see A. Jones 1940, 141–42); see a third-century inscription of tenants to the Emperor Philip, who complain that they are taken from their work and pressed into service of traveling military and dignitaries (Broughton 1938, 659–61).

82. Here is another clear example of continuity between Vespasian and Domitian. Note also the respect with which he refers to Vespasian in this public document. That does not square with the comments in the standard sources for Domitian's life (see chapter 6). For another example of Domitian's care for the cities, note his concern to restrict endowments to cities to the purposes intended (see Oliver 1953, 970–71).

83. For another example of Domitian's policy toward the land, note his handling of squatter's rights to land that was not being tilled (see McCrum and Woodhead, 1961, 462) and possibly the breaking up of the estate of Claudius Hipparchus, an Athenian millionaire (Oliver 1953, 954; Pleket 1961, 306).

84. Cf. Frontin. *Str.* 2.11.7, Sil. *Pun.* 14.686–88; McDermott and Orentzel 1977, 30–31.

Chapter 10

1. Since the definite article occurs only once in the Greek text (ἐν τῇ θλίψει καὶ βασιλείᾳ καὶ ὑπομονῇ Ἰησοῦ), all three nouns are linked to the state of being "in Jesus."

2. For the meaning "arrived," cf. Acts 13:5; 2 Tim. 1:17.

3. The genitive Ἰησοῦ is subjective, referring to the witness made by Jesus, which John and the others also proclaim (cf. Rev. 1:2).

4. Pliny the elder says simply *Patmus circuitu xxx* (*HN* 4.12.69). For a list of islands of deportation mentioned by ancient authors, see Saffrey 1975, 398.

5. The grammar and text are problematic in the phrase *in the days of Antipas*.

6. "Hold fast" and "did not deny" (Rev. 2:13) are general encouragements and not technical terms for response to political persecution (cf. Rev. 2:3, 3:8, 10, 14:12, 16:15).

7. The verbal stem is used to describe men "blaspheming" God (16:9, 11, 21).

8. The seer may here be playing on the postexilic tradition that gentiles are expected to come and bow down before the Jews; cf. Isa. 49:23, Zech. 8:20–23.

9. Political Rome is fully identified with demonic forces, but the demonic forces are "larger," and more comprehensive, than Rome.

10. πᾶς τεχνίτης πάσης τέχνης (Rev. 18.22). Is the seer here displaying his knowledge of culture?

11. There is a *cognitive* dimension to social institutions, just as there is a *social* dimension to revelatory genres.

12. There was some ambiguous status; see Meeks 1983a, 72–73.

13. McCrum and Woodhead 1961, nos. 121, 142, 148; *IG* 3.1091. As can be seen from those inscriptions, homologues were formed in these cults between Caesar and the divinity rather than Caesar and Satan. See also pp. 158–64.

14. These relations were expressed most explicitly in processions; see Price 1984, 111.

15. There is no evidence at this time so far as I can tell for widespread anxiety and insecurity; if anything, the reverse seems more correct.

16. This apocalypse is also called prophecy (*propheteia*, Rev. 1:3), mystery (*mysterion*, 10:7), and eternal gospel (*euangelion*, 14:6). See Schüssler Fiorenza 1981, 36–38 for connections between "words of prophecy" and "apocalypse." On "mystery" see also Rev. 1:20. The seer can also use different languages and incomprehensible names to give an air of esoteric mystery and profundity to his message (cf. Rev. 9:11; 13:17; 16:16; 17:5, 7; 19:12).

17. Note the parallelism between the "I" statement of God (Rev. 1:8) and that of John (1:9). God's statement provides the transition to first person.

18. The seven stars on the voice's right hand (Rev. 1:16) are the messengers (*angeloi*, cf. 1:1) of the seven churches of Asia Minor; and the seven gold lampstands (1:12) are the seven churches (1:20).

19. Note how the language of Rev. 3:20 is picked up at 4:1.

20. Cf. Schüssler Fiorenza 1985, 23, 51–52, 170; Court 1979, 20; Collins 1977, 340–41.

21. Cf. Sanders' emphasis on "restoration and reversal" (1983, 456).

22. See MacMullen 1966, 13–22, 32–34, 65, 68–69.

23. For details see MacMullen 1966, 34, 36, 38, 50, 53, 56, 63, 71, 76–77, 80–81. Martyrdom is also, of course, a theme in the Book of Revelation.

24. Tacitus' *Agricola*, 4 Maccabees, the Alexandrian *Acts of the Pagan Martyrs*, and the later genre of Christian martyrdom illustrate the spread of this literature of opposition (see MacMullen 1966, 79–93).

25. See MacMullen 1966, 50–51, 306. For similarities between the Cynic κοσμοπολιτεία and the Christian citizenship in "Jerusalem above," see p. 315.

26. See MacMullen 1966, 243: "It is striking how interchangeable and ambiguous were the attitudes of the different groups in the aristocracy, how Janus-faced they were, looking toward the past, libertas, and senate, and at the same time toward the future, stability, and the emperor."

27. "The mind, in fact, from the second century on, comes under increasingly open, angry, and exasperated attack" (MacMullen 1966, 109).

28. See MacMullen 1966, 97, 99–102, 106, 110–115.

29. See MacMullen 1966, 102–3, 106, 111–12, 143–46, 149, 151, 158.

30. MacMullen calls attention to Apuleius' *Apology* as a nice example of the legal status of magic in the empire (1966, 121–24).

31. See MacMullen 1966, 121–125, 130–31, 144.

32. Sherwin-White 1966, 287, 641; MacMullen 1966, 55. Quintilian provides evidence that Domitian allowed speech against evil tyranny (*Inst.* 12.1.40. Contrast to Philostr. *VA* 7–8). Even speaking favorably about past subversives was tolerated under Domitian (see Mart. 1.8, 13; 4.54.7). Scholarly debate continues over whether philosophers received special privileges under the Flavians (see Bonner 1977, 160–62).

33. See Plin. *Ep.* 3.11, 7.19. For Pliny's rewriting of his own political situation at that time see chapter 6 above.

34. See Suet. *Dom.* 10; Dio Cass. 67.13.

35. Rogers 1960, 22. Cf. Sherwin-White 1966, 243; MacMullen 1966, 37, 77, 79; Syme 1958, 1:76. Waters 1964, 76.

36. See MacMullen 1966, 126–27, 138–41.

37. Cf. the following comments by Leo Curran cited by Irving Massey (1976, 23): "The Roman passion for order, expressed in the desire for the strict preservation of boundaries between individuals, is repeatedly violated in the flight-pursuit sequences of the Metamorphoses. . . . The anti-Augustanism of Ovid is conveyed in part by his dwelling on irresistible and largely meaningless violent change."

Chapter 11

1. For example, there is no center or periphery to the surface of a sphere.

2. Lohmeyer comments, "The special highlighting of the name indicates how uncommon and impressive was the death 'for the sake of the faith'" (1953, 25).

3. For a general discussion of irony in the New Testament, see Thompson 1978, 221–31, 299.

4. The preposition ἐπί indicates that the mourning is directed at the one who has been pierced (see John 19:37).

5. Barnabas also plays on the irony of the "tree": "the kingdom of Jesus is on the cross [ξύλου]" (*Barn.* 8.5).

6. See Tert. *Ad Scap.* 5: "Your cruelty is our glory."

7. See Booth 1974, 28, 42. On suffering and social boundaries in the writings of Paul, see Meeks 1979, 11; for suffering in Ignatius see Schoedel 1980, 32–33.

8. A similar transformation from history to world structure occurs with respect to the resurrection, when the "firstborn from the dead and the *archon* of the kings of the earth" (Rev. 1:5) becomes "the *arche* of God's creation" (3:14).

9. Apart from the Book of Revelation, "explicit comments made in the New Testament about non-Christians are remarkably free from condemnation" (Malherbe 1983, 21, cf. Colwell 1939, 59).

10. Here *crisis*—if the word must be used—points to a specific social location in the Roman Empire, not to the whole of social existence at that time; it does not characterize the Roman era as a whole. Amos Wilder has written an essay demonstrating "that the sociological, motivating occasions for the early Christian sense of crisis and fulfillment were not social in the sense of property, slavery, persecution, etc., but in a deeper and more comprehensive sense were appropriate to ancient life-patterns. Thus understood, 'hope for men living in this impermanent and not too secure world' can be indistinguishably social-political and spiritual" (Wilder 1961, 74). His point fits well with the present study. But he assumes that the whole of life in the first century CE was "seen as at a point of radical crisis. Apocalyptic or highly dualistic eschatology as world-view and timeview arose out of such a general crisis in the inherited total way of life, . . . a crisis in the tradition" (p. 70). Here he confuses the viewpoint of those in a particular social location in the empire with Roman culture as a whole. So far as I can tell, there was no "general crisis in the inherited total way of life." First-century Roman life was rather one of the most integrated, peaceful, meaningful periods of history for most of those who lived in the empire. This confusion of a particular social location with society as a whole is not uncommon in the study of early Christianity.

11. "At best, a minority viewpoint is forced to be defensive. At worst, it ceases to be plausible to anyone" (Berger 1970, 7).

12. There is an analogous instability in the literary dimension of the genre. Aune points out that an apocalypse "can exist as an independent text or as a constituent part of a host genre" (1986a, 80).

13. An apocalypse can just as likely be a "cognitive base" for social change as a response to change. In different but, I think, compatible terms, Jonathan Smith characterizes social change as involving "symbolic-social questions: what is the place on which I stand? what are my horizons? what are my limits? . . . It is through an understanding and symbolization of place that a society or individual creates itself" (1973, 140). Berger and Luckmann make the point that knowledge is always in a dynamic relation with the social order: "Knowledge is a social product *and* knowledge is a factor in social change" (Berger and Luckmann 1967, 87). Knowledge within the genre "apocalypse" can be an especially effective factor in social change. See also Hopkins 1978, 198.

Appendix

1. Only by means of such a theory can data be transformed into evidence; moreover, different theories will look for different kinds of evidence in different sets of data. All that has to be sorted out if conversation among competing theories is to be fruitful.

2. Hemer expresses the intention "to view the book historically without a judgment coloured unduly by the supposition that it may be most fruitfully explained as representative of a certain type of apocalyptic, epistolary or dramatic composition. . . . to insist on the comparative importance of the detailed historical criticism of background and content in a work which transcends its literary models" (1986, 15, cf. 30).

3. Court criticizes Minear's interpretation on those grounds (1979, 10).

4. See chapter 9 on this incident.

5. Many literary critics would assume the same for all literature, i.e., that literary worlds are never completely separable from the "real" world of the everyday. The question is how they are related.

6. On plausibility structure see Berger 1969, 45–47. That "symbolic universe" divides the life of ordinary Christians into social realities of the everyday and imaginative constructs supporting religious beliefs.

7. Rhetorical criticism need not reduce symbolic-poetic elements in such a thoroughgoing manner. In fact, rhetorical criticism can offer just as rich fare as the symbolic-poetic.

8. "I propose to look for the integrating center, that is, the distinct historical-social-religious experience and resulting theological perspective that have generated the particular form-content configuration (*Gestalt*) of Rev[elation]" (Schüssler Fiorenza 1985, 2, 183).

9. "Any change in theological ideas and literary forms is preceded by a change in social function and perspective" (Schüssler Fiorenza 1983, 311). "It must be kept in mind that it is the rhetorical situation that calls forth a *particular* rhetorical response and not vice versa" (Schüssler Fiorenza 1985, 192, my emphasis). So John's prophetic interpretation contrasts with those of his prophetic rivals (the Nicolaitans, Jezebel, Balaamites) because it "is rooted in an experience different" from theirs (1985, 5).

10. Other ways include the sacraments, meditation, asceticism, and mystical visions. Less individual forms involve community organization, ethical standards that prefigure the coming kingdom, and apocalyptic mythology (Gager 1975, 49–50).

11. For a critique of this approach see Thompson 1985.

12. The destruction of Jerusalem, the Neronian persecution, imperial worship, the martyrdom of Antipas, and John's own *relegatio in insulam* also factor into the book.

13. See Talmon 1962, 136–37.

14. See Rowland: "The antithesis between theological affirmation and historical reality could not have been more starkly put. The Christians were living in a world where the dominion of the creator was barely acknowledged, yet the seer was telling them about heavenly choirs which sang the praises of God as the king of the universe" (1982, 425).

15. Yarbro Collins here refers to the theory of cognitive dissonance as a way of understanding this tension or conflict, but she does not explore that theory's possibilities for interpreting apocalyptic. Wayne Meeks does appropriate that theory and — I think rightly — observes that "an 'explanation' of apocalyptic beliefs therefore needs to take account of 'cognitive dissonance' theory at least as much as 'relative deprivation' theory" (1983b, 688). An apocalyptic myth reduces cognitive dissonance by "providing a comprehensive cognitive map, an alternative vision of reality," which in turn "offers access to social power, if only within that counter-cultural community" (p. 701).

16. For another treatment of catharsis, cf. Schüssler Fiorenza 1985, 198; Barr 1984, 49.

WORKS CITED

Abbot, F. W. and A. C. Johnson. 1926. *Municipal Administration in the Roman Empire*. Princeton: Princeton University Press.

Abrams, M. H. 1984. "Apocalypse: Theme and Variations." In *The Apocalypse in English Renaissance Thought and Literature*, ed. C. A. Patrides and Joseph Wittreich, 342–68. Ithaca: Cornell University Press.

Applebaum, S. 1974a. "The Legal Status of Jewish Communities in the Diaspora." In *The Jewish People in the First Century*, ed. S. Saffrai and M. Stern, 1:420–63. Assen, Netherlands: Van Gorcum.

———. 1974b. "The Organization of the Jewish Communities in the Diaspora." In *The Jewish People in the First Century*, ed. S. Saffrai and M. Stern, 1:464–503. Assen, Netherlands: Van Gorcum.

———. 1976. "The Social and Economic Status of the Jews in the Diaspora." In *The Jewish People in the First Century*, ed. S. Saffrai and M. Stern, 2:701–27. Philadelphia: Fortress.

Arndt, William F. and F. Wilbur Gingrich. 1979. *A Greek-English Lexicon of the New Testament and Other Early Christian Literature*. Chicago: University of Chicago Press.

Aune, David E. 1981. "The Social Matrix of the Apocalypse of John." *Biblical Research* 24:16–32.

———. 1983. "The Influence of Roman Imperial Court Ceremonial on the Apocalypse of John." *Biblical Research* 28:5–26.

———. 1986a. "The Apocolypse of John and the Problem of Genre." *Semeia* 36:65–96.

———. 1986b. "Now You See It, Now You Don't: Ancient Magic and the Apocalypse of John." SBL Seminar on Early Christian Apocalypticism. Photocopy.

Austin, John L. 1962. *How To Do Things with Words*. Cambridge: Harvard University Press.

Barnard, L. W. 1963–64. "Clement of Rome and the Persecution of Domitian." *New Testament Studies* 10:251–60.

Barr, David L. 1984. "The Apocalypse As a Symbolic Transformation of the World: A Literary Analysis." *Interpretation* 38:39–50.

Beardslee, William A. 1970. *Literary Criticism of the New Testament*. Philadelphia: Fortress.

———. 1979. "Whitehead and Hermeneutic." *Journal of the American Academy of Religion* 47:31–37.

Beasley-Murray, G. R. 1974. *The Book of Revelation*. London: Marshall, Morgan, & Scott.

Benario, Herbert. 1975. *An Introduction to Tacitus*. Athens: University of Georgia Press.

Benko, Stephen, and John J. O'Rourke. 1977. *The Catacombs and the Colosseum*. Valley Forge: Judson.

———. 1986. *Pagan Rome and the Early Christians*. Bloomington: Indiana University Press.

Benveniste, Emil. 1971. *Problems in General Linguistics*. Coral Gables, Fla.: University of Miami Press.

Berger, Peter. 1969. *The Sacred Canopy*. Garden City: Anchor.

———. 1970. *A Rumor of Angels*. Garden City: Anchor.

Berger, Peter, and Thomas Luckmann. 1967. *The Social Construction of Reality*. Garden City: Anchor.

Bickerman, Elias. 1973. "Consecratio." In *Le culte des souverains dans l'Empire Romain*, W. den Boer, ed., *Foundation Hardt pour l'Étude de l'Antiquité Classique, Entretiens 19*, 3–25. Geneva: Vandoeuvres.

Billings, Robert S., et al. 1980. "A Model of Crisis Perception: A Theoretical and Empirical Analysis." *Administrative Science Quarterly*, 25:300–316.

Bloch, Joshua. 1952. *On the Apocalyptic in Judaism*. Philadelphia: Dropsie College for Hebrew and Cognate Learning.

Bogart, R. 1976. *Epigraphica* 3. Leiden: E. J. Brill.

Bohm, David. 1983. *Wholeness and the Implicate Order*. London: Ark.

Bonner, Stanley F. 1977. *Education in Ancient Rome*. Berkeley: University of California Press.

Bonz, Marianne. "The Wealth and Social Status of Western Diaspora Jews: A Reassessment of the Sardis Evidence." Harvard University. Photocopy.

Booth, Wayne. 1974. *A Rhetoric of Irony*. Chicago: University of Chicago Press.

Bowersock, G. W. 1973. "Greek Intellectuals and the Imperial Cult in the Second Century A.D." In *Le culte des souverains dans l'Empire Romain*, ed. W. den Boer, *Foundation Hardt pour L'Étude de L'Antiquité Classique: Entretiens 19*, 179–206. Geneva: Vandoeuvres.

Broughton, T. R. S. 1938. "Roman Asia Minor." In *An Economic Survey of Ancient Rome*, ed. Tenney Frank, 499–918.

———. 1951. "New Evidence on Temple-Estates in Asia Minor." In *Studies in Roman Economic and Social History in Honor of Allan Chester Johnson*, ed. P. R. Coleman-Norton, 236–50. Princeton: University Press.

Bruner, Jerome. 1986. *Actual Minds, Possible Worlds*. Cambridge: Harvard University Press.

Brunt, Peter. 1961. "Charges of Provincial Maladministration under the Early Principate." *Historia* 10:189–227.

Burford, Alison. 1972. *Craftsmen in Greek and Roman Society*. Ithaca: Cornell University Press.

Burridge, Kenelm O. L. 1982. "Reflections on Prophecy and Prophetic Groups." *Semeia* 21:99–102.

Burton, G. P. 1975. "Procon Assizes and the Administration of Justice under the Empire." *Journal of Roman Studies* 65:92–106.

Caird, G. B. 1966. *The Revelation of St. John the Divine*. New York: Harper & Row.

Calder, W. M. 1922–23. "Philadelphia and Montanism." *John Rylands Library* 7:309–54.

Canfield, Leon Hardy. 1913. *The Early Persecutions of the Christians*. New York: Columbia University Press.

Charles, R. H. 1920. *A Critical and Exegetical Commentary on the Revelation of St. John*. Edinburgh: T. & T. Clark.

Charlesworth, James H., ed. 1983. *The Old Testament Pseudepigrapha*. Garden City: Doubleday.

Cogley, Richard W. 1987. "Seventeenth-century English Millenarianism." *Religion* 17: 379–96.

Collins, John J. 1977. "Pseudonymity, Historical Reviews, and the Genre of the Revelation of John." *Catholic Biblical Quarterly* 39:329–43.

———, ed. 1979. *Semeia 14: Apocalypse: The Morphology of a Genre*. Missoula: Scholars.

———. 1983. "Sybilline Oracles." In *The Old Testament Pseudepigrapha*, ed. James Charlesworth, 1:317–472. Garden City: Doubleday.

———. 1984a. *The Apocalyptic Imagination*. New York: Crossroad.

———. 1984b. *Daniel with an Introduction to Apocalyptic Literature*. Grand Rapids: William B. Eerdmans.

Colwell, Ernest Cadman. 1939. "Popular Reactions against Christianity in the Roman Empire." In *Environmental Factors in Christian History*, ed. John Thomas McNeill, et al. Chicago: University of Chicago Press. Pp. 53–71.

Coover, Robert. 1971. *The Origin of the Brunists*. New York: Ballantine.

Court, John. 1979. *Myth and History in the Book of Revelation*. Atlanta: John Knox.

Culley, Robert C., and Thomas W. Overholt. 1982. *Semeia 21: Anthropological Perspectives on Old Testament Prophecy*. Chico, CA: Scholars.

Deichgräber, Reinhard. 1967. *Gotteshymnus und Christushymnus in der frühen Christenheit*. Göttingen: Vandenhöck & Ruprecht.

Deissmann, Adolf. [1922] 1978. *Light from the Ancient East*. Reprint. Grand Rapids: Baker.

Dilke, O. A. W. 1954. *Statius: Achilleid*. Cambridge: Cambridge University Press.

Duling, D. C. 1983. Introduction to "Testament of Solomon." In *The Old Testament, Pseudepigrapha*, 1:935–87. Garden City: Doubleday.

Eco, Umberto. 1984. *The Name of the Rose*. London: Picador.

Ehrenberg, V. and A. H. M. Jones. 1955. *Documents Illustrating the Reigns of Augustus and Tiberius*. Oxford: Oxford University Press.

Eliade, Mircea. 1959. *The Sacred and the Profane*. New York: Harcourt, Brace & World.

———. 1963. *Patterns in Comparative Religion*. Cleveland: World.

Elliott, John H. 1981. *A Home for the Homeless: A Sociological Exegesis of 1 Peter*. Philadelphia: Fortress.

Ferguson, John. 1970. *The Religions of the Roman Empire*. Ithaca: Cornell University Press.

Feuillet, André. 1965. *The Apocalypse*. Staten Island: Alba.

French, D. H. 1980. "The Roman Road-system of Asia Minor." *Aufstieg und Niedergang der römischen Welt* 2.7.2:698–729.

Frend, W. H. C. 1981. *Martyrdom and Persecution in the Early Church*. Grand Rapids: Baker.

Frey, R. P. Jean-Baptiste. 1952. *Corpus Inscriptionum Iudaicarum* vol 2. Roma: Pontificio Istituto Di Archeologia Cristiana.

Frye, Northrop. 1957. *Anatomy of Criticism*. Princeton: Princeton University Press.

Gager, John. 1975. *Kingdom and Community*. Englewood Cliffs: Prentice-Hall.

Geffcken, J. 1902. "Römische Kaiser im Volksmunde der Provinz." In *Nachrichten von der Koenigl. Gesellschaft der Wissenschaften zu Göttingen, Philologisch-historische Klasse aus dem Jahre 1901.* 183–95. Göttingen: Lüder Horstmann.

Giglioli, Pier. 1982. *Language and Social Context.* New York: Viking Penguin.

Grant, Frederick C. 1953. *Hellenistic Religions.* Indianapolis: Library of Liberal Arts.

Grant, Robert M. 1977. *Early Christianity and Society.* New York: Harper & Row.

———. 1980. "The Social Setting of Second-century Christianity." In *Jewish and Christian Self-definition* volume 1, ed. E. P. Sanders. 16–29. Philadelphia: Fortress.

Gruenwald, Ithamar. 1980. *Apocalyptic and Merkavah Mysticism.* Leiden: E. J. Brill.

Guiraud, Pierre. 1975. *Semiology.* London: Routledge & Kegan Paul.

Habicht, Christian. 1975. "New Evidence on the Province of Asia." *Journal of Roman Studies* 65:64–91.

Hammond, Nicholas G. L. 1981. *Atlas of the Greek and Roman World in Antiquity.* Park Ridge, N.J.: Noyes.

Hanfmann, G. M. A. 1972. *Letters from Sardis.* Cambridge: Harvard University Press.

———. 1975. *From Croesus to Constantine.* Ann Arbor: University of Michigan Press.

Hanson, Paul D. 1976. "Apocalypticism." In *Interpreter's Dictionary of the Bible.* Suppl. 29–31. Nashville: Abingdon.

———. 1979. *The Dawn of Apocalyptic.* Philadelphia: Fortress.

Harnack, Adolf. 1961. *The Mission and Expansion of Christianity in the First Three Centuries.* New York: Harper & Row.

Hartman, Lars. 1983. "Survey of the Problem of Apocalyptic Genre." In *Apocalypticism in the Mediterranean World and the Near East*, ed. Hellholm, 329–43. Tübingen: J. C. B. Mohr.

Hellholm, David, ed. 1983. *Apocalypticism in the Mediterranean World and the Near East.* Tübingen: J. C. B. Mohr.

———. 1986. "The Problem of Apocalyptic Genre and the Apocalypse of John." *Semeia* 36:13–64.

Hemer, Colin J. 1986. *The Letters to the Seven Churches of Asia in Their Local Setting.* Sheffield: Journal for the Study of the Old-Testament.

Henderson, Bernard. 1927. *Five Roman Emperors.* Cambridge: Cambridge University Press.

Hock, Ronald F. 1978. "Paul's Tentmaking and the Problem of His Social Class." *Journal of Biblical Literature* 97:555–64.

———. 1980. *The Social Context of Paul's Ministry: Tentmaking and Apostleship.* Philadelphia: Fortress.

Hopkins, Keith. 1978. *Conquerors and Slaves.* Cambridge: Cambridge University Press.

Howell, Peter. 1980. *A Commentary on Book One of Epigrams of Martial.* London: Athlone.

Isaac, E. 1983. "1 (Ethiopic Apocalypse of) Enoch." In *The Old Testament Pseudepigrapha* ed. James Charlesworth, 1:5–89. Garden City: Doubleday.

Jaspers, Karl. 1970. *Philosophy* volume 2. Chicago: University of Chicago Press.

Johnson, Sherman E. 1958. "Early Christianity in Asia Minor." *Journal of Biblical Literature* 77:1–17.

———. 1972. "Unsolved Questions about Early Christianity in Anatolia." In *Studies in New Testament and Early Christian Literature*, ed. David E. Aune, 181–93. Leiden: E. J. Brill.

———. 1975. "Asia Minor and Early Christianity." In *Christianity, Judaism, and Other Greco-Roman Cults* Pt 2, *Early Christianity*, ed. Jacob Neusner, 77–145. Leiden: E. J. Brill.

Jones, A. H. M. 1940. *The Greek City from Alexander to Justinian*. Oxford: Clarendon.

_____. 1983. *The Cities of the Eastern Roman Provinces*. Amsterdam: Adolf M. Hakkert.

Jones, Brian W. 1973. "Domitian's Attitude to the Senate." *American Journal of Philology* 94:79–91.

_____. 1979. *Domitian and the Senatorial Order*. Philadelphia: American Philosophical Society.

_____. 1984. *The Emperor Titus*. London: Croom Helm.

Jones, C. P. 1978. *The Roman World of Dio Chrysostom*. Cambridge: Harvard University Press.

Jung, C. G. 1973. *Answer to Job*. Princeton: Princeton University Press.

Keresztes, Paul. 1973. "The Jews, the Christians, and Emperor Domitian." *Vigiliae Christianae* 27:1–28.

Kermode, Frank. 1979. *The Genesis of Secrecy*. Cambridge: Harvard University Press.

Knierim, Rolf. 1973. "Old Testament Form Criticism Reconsidered." *Interpretation* 27: 435–68.

Koch, Klaus. 1969. *The Growth of the Biblical Tradition*. New York: Charles Scribner's Sons.

_____. 1972. *The Rediscovery of Apocalyptic*. London: SCM.

Köster, Helmut. 1971. "GNOMAI DIAPHOROI: The Origin and Nature of Diversification in the History of Early Christianity." In *Trajectories through Early Christianity*, ed. James M. Robinson & Helmut Köster, 114–57. Philadelphia: Fortress.

Kraabel, Alf Thomas. 1968. *Judaism in Western Asia Minor under the Roman Empire, with a Preliminary Study of the Jewish Community at Sardis, Lydia*. Thesis, Harvard University.

_____. 1971. "Melito the Bishop and the Synagogue at Sardis: Text and Context." In *Studies Presented to George M. A. Hanfmann*, 77–85. Mainz: Verlag Philipp von Zabern.

_____. 1978. "Paganism and Judaism: The Sardis Evidence." In *Paganisme, Judaïsme, Christianisme*. 13–33. Paris: Editions E. de Boccard.

Kümmel, Werner Georg. 1975. *Introduction to the New Testament*. Nashville: Abingdon.

Lacocque, Andre. 1979. *The Book of Daniel*. Atlanta: John Knox.

Latyschev, Basilius. 1965. *Inscriptiones Antiquae Orae Septentrionalis Ponti Euxini Graecae et Latinae Per Annos 1885–1900 Reertae*. Hildesheim: Georg Olms.

Leeuw, G. van der. 1963. *Religion in Essence and Manifestation*. New York: Harper Row.

Lévi-Strauss, Claude. 1967. *Structural Anthropology*. Garden City: Anchor.

Lewis, N. 1968. "Domitian's Order on Requisitioned Transport and Lodgings." *Revue Internationale des Droits de l'Antiquité* 15:135–42.

Lohmeyer, Ernst. 1953. *Die Offenbarung des Johannes*. Tübingen: J. C. B. Mohr.

McCrum, Michael, and A. G. Woodhead. 1961. *Select Documents of the Principates of the Flavian Emperors*. Cambridge: Cambridge University Press.

McDermott, W. C., and A. Orentzel. 1977. "Silius Italicus and Domitian." *American Journal of Philology* 98:23–34.

McGinn, Bernard. 1979. *Apocalyptic Spirituality*. New York: Paulist.

_____. 1984. "Early Apocalypticism: The Ongoing Debate." In *The Apocalypse in English Renaissance Thought and Literature*, ed. C. A. Patrides and Joseph Wittreich, 2–39. Ithaca: Cornell University Press.

MacMullen, Ramsay. 1966. *Enemies of the Roman Order*. Cambridge: Harvard University Press.

———. 1970. "Market-Days in the Roman Empire." *Phoenix* 24:333–41.

Macro, Anthony D. 1980. "The Cities of Asia Minor under the Roman Imperium." *Aufstieg und Niedergang der römischen Welt* 2.7.2:658–97.

Magie, David. 1950. *Roman Rule in Asia Minor*. Princeton: Princeton University Press.

Malherbe, Abraham J. 1983. *Social Aspects of Early Christianity*. Philadelphia: Fortress.

Massey, Irving. 1976. *The Gaping Pig: Literature and Transformation*. Berkeley: University of California Press.

Meeks, Wayne A. 1979. "'Since Then You Would Need To Go out of the World': Group Boundaries in Pauline Christianity." In *Critical History and Biblical Faith: New Testament Perspectives*, ed. Thomas J. Ryan, 4–58. Villanova: Villanova University Press.

———. 1983a. *The First Urban Christians*. New Haven: Yale University Press.

———. 1983b. "Social Functions of Apocalyptic Language in Pauline Christianity." In *Apocalypticism in the Mediterranean World and the Near East*, ed. David Hellholm, 687–705. Tübingen: J. C. B. Mohr.

Metzger, Bruce M. 1983. "The Fourth Book of Ezra." In *The Old Testament Pseudepigrapha*, 1:517–59. Garden City: Doubleday.

Millar, Fergus. 1964. *A Study of Cassius Dio*. Oxford: Clarendon.

———. 1966. "The Emperor, the Senate, and the Provinces." *Journal of Roman Studies* 56:156–66.

———. 1972. "The Imperial Cult and the Persecutions." In *Le culte des souverains dans l'Empire Romain*, Foundation Hardt pour L'Étude de L'Antiquité Classique; Entretiens Tome 19, ed. W. den Boer, 145–75. Geneva: Vandoeuvres.

———. 1977. *The Emperor in the Roman World*. Ithaca: Cornell University Press.

Minear, Paul. 1962. "The Cosmology of the Apocalypse." In *Current Issues in New Testament Interpretations*, ed. William Klassen and Graydon Snyder, 23–37. New York: Harper & Row.

Mounce, R. H. 1977. *The Book of Revelation*. Grand Rapids: Eerdmans.

Mouterde, René, and Claude Mondesert. 1957. "Deux inscriptions Grecques de Hama." *Syria* 34:278–87.

Mowry, Lucetta. 1952. "Revelation 4–5 and Early Christian Liturgical Usage." *Journal of Biblical Literature* 71:75–84.

Newton, H. C. 1901. *The Epigraphical Evidence for the Reigns of Vespasian and Titus*. New York. Macmillan.

Nicholson, E. W. 1979. "Apocalyptic." In *Tradition and Interpretation*, ed. G. W. Anderson, 189–213. Oxford: Clarendon.

Nickelsburg, George W. E. 1981. *Jewish Literature between the Bible and the Mishnah*. Philadelphia: Fortress.

———. 1983. "Social Aspects of Palestinian Jewish Apocalypticism." In *Apocalypticism in the Mediterranean World and the Near East*, ed. David Hellholm, 641–65. Tübingen: J. C. B. Mohr.

Norden, Eduard. 1956. *Agnostos Theos: Untersuchungen zur Formengeschichte Religiöser Rede*. Stuttgart: B. G. Teubner.

Ogilvie, Robert M., and I. Richmond. 1967. *Cornelii Taciti: De Vita Agricolae*. Oxford: Clarendon.

Oliver, James H. 1953. *The Ruling Power*. Philadelphia: American Philosophical Society.

Olsson, Tord. 1983. "The Apocalyptic Activity: The Case of Jamasp Namag." In *Apocalypticism in the Mediterranean World and the Near East*, ed. David Hellholm, 21–49. Tübingen: J. C. B. Mohr.

Oster, Richard. 1976. "The Ephesian Artemis As an Opponent of Early Christianity." *Jahrbuch für Antike und Christentum* 19:24–44.

Petersen, Norman R. 1985. *Rediscovering Paul: Philemon and the Sociology of Paul's Narrative World*. Philadelphia: Fortress.

Peterson, Erik. 1926. ΕΣ ΘΕΟΣ: *Epigraphische, formgeschichtliche, und religionsgeschichtliche Untersuchungen*. Göttingen: Vandenhöck & Ruprecht.

———.1964. *The Angels and the Liturgy*. New York: Herder & Herder.

Pleket, Henri. 1961. "Domitian, the Senate, and the Provinces." *Mnemosyne* 7:296–315.

———. 1965. "An Aspect of the Emperor Cult: Imperial Mysteries." *Harvard Theological Review* 58:331–47.

Price, S. R. F. 1984. *Rituals and Power: The Roman Imperial Cult in Asia Minor*. Cambridge: Cambridge University Press.

Prigent, P. 1974. "Au temps de l'Apocalypse." Pt. 1, "Domitien." *Revue d'histoire et de philosophie religieuse* 54:455–83.

Radice, Betty. 1975. "The Letters of Pliny." In *Empire and Aftermath*, ed. T. A. Dorey. London: Routledge & Kegan Paul.

Rajak, Tessa. 1984. *Josephus: The Historian and His Society*. Philadelphia: Fortress.

Ramsay, W. M. 1895. *The Cities and Bishoprics of Phrygia*. Volume 1, *The Lycos Valley and South-western Phrygia*. Oxford: Clarendon.

———. 1924. "Studies in the Roman Province Galatia." *Journal of Roman Studies* 14: 179–84.

Reeves, Majorie. 1984. "The Development of Apocalyptic Thought: Medieval Attitudes." In *The Apocalypse in English Renaissánce Thought and Literature*. ed. C. A. Patrides and Joseph Wittreich. Tübingen: J. C. B. Mohr.

Ricoeur, Paul. 1976. *Interpretation Theory*. Fort Worth: Texas Christian University Press.

Rissi, Mathias. 1966. *Time and History*. Richmond, Va.: John Knox.

Robert, Louis. 1937. *Études Anatoliennes*. Paris: E. de Boccard.

———. 1940. "Inscriptions Gréco-Juives." *Hellenica* 1:25–29.

———. 1964. *Nouvelles Inscriptions de Sardes*. Paris: Librairie d'Amerique et d'Orient.

Robinson, David M. 1924. "I. – A New Latin Economic Edict from Pisidian Antioch." *Transactions of the American Philological Association* 55:5–20.

Robinson, John A. T. 1976. *Redating the New Testament*. Philadelphia: Westminster.

Rogers, Robert. 1960. "A Group of Domitianic Treason Trials." *Classical Philology* 55: 19–23.

Rostovtzeff, Michael. 1957. *The Social and Economic History of the Roman Empire*. Oxford: Clarendon.

Rowland, Christopher. 1979. "The Visions of God in Apocalyptic Literature." *Journal for the Study of Judaism* 10:137–54.

———. 1982. *The Open Heaven*. New York: Crossroad.

Rowley, H. H. [1963] 1980. *The Relevance of Apocalyptic*. Reprint. Greenwood, S.C.: Attic.

Saffrey, H. D. 1975. "Relire L'apocalypse à Patmos." *Revue Biblique* 82:385–417.

Sanders, E. P. 1983. "The Genre of Palestinian Jewish Apocalypses." In *Apocalypticism in the Mediterranean World and the Near East*, ed. David Hellholm, 447–59. Tübingen: J. C. B. Mohr.

Scherrer, Steven J. 1984. "Signs and Wonders in the Imperial Cult." *Journal of Biblical Literature* 103:599–610.

Schmithals, Walter. 1975. *The Apocalyptic Movement*. New York: Abingdon.

Schoedel, William R. 1980. "Theological Norms and Social Perspectives in Ignatius of Antioch." In *Jewish and Christian Self-Definition*, ed. E. P. Sanders, 1:30–56. Philadelphia: Fortress.

Schürer, Emil. 1986. *The History of the Jewish People in the Age of Jesus Christ*. Vol. 3, pt. 1. Edinburgh: T. & T. Clark.

Schüssler Fiorenza, Elisabeth. 1981. *Invitation to the Book of Revelation*. Garden City, N.Y.: Doubleday.

_____. 1983. "The Phenomenon of Early Christian Apocalyptic: Some Reflections on Method." *Apocalypticism in the Mediterranean World and the Near East*, ed. David Hellholm, 295–316. Tübingen: J. C. B. Mohr.

_____. 1985. *The Book of Revelation: Justice and Judgment*. Philadelphia: Fortress.

Scott, Kenneth. 1932. "The Elder and Younger Pliny on Emperor Worship." *Transactions of the American Philological Association* 63:156–65.

_____. 1933. "Statius' Adulation of Domitian." *American Journal of Philology* 54: 247–59.

_____. [1936] 1975. *The Imperial Cult under the Flavians*. Reprint. New York: Arno.

Sherwin-White, A. N. 1966. *The Letters of Pliny*. Oxford: Clarendon.

Skinner, Quentin. 1969. "Meaning and Understanding in the History of Ideas." *History and Theory* 8:3–53.

_____. 1970. "Conventions and the Understanding of Speech Acts." *Philosophical Quarterly* 20:118–38.

_____. 1974. "'Social Meaning' and the Explanation of Social Action." In *The Philosophy of History*, ed. Patrick Gardiner, 106–26. Oxford: Oxford University Press.

Smallwood, E. Mary. 1956. "Domitian's Attitude toward the Jews and Judaism." *Classical Philology* 51:1–13.

_____. 1966. *Documents Illustrating the Principates of Nerva, Trajan, and Hadrian*. Cambridge: Cambridge University Press.

_____. 1981. *The Jews under Roman Rule*. Leiden: E. J. Brill.

Smith, Jonathan Z. 1973. "The Influence of Symbols upon Social Change: A Place on Which To Stand." In *The Roots of Ritual*, ed. James D. Shaughnessy, 121–43. Grand Rapids: William B. Eerdmans.

_____. 1978. "Wisdom and Apocalyptic." *Map Is Not Territory*. Leiden: E. J. Brill.

Smith, Morton. 1983. "On the History of ΑΠΟΚΑΛΥΠΤΩ and ΑΠΟΚΑΛΥΨΙΣ." In *Apocalypticism in the Mediterranean World and the Near East* ed. David Hellholm, 9–20. Tübingen: J. C. B. Mohr.

Stalnaker, Robert C. 1978. "Assertion." *Syntax and Semantics* 9:315–32.

Stern, Menahem. 1974. "The Jewish Diaspora." In *The Jewish People in the First Century*, ed. S. Safrai and M. Stern, 117–83. Assen, Netherlands: Van Gorcum.

_____. 1980. *Greek and Latin Authors on Jews and Judaism*. Volume 2, *From Tactitus to Simplicius*. Jerusalem: Israel Academy of Sciences & Humanities.

Stone, Michael. 1980. *Scriptures, Sects, and Visions*. Philadelphia: Fortress.

Sutherland, C. H. V. 1935. "The State of the Imperial Treasury at the Death of Domitian." *Journal of Roman Studies* 25:150–62.

Sweet, J. P. M. 1979. *Revelation*. Philadelphia: Westminster.

Syme, Ronald. 1930. "The Imperial Finances under Domitian, Nerva, and Trajan." *Journal of Roman Studies* 20:55–70.

_____. 1936. *Cambridge Ancient History*, vol. 2. Cambridge: Cambridge University Press.

———. 1958. *Tacitus*. Oxford: Clarendon.

Szelest, Hanna. 1974. "Domitian und Marital." *Eos* 62:105–14.

Talmon, Yonina. 1962. "Pursuit of the Millennium: The Relation between Religious and Social Change." *Archives Europeenes de sociologie* 3:125–48.

Taylor, Lily Ross. [1931]. n.d. *The Divinity of the Roman Emperor*. Reprint. Middletown: Scholars.

Thackeray, H. St. John. 1967. *Josephus, the Man and the Historian*. New York: KTAV.

Theissen, Gerd. 1977. *Sociology of Early Palestinian Christianity*. Philadelphia: Fortress.

———. 1982. *The Social Setting of Pauline Christianity*. Philadelphia: Fortress.

Thompson, Leonard L. 1969. "Cult and Eschatology in the Apocalypse of John." *Journal of Religion* 49:330–50.

———. 1973. "Hymns in Early Christian Worship." *Anglican Theological Review* 55: 458–72.

———. 1978. *Introducing Biblical Literature*. Englewood Cliffs: Prentice-Hall.

———. 1984. "*Domitianus Dominus*: A Gloss on Statius *Silvae* 1.6.84." *American Journal of Philology* 105:469–75.

———. 1985. "The Mythic Unity of the Apocalypse." In *Society of Biblical Literature 1985 Seminar Papers* ed. Kent Harold Richards. 13–28. Atlanta: Scholars.

———. 1986. "A Sociological Analysis of Tribulation in the Apocalypse of John." *Semeia* 36:147–74.

Tilly, Charles. 1984. *Big Structures, Large Processes, Huge Comparisons*. New York: Russell Sage.

Tracy, David. 1974. "Religious Language As Limit-Language." *Theology Digest* 22:291–307.

Tucker, Gene M. 1971. *Form Criticism of the Old Testament*. Philadelphia: Fortress.

Vessey, David. 1974. "Pliny, Martial, and Silius Italicus." *Hermes* 102:109–16.

Vielhauer, Philipp. 1965. "Introduction to Apocalypses and Related Subjects." In *New Testament Apocrypha*, ed. Edgar Hennecke, Wilhelm Schneemelcher, and R. M. Wilson, 2:579–607. Philadelphia: Westminster.

Viscusi, Peter. 1973. "Studies on Domitian." Ann Arbor: University Microfilms.

Waters, Kenneth. 1964. "The Character of Domitian." *Phoenix* 18:49–77.

———. 1969. "*Traianus Domitiani Continuator*." *American Journal of Philology* 90: 385–405.

Weinstock, S. 1971. *Divus Julius*. Oxford: Oxford University Press.

Weiss, Johannes. 1959. *Earliest Christianity*. New York: Harper & Row.

Werner, Eric. 1939. *The Sacred Bridge*. New York: Columbia University Press.

Wilder, Amos N. 1961. "Social Factors in Early Christian Eschatology." In *Early Christian Origins*, ed. Allen P. Wikgren, 67–76. Chicago: Quadrangle.

Wilken, Robert L. 1976. "Melito, the Jewish Community at Sardis, and the Sacrifice of Isaac." *Theological Studies* 37:53–69.

———. 1984. *The Christians As the Romans Saw Them*. New Haven: Yale University Press.

Wilson, Robert R. 1982. "From Prophecy to Apocalyptic: Reflections on the Shape of Israelite Religion." *Semeia* 21:79–95.

Yamauchi, Edwin M. 1980. *The Archaeology of New Testament Cities in Western Asia Minor*. Grand Rapids: Baker.

Yarbro Collins, Adela. 1979. *The Apocalypse*. Wilmington: Michael Glazier.

———. 1980. "Revelation 18: Taunt-Song or Dirge?" In *L'Apocalypse johannique et l'apocalyptique dans le Nouveau Testament*, ed. J. Lambrecht. Gembloux: J. Duculot.

_____. 1981. "Myth and History in the Book of Revelation: The Problem of Its Date." In *Traditions in Transformation*, ed. Baruch Halpern and Jon D. Levenson, 377–403. Winona Lake, Ind.: Eisenbrauns.

_____. 1983. "Persecution and Vengeance in the Book of Revelation." In *Apocalypticism in the Mediterranean World and the Near East*, ed. David Hellholm, 729–49. Tübingen: J. C. B. Mohr.

_____. 1984. *Crisis and Catharsis*. Philadelphia: Westminster.

_____. 1986. "Introduction." *Semeia* 36:1–11.

ADDITIONAL SOURCES

Akurgal, Ekrem. 1983. *Ancient Civilizations and Ruins of Turkey.* Istanbul: Mobil Oil Türk A. S.

Aune, David E. 1983. *Prophecy in Early Christianity and the Ancient Mediterranean World.* Grand Rapids: William B. Eerdmans.

Bauer, Walter. 1971. *Orthodoxy and Heresy in Earliest Christianity.* Philadelphia: Fortress.

Baus, Karl. 1980. *From the Apostolic Community to Constantine.* New York: Seabury.

Beardslee, William A. 1970. "Hope in Biblical Eschatology and in Process Theology." *Journal of the American Academy of Religion* 38:227–39.

Bruce, I. A. F. 1964. "Nerva and the *Fiscus Iudaicus.*" *Palestine Exploration Quarterly* 96:34–45.

Buttrey, T. V. 1975. "Domitian's Perpetual Censorship and the Numismatic Evidence." *Classical Journal* 71:26–34.

Casson, Lionel. 1974. *Travel in the Ancient World.* London: George Allen & Unwin.

Charlesworth, M. P. 1937. "Flaviana." *Journal of Roman Studies* 27:54–62.

Cousins, Mark, and Athar Hussain. 1984. *Michel Foucault.* New York: St. Martin's.

Devreker, J. 1977. "La Continuité dans le *Consilium Principis* sous les Flaviens." *Ancient Society* 8:223–43.

Dorey, T. A. 1960. "Agricola and Domitian." *Greece and Rome* 7:66–71.

Duncan-Jones, Richard. 1982. *The Economy of the Roman Empire.* Cambridge: Cambridge University Press.

Evans, John Karl. 1976. "Tacitus, Domitian and the Proconsulship of Agricola." *Rheinisches Museum für Philologie* 119:79–84.

Finley, M. I. 1973. *The Ancient Economy.* Berkeley: University of California Press.

Fishwick, Duncan. 1964. "The Institution of the Provincial Cult in Africa Proconsularis." *Hermes* 92:342–63.

Foakes Jackson, F. J., and Kirsopp Lake. 1979. *The Beginnings of Christianity.* Grand Rapids: Baker.

Francis, Fred O., and Wayne A. Meeks. 1975. *Conflict at Colossae.* Missoula, Mont.: Society of Biblical Literature.

Fritz, Kurt von. 1957. "Tacitus, Agricola, Domitian, and the Problem of the Principate." *Classical Philology* 52:73–97.

Funk, Robert W. 1969. *Apocalypticism.* New York: Herder & Herder.

Garnsey, Peter, and Richard Saller. 1982. *The Early Principate: Augustus to Trajan.* Oxford: Clarendon.

Ginsburg, Michael S. 1930. "Fiscus Judaicus." *Jewish Quarterly Review* 21:281–91.

Goodspeed, Edgar J. 1966. *A History of Early Christian Literature*. Rev. Robert M.
    Grant. Chicago: University of Chicago Press.

Grant, Robert M. 1970. *Augustus to Constantine*. New York: Harper & Row.

Hall, Stuart George. 1979. *Melito of Sardis: On Pascha and Fragments*. Oxford:
    Clarendon.

Hammond, Nicholas G. L., and H. H. Scullard. 1970. *Oxford Classical Dictionary*.
    Oxford: Clarendon.

Hengel, Martin. 1974. *Judaism and Hellenism*. Philadelphia: Fortress.

Jarvie, I. C. 1969. *The Revolution in Anthropology*. Chicago: Henry Regnery.

Johnson, Sherman E. 1961. "Christianity in Sardis." *Early Christian Origins*, ed. Allen
    P. Wikgren, 81–90. Chicago: Quadrangle.

_____. 1964. "Laodicea and its Neighbors." In *The Biblical Archaeologist Reader*, vol.
    2, ed. David Noel Freedman and Edward F. Campbell, Jr., 352–68. Garden City:
    Anchor.

Judge, E. A. 1960. *The Social Pattern of the Christian Groups in the First Century*.
    London: Tyndale.

Kraabel, Alf Thomas. 1981. "Social Systems of Six Diaspora Synagogues." In *Ancient
    Synagogues: The State of Research*, ed. Joseph Gutmann. Chico, Calif.: Schol-
    ars.

Lebram, J. C. H. 1983. "The Piety of the Jewish Apocalyptists." In *Apocalypticism in
    the Mediterranean World and the Near East*, ed. David Hellholm, 171–210.
    Tübingen: J. C. B. Mohr.

Lévi-Strauss, Claude. 1966. *The Savage Mind*. Chicago: University of Chicago Press.

Levy, Jean-Philippe. 1967. *The Economic Life of the Ancient World*. Chicago: Univer-
    sity of Chicago Press.

MacDonald, Dennis Ronald. 1983. *The Legend and the Apostle*. Philadelphia: West-
    minster.

McGinn, Bernard. 1979. *Visions of the End: Apocalyptic Traditions in the Middle
    Ages*. New York: Columbia University Press.

MacMullen, Ramsay. 1959. "Roman Imperial Building in the Provinces." *Harvard Stud-
    ies in Classical Philology* 64:207–35

_____. 1974. *Roman Social Relations*. New Haven: Yale University Press

Mellor, Ronald. 1975. ΘΕΑ ΡΩΜΗ: *The Worship of the Goddess Roma in the Greek
    World*. Göttingen: Vandenhoeck & Ruprecht.

Müller, Ulrich B. 1976. *Zur frühchristlichen Theologiegeschichte*. Gütersloh: Gütersloh-
    her Verlagshaus Mohn.

Pedley, John Griffiths. 1972. *Ancient Literary Sources on Sardis*. Cambridge: Harvard
    University Press.

Quasten, Johannes. 1984. *Patrology*. Vol. 1. Westminster, Md.: Christian Classics.

Ramage, Andrew, and Nancy H. Ramage. 1983. *Twenty-five Years of Discovery at
    Sardis: 1958–1983*. Archaeological Exploration of Sardis.

Seager, A. 1982. "The Synagogue at Sardis." In *Ancient Synagogues Revealed*, ed. Lee
    I. Levine. Jerusalem: The Israel Exploration Society, 178–84.

Sevenster, J. N. 1975. *The Roots of Pagan Anti-Semitism in the Ancient World*. Leiden:
    E.J. Brill.

Skinner, Quentin. 1973. "Motives, Intentions and the Interpretation of Texts." *The New
    Literary History* 5:393–408.

Snyder, Graydon F. 1985. *Ante Pacem*. Macon, Georgia: Mercer University Press.

Stambaugh, John E., and David L. Balch. 1986. *The New Testament in its Social Environment*. Philadelphia: Westminster.

Stark, Rodney. 1986. "The Class Basis of Early Christianity: Inferences from a Sociological Model." *Sociological Analysis* 47:216–25.

Stern, Menahem. 1976. Greek and Latin Authors on Jews and Judaism. Vol. 1: *From Herodotus to Plutarch*. Jerusalem: The Israel Academy of Sciences and Humanities.

Sternberg, Meir. 1985. *The Poetics of Biblical Narrative*. Bloomington: Indiana University Press.

Stevenson, J. 1987. *A New Eusebius*. London: SPCK.

Traub, Henry W. 1954. "Agricola's Refusal of a Governorship." *Classical Philology* 49:255–57.

Tucker, Gene M. 1976. "Form Criticism, OT." In *The Interpreter's Dictionary of the Bible: Supplementary Volume*, 342–45. Nashville: Abingdon.

Verner, David C. 1983. *The Household of God: The Social World of the Pastoral Epistles*. Chico, Calif.: Scholars.

Walton, C. S. 1929. "Oriental Senators in the Service of Rome: A Study of Imperial Policy Down to the Death of Marcus Aurelius." *Journal of Roman Studies* 19:38–66.

SUBJECT INDEX

Abyss, 79, 83, 220 n.14; beast of, 80; ruler of, 79. *See also* Bottomless pit
Acclamation, 53, 60, 62; new song, 59; worthy, 58
Accommodation (cultural): Christian, 120, 129; in Tertullian, 129
Acmonia: Jews at, 140; Julia Severa, 140
Ages: in 4 Ezra, 22; new, 115, 134, 225 n.65; no hard division between, 84, 221 n.33; present, 19, 63; to come, 19, 21, 63; two, 19
Agricola, 98; Tacitus's father-in-law, 98
Alce, 126; Herod's aunt, 126; sister of Nicetes, 127; at Smyrna, 126
Alienation in apocalyptic, 24, 196
Alpha and omega, 45
Altar, 62
Amulets. *See* Magic
Ancient of Days, 21
Angel of waters, 61
Animal symbolism, 20, 21
Antichrist, 15
Antioch of Pisidia, 165
Antiochus Epiphanes, 21
Antipas, 17, 173
Antoninus Pius, xii
Apamea: Jews in, 139; Noah story, 139
Apocalypse, 23; characteristics of genre, 18, 175, 192; content of, 31; defined, 24; element in worship, 72; eschatology, 23; formal elements, 175; language of, 91, 199; spirituality, 198; themes in, 175; world creating, 86; world expanding, 31
Apocalypticism, 23, 198
Apollos, 118; educated in rhetoric, 118; independent wealth, 118
Apollyon, 79; and apoleian, 79
Apphia, 121, 227 n.18
Aquila, 117. *See also* Priscilla
Archetypes, 205; heavenly, 20, 21; themes, 6
Archippus, 227 n.18; Christian leader at Colossae, 121

Archon of Philippi, 124
Aristarchus, 119
Armageddon, 51
Artemis of Ephesus, 118; economic significance, 147; silversmiths, 119
Artemis of Sardis, 124
Asebeia, 134, 135
Asia, province of, 11–12; agricultural resources, 146; Christians and Jews, 130, 133, 172; churches of, 117, 178, 227 n.11; home base in Revelation, 180; Jewish accommodation, 134; Jews, 137–145; John's churches in, 116; land control, 147; natural resources, 146; opposition to Christianity, 130; proconsul of, 12; public knowledge in, 176–77; river valleys, 146; and Rome, 156; senatorial province, 156; under Augustus, 156; under Flavians, 156–57; urban life (*see* Cities of Asia); wealth of, 154. *See also* Map of Asia Minor
Asia Minor, 11, 146; legal action against Christians rare, 132; mixture of people, 147; renewal of local rites, 131; turning from Christianity, 131; suspicion of Christianity, 132. *See also* Map of Asia Minor
Asiarchs, 161
Assembly, 149–50
Assize districts, 149
Astronomy, 19, 182–83
Augustales, 233 n.52. *See also* Imperial cult
Augustus, xi, 159

Babylon and Rome, 14
Babylon the Whore, 39, 68; and heavenly worship, 68; judgment of, 62; sexual expressions, 90. *See also* Harlot
Babylonian exile, 26
Balaam, 80; food offered to idols, 122; at Pergamum, 122; practiced immorality, 122. *See also* Jezebel
Baldness, 108

255

INDEX OF
ANCIENT SOURCES

INDEX OF
MODERN AUTHORS

228.06
T473

42507

LINCOLN CHRISTIAN COLLEGE AND SEMINARY